Turkey in the European Union

Implications for Agriculture, Food and Structural Policy

Turkey in the European Union

Implications for Agriculture, Food and Structural Policy

Edited by
Alison Burrell and Arie Oskam
Wageningen University, Wageningen, The Netherlands

CABI Publishing

CABI Publishing is a division of CAB International

CABI Publishing
CAB International
Wallingford
Oxfordshire OX10 8DE
UK

Tel: +44 (0)1491 832111
Fax: +44 (0)1491 833508
E-mail: cabi@cabi.org
Website: www.cabi-publishing.org

CABI Publishing
875 Massachusetts Avenue
7th Floor
Cambridge, MA 02139
USA

Tel: +1 617 395 4056
Fax: +1 617 354 6875
E-mail: cabi-nao@cabi.org

A catalogue record for this book is available from the British Library, London, UK.

Library of Congress Cataloging-in-Publication Data
Turkey in the European Union : implications for agriculture, food and structural policy / edited by Alison Burrell and Arie Oskam.
p. cm.
Includes bibliographical references and index.
ISBN-10: 0-85199-094-0 (alk. paper)
ISBN-13: 978-0-85199-094-1 (alk. paper)
1. Agriculture and state--Turkey. 2. Agriculture--Economic aspects--Turkey. 3. European Union--Turkey. I. Burrell, Alison. II. Oskam, A. J. III. Title.

HD2056.7.Z8T873 2005
338.1'8561--dc22

2005008081

ISBN 0 85199 094 0
978 0 85199 094 1

Printed and bound in the UK by Cromwell Press, Trowbridge, from copy supplied by the editors.

Contents

List of Tables, Figures and Boxes

Tables

Figures

Boxes

Acknowledgements

This book is based on a report commissioned by the Dutch Ministry of Agriculture, Nature and Food Quality and prepared under the leadership of Arie Oskam. It was delivered in early December 2004. We thank the Guiding Committee composed of members from the Ministry of Agriculture, Nature and Food Quality, the Ministry of Finance, the Ministry of Economic Affairs, the Ministry of Foreign Affairs, the Netherlands Scientific Council for Government Policy, the Netherlands Bureau for Economic Policy Analysis, and the Agricultural Economics Research Institute of Wageningen University, who studied our report in detail and provided useful guidance, both during our regular meetings in The Hague and bilaterally.

In particular we would like to express our gratitude to Gerrit Meester (Chair of the Guiding Committee) for his sustained interest and professional support of the total project, as well as to Wendy Asbeek Brusse, Homa Ashtari, Sander de Does, Doekele Haagsma, Sander Kers, Jouke Knol, Regine Neyt, Huib Silvis, Herman Stolwijk, Frank van Tongeren, Albert Vermue and Mark Woldberg. Carla Konsten and Gerard Terberg (the present and previous Agricultural Counsellors at the Embassy of the Netherlands in Turkey, also members of the Guiding Committee) gave constant support to our work by providing information and advice.

Various other people contributed to the final outcome. At a 2-day seminar organised in Wageningen in mid-October 2004 to discuss the draft report, we benefited greatly from fruitful exchanges of ideas with Erol Cakmak, Eduardo Díez Patier, Harald Grethe and Gerrit Meester, and also from helpful written comments received from Ahmet Bayaner, Sophia Davidova, Ulrich Koester and Sübidey Togan.

The authors of the chapters published here were generous in commenting on the chapters of their colleagues. In addition, each chapter was reviewed by several external referees. We acknowledge the constructive remarks kindly provided by Uygun Aksoy, Floor Brouwer, Asim Karaomerlioglu, Roald Laperre, Mark Lundell, Peter van der Meer, Jos van Meggelen, Susanna Perachino, Serdar Sayan, Rob Vijftigschild, Marie-Josée Mangen, David Ward, Jo Wijnands and Erdal Yilmaz.

We are also thankful to everyone who directly or indirectly contributed data, information, suggestions and moral support throughout the project, not least the members of the Agricultural Economics and Rural Policy Group at Wageningen University. During the first stage of the project Irfan Ozcan and Sibel Kacan made a constructive contribution to our research activities. Natasha Longworth took care of the layout of the book and made many suggestions for improving the final text.

The authors of the chapters and the editors of the book are responsible for its content, as well as for any errors and omissions.

Alison Burrell and Arie Oskam, Editors
Wageningen, 15 January 2005

Acronyms and Abbreviations

AAF Canada	Agriculture and Agri-Food Canada
AARI	Aegean Agricultural Research Institute
AC-2	Bulgaria and Romania
ACC	Agricultural Credit Cooperative
ACP	African, Caribbean and Pacific Countries
ADC	Agricultural Development Cooperative
AEARP	Agricultural Extension and Applied Research Project
AKIS	Agricultural knowledge and information system
AMS	Aggregate Measure of Support
ARIP	Agriculture Reform Implementation Project
ASC	Agricultural Sales Cooperative
ASCU	Agricultural Sales Cooperatives Union
ASP	Annual Seed Programme
BK	Bağ-Kur
bn	billion
BSE	Bovine Spongiform Encephalopathy
CAP	Common Agricultural Policy
ÇAYKUR	General Directorate of Tea Establishments
CEECs	Central and Eastern European Countries
CET	Common External Tariff
CIMMYT	International Maize and Wheat Improvement Centre
CIS	Commonwealth of Independent States[1]
CITES	Convention on International Trade of Endangered Species of Flora and Fauna
CPB	Centraal Plan Bureau (Netherlands Bureau for Economic Policy Analysis)
CPI	Consumer Price Index
ÇSGB	Ministry of Labour and Social Security
CU	customs union
DIS	Direct Income Support
DOKAP	Eastern Black Sea Regional Development Plan
DSİ	General Directorate of State Hydraulic Works
DTM	Undersecretariat of the Prime Ministry for Foreign Trade
EAGGF	European Agricultural Guidance and Guarantee Fund
EBA	Everything but Arms
EBK	Meat and Fish Production Inc.

[1] Armenia, Azerbaijan, Belarus, Kazakhstan, Krygyzstan, Republic of Moldova, Russian Federation, Tajikistan, Turkmenistan, Ukraine and Uzbekistan.

ECO	Economic Cooperation Organisation
ECOTA	Economic Cooperation Organisation Trade Agreement
EEA	European Environmental Agency
EEC	European Economic Community
EFTA	European Free Trade Association
EIA	Environmental impact assessment
EP	European Parliament
EPPO	European and Mediterranean Plant Protection Organisation
ERDF	European Regional Development Fund
ESF	European Social Fund
ETO	Turkish Association on Organic Agriculture
EU	European Union
EU-15	European Union of 15 Member States[2]
EU-25	European Union of 25 Member States (EU-15 + NMS)
EU-27	European Union of 27 Member States (EU-25 + AC-2)
EU-28	European Union of 28 Member States (EU-27 + Turkey)
EUFMD	European Commission for the Control of Foot and Mouth Disease
EUR	Euro
EUREPGAP	Euro-Retailer Produce Working Group's Good Agricultural Practices
FAO	Food and Agriculture Organisation of the United Nations
FAPAS	Food Analyses Performance Assessment Scheme
FDI	Foreign Direct Investment
FFV	Fresh Fruits and Vegetables
FMD	Foot and mouth disease
FTA	Free Trade Agreement
GAIN	Global Agriculture Information Network
GAP	Southeastern Anatolia Project
GATT	General Agreement on Tariffs and Trade
GCE	Computable General Equilibrium
GDAR	General Directorate of Agricultural Research
GDP	Gross Domestic Product
GDPC	General Directorate of Protection and Control
GDRS	General Directorate of Rural Services
GNI	Gross National Income
GNP	Gross National Product
GTAP	Global Trade Analysis Project
GVA	Gross Value Added
HACCP	Hazard Analysis at Critical Control Points
HLFS	Household Labour Force Survey
HPAI	Highly pathogenic avian influenza

[2] Belgium, The Netherlands, Luxembourg, France, Italy, Germany, United Kingdom, Ireland, Spain, Portugal, Greece, Austria, Sweden, Finland and Denmark.

IBA	Important Bird Area
ICARDA	International Centre for Agricultural Research in the Dry Areas
ICT	Information and Communication Technologies
İGEME	Export Promotion Centre of Turkey
ILO	International Labour Organisation
IMF	International Monetary Fund
İMKB	Istanbul Stock Exchange
IPEC	International Program on the Elimination of Child Labour
IPM	Integrated pest management
IPPC	International Plant Protection Convention
ISIC	International Standard Industrial Classification System
ISO	International Standards Organisation
IUCN	International Union for Nature Conservation
JICA	Japan International Cooperation Agency
LEADER	*Liaisons Entre Actions de Développement de l'Économie Rurale*
LFA	Less Favoured Area
MARA	Ministry of Agriculture & Rural Affairs
MBM	Meat and bone meal
MeBr	Methyl bromide
MFN	Most favoured nation
mn	million
MoE	Ministry of Environment
MoEF	Ministry of Environment and Forestry
NARS	National Agricultural Research System
NDP	National Development Project
NEAP	National Environmental Action Plan
NGO	Non-governmental organisation
NMS	New Member States[3]
NPAA	National Plan for the Adoption of the *Acquis*
NTB	Non-tariff Trade Barriers
NUTS	Nomenclature of Territorial Units for Statistics
OAC	Organic Agriculture Committee
OECD	Organisation for Economic Cooperation and Development
OIE	World Animal Health Organisation
PALiLZ	Polish Information and Foreign Investment Agency
PPP	Purchasing Power Parity
PPR	*Peste des petits ruminants*
PPS	Purchasing Power Standard
PPV	Plum pox potyvirus
PSE	Producer Support Estimate
R&D	Research and Development

[3] Poland, Czech Republic, Hungary, Slovak Republic, Slovenia, Estonia, Latvia, Lithuania, Malta and Cyprus.

SARS	Severe Acute Respiratory Syndrome
SEE	State Economic Enterprise
SIS	State Institute of Statistics
SPA	Specially Protected Area
SPO	State Planning Organisation
SPS	Sanitary and Phytosanitary
SRM	Specific Risk Material
SSK	Social Insurance Organisation
TBT	Technical Barriers to Trade
TCMB	Central Bank of the Republic of Turkey
TEKEL	Directorate General for Tobacco and Tobacco Products, Salt and Alcohol Industry
TİGEM	General Directorate of Agricultural Enterprises
TİSK	Turkish Confederation of Employers' Unions
TMO	Turkish Grain Board
TOBB	Union of Chambers of Commerce
TRIPS	Trade-Related Aspects of Intellectual Property Rights
TRL	Turkish Lira
TRQ	Tariff-rate quota
TSE	Turkish Standards Organisation
TTA	Turkish Exporters' Association
TTGV	Technology Development Foundation of Turkey
TÜBİTAK	Scientific and Technical Research Council of Turkey
TÜGEM	General Directorate of Agricultural Production and Improvement
TÜGSAŞ	Turkish Fertiliser Industry Co.
TÜRKŞEKER	Turkish Sugar Company
TÜRK-TED	Turkish Seed Industry Association
TÜSİAD	Turkish Industrialists' and Businessmen's Association
TZOB	Union of Turkish Chambers of Agriculture
UAA	Utilised Agricultural Area
UN	United Nations
UNCTAD	United Nations Conference on Trade and Development
UNDP	United Nations Development Programme
URA	Uruguay Round Agreement
URAA	Uruguay Round Agreement on Agriculture
USD	United States Dollars
USDA	United States Department of Agriculture
VAT	Value Added Tax
WB	World Bank
WHO	World Health Organisation
WPI	Wholesale Price Index
WTO	World Trade Organisation
YASED	Foreign Investors Association
YTL	New Turkish Lira

Reference Exchange Rates

Year	1 ECU/EUR = … USD	1 ECU/EUR = … TRL
1980	1.39233	54.58
1981	1.11645	98.71
1982	0.97972	164.03
1983	0.89022	251.29
1984	0.78903	460.69
1985	0.76309	680.92
1986	0.98417	679.76
1987	1.15444	740.27
1988	1.18248	1,200.36
1989	1.10175	1,924.21
1990	1.27343	3,329.06
1991	1.23916	5,153.29
1992	1.29810	8,930.95
1993	1.17100	12,879.30
1994	1.18952	35,535.30
1995	1.30801	59,912.10
1996	1.26975	103,214.00
1997	1.13404	171,848.00
1998	1.12109	293,736.00
1999	1.06580	447,238.00
2000	0.92360	574,816.00
2001	0.89560	1,102,425.00
2002	0.94560	1,439,680.00
2003	1.13120	1,694,851.00

1. USD/(ECU)EUR exchange rates are from European Central Bank reference rates as published by Eurostat. TRL(ECU)/EUR exchange rates are based on TRL/USD exchange rates published by OECD, for 1980-1989. From 1990 onwards these are European Central Bank reference rates as published by Eurostat.
2. All exchange rates are yearly averages.

1

Introduction

Arie Oskam and Alison Burrell

The relationship between the European Economic Community – later the European Union – and Turkey has developed in phases. The process began in 1963 when Turkey and the EEC signed the Ankara Agreement, an association agreement covering the liberalisation of markets for goods and financial aid. Turkey applied for membership of the European Community in 1987 and in 1996 the EU-Turkey Customs Union took effect. Agricultural products were not included in the Customs Union, although a significant part of agricultural trade takes place under preferential agreements.

At the Copenhagen summit of June 1993, the EU member states agreed that:

> "Accession will take place as soon as an associated country is able to assume the obligations of membership by satisfying the economic and political conditions required. Membership requires that the candidate country has achieved stability of institutions guaranteeing democracy, the rule of law, human rights and respect for and protection of minorities, the existence of a functioning market economy as well as the capacity to cope with competitive pressure and market force within the Union. Membership presupposes the candidate's ability to take on the obligations of membership including adherence to the aims of political, economic and monetary union" (Bulletin, EC 6–1993: 13).

These criteria have from then on been referred to as the Copenhagen criteria.

Turkey achieved candidate status at the 1999 Helsinki summit, but in December 2002 the European Council postponed the decision regarding a starting date for Turkey's accession negotiations. The decision was shifted to December 2004, and it was agreed that it would depend on Turkey's ability to satisfy only that part of the Copenhagen criteria covering political conditions:

> "The Union recalls that membership requires that a candidate country has achieved stability of institutions guaranteeing democracy, the rule of law, human rights and respect for and protection of minorities .... If the European Council in December 2004, on the basis of a report and a recommendation from the Commission, decides that Turkey fulfils the Copenhagen political criteria, the European Union will open accession negotiations with Turkey without delay" (European Council, 2003: 5).

At the European Council meeting in Amsterdam on 16-17 December, the Council decided that "...Turkey sufficiently fulfils the Copenhagen political criteria to open accession negotiations" (European Council, 2004: 4-8), conditional on Turkey's implementation of certain legislation identified in the European Commission's October 2004 report (European Commission, 2004a). The Commission was invited to prepare a framework with a view to opening negotiations on 3 October 2005. As in previous negotiations, the *acquis communautaire* will be divided into a number of chapters[1]. Regarding future negotiations with candidate states, the Council decision specifies that "long transition periods, derogations, specific arrangements or permanent safeguard clauses ... may be considered". The areas mentioned in which such arrangements might be appropriate include structural policy and agriculture. Although the shared objective of the negotiations is accession, they are "an open-ended process, the outcome of which cannot be guaranteed beforehand". Negotiations may be suspended in the case of "a serious and persistent breach in a candidate state of the principles of liberty, democracy, respect for human rights and fundamental freedoms and the rule of law on which the Union is founded". Finally, the Council announced that, in parallel to accession negotiations, "the Union will engage with every candidate state in an intensive political and cultural dialogue" that will also involve the civil society.

The study published in this book was completed at the end of November 2004. It focuses on just one dimension of the Copenhagen criteria: the existence of a functioning market economy, including the existence of institutions required to support such a market economy within the EU. Specifically, we examine Turkey's readiness to meet EU requirements in relation to agriculture, food and the rural economy. The aim of the study was to provide useful background information in the event that it was decided to begin accession negotiations with Turkey. Now that this decision has been made, the full set of Copenhagen criteria is again relevant. In particular, information contained in this study constitutes a benchmark against which progress achieved by Turkey's agrifood sector during the pre-accession period can be gauged.

Previous studies have examined the existence in Turkey of a functioning market economy, suitable for entering the European Union[2]. These studies have, however, paid little attention to agriculture, food, rural development and structural policy. Yet these areas cover 80 per cent of the present budget expenditure of the European Union and account for a large share of the *acquis communautaire*. Our study examines in detail these areas, which will be very important in the forthcoming negotiations with Turkey.

Objective and research questions

Our main objective is to provide a balanced overview of the Turkish agrifood system and rural areas. Taking Turkey's accession to the European Union in 2015 as a working hypothesis, consequences for both Turkey and the EU are explored on the basis of available literature and information, using an approach that places strong emphasis on long-term processes and institutional developments.

---

[1] In the negotiations of the NMS, the chapter relating to agriculture was the largest and one of the last to be finalised.

[2] See for example, European Commission (2003a) and (2004a and b); Hughes (2004); Quaisser and Reppegather (2004); Togan (2004).

The various chapters attempt to answer the following questions:

- What are the main problems facing the economic institutions involved with agriculture and food in the context of Turkey's accession to the European Union?
- What is the economic situation of Turkey and its regional income structure?
- What is the current performance of the Turkish agricultural sector?
- What are the main differences between EU and Turkish agricultural policy?
- How are rural, structural and food policies organised in Turkey?
- How is the Turkish agrifood sector organised and what is its performance?
- What is the situation regarding technology and human capital in Turkish agriculture?
- What is the trade position of Turkey, with respect to the EU, in a regional context and within the WTO?
- What is the relationship between agriculture and the environment, and how does this compare with EU standards?
- What standards apply in Turkey with respect to food safety, plant health, animal health and biosecurity?
- What are the implications, for both Turkey and the EU member states, of applying EU agricultural, food, rural and structural policy rules to Turkey?

In the following chapters, we focus on Turkey and EU-27, which is sometimes divided into EU-15, NMS (the ten new member states from 1 May 2004) and AC-2 (Bulgaria and Romania). Although Croatia is likely to be an EU member before 2015, we have not included this country in our analysis.

Macro-economic context

Turkey has some important distinguishing characteristics, as an economy and a society. It is a country with large differences in economic and social development between regions in the east and the west. Turkey is classified as a developing country, and has experienced significant economic instability in the recent past. The domestic labour market shows some peculiarities. The formal labour market is quite rigid but the informal sector, accounting for more than 50 per cent of total employment, is very flexible. This large informal sector is seen as unsustainable because it avoids taxes and social security payments (OECD, 2004; Togan and Ersel, 2005). Although Turkey has a longer tradition as a market economy than central and eastern European countries, the role of the state and state-owned enterprises has been strong for a long period. Competition and competition rules have a short history. It is a country with very high, although declining, inflation, a low level of investment and population growth of about 1.7 per cent per year.

Per capita income growth has been declining over the last two decades: the annual average growth rate was 2.8 per cent over the period 1983-93 and 0.9 per cent during 1993-2003. Compared to other lower-middle income countries at the end of the twentieth century, this indicates a rather stagnant economy (World Bank, 2004). Turkey's GDP in 2003 was EUR 212 billion, which is about 2 per cent of the GDP of EU-25. Total public debt relative to GDP increased from 33 per cent in 1983 to 61 per cent in 2002.

Within the Turkish economy as a whole, the agricultural sector is also rather stagnant, with long-term growth rates of 1.5 and 1.0 per cent per year during the periods 1982-1992 and 1992-2002 respectively (World Bank, 2004).

The share of the agricultural sector in the Turkish economy is in relative decline, but it is still large compared to the EU-15 and NMS. Approximately 90 per cent of the labour force at the beginning of the Republic was employed in agriculture. This had fallen to 34 per cent in 2003. The future of both the agriculture and food sectors is linked to the development of the total Turkish economy, since agricultural incomes can only keep pace with incomes outside agriculture if other sectors of the economy grow fast enough to absorb agriculture's labour surplus. Moreover, developments in the food sector will depend on growing consumer awareness and ability to pay for food safety and quality.

Evidence from other studies

A number of studies have analysed the consequences of Turkish accession to the EU. Expected consequences differ significantly between these reports. The expected economic developments of both Turkey and the EU are important for determining EU budget allocations, Turkey's absorption capacity for structural funds and opportunities for foreign direct investment and trade. This book deals with these issues in detail, in relation to agriculture and food, in Chapters 11 and 12. A brief summary only is given here.

CPB (Lejour *et al.*, 2004) estimated the effect of accession as an increase in Turkey's GDP of 0.8 per cent without institutional improvement, and 5.6 per cent with institutional improvement, where 'institutional improvement' means a decrease in corruption. If these numbers are accepted, they imply: (a) negligible economic effects of accession *per se* and (b) large effects of 'institutional factors'.

ABN-AMRO (Kalshoven and Küçükakin, 2004) explored the long-term effects of accession on the Turkish economy. Their assumptions about economic growth (4.9 per cent per year up to 2014 and then 5.0 to 6.2 per cent up to 2024) are introduced exogenously, and growth in foreign direct investment and trade is linked directly to this assumed economic growth. Needless to say, given these assumptions, accession has some substantial positive impacts. However, it has to be borne in mind that the outcomes in the ABN-AMRO report depend on a number of sensitive assumptions. A report from the OstEuropa-Institut in Munich (Quaisser and Reppegather, 2004) is much less optimistic with respect to the development possibilities of the Turkish economy in the period before and after accession. These three studies span the range of the many studies devoted to the consequences of accession for Turkey and the EU.

Areas of uncertainty

In preparing this study, we encountered some statistical puzzles, which other studies tend to circumvent but which are essential for our conclusions. Some of these puzzles at first sight look rather simple: for example, ascertaining the relevant area of agriculture land, the precise number of agricultural workers (and their income levels) and the size and production of the livestock herd. More generally, after intensive searching and better familiarity with Turkish agricultural statistics, we can say that there is a considerable body of statistical material available for Turkey, but its quality and reliability are much less clear. In certain areas that are important for the implementation and monitoring of EU legislation, information is completely lacking. The same holds for particular topics that are relevant to the general subject area of our study, namely animal welfare, the performance of the food industry, food quality and so on.

We have had to deal with other sources of uncertainty when trying to answer questions with respect to the future level of competition, the size and budget of structural, rural policy and agricultural policy, and the budget costs of the accession. Many estimates depend on the future rate of economic growth of Turkey and the exchange rate between the Turkish lira and the euro at the moment of accession and beyond. Other reports work with quite different growth rates from those assumed in our study. These elements are crucial, however, since the projected level of Turkey's GDP in 2015 determines Turkey's estimated contribution to the EU budget, as well as the level of EU structural and rural development spending in Turkey.

Approach and structure of the study

Our study adopts an institutional approach. Institutions are particularly important in the case of Turkey, and especially because of large differences between Turkey and the EU in *informal* institutional rules (that is, the behavioural traditions and expectations underlying political, economic and social interactions). Informal rules, however, generally take more than one generation to change. This will be a recurrent theme in the following chapters. Here we provide a few examples of the importance of institutions. First, the CPB report (Lejour *et al.*, 2004) indicated that the growth effect of Turkish accession would be 7 times larger if corruption decreased to the level of Portugal. This is has strong implications for the institutional context in Turkey. A second example concerns the three 'traps' for the Turkish economy identified by OECD (2004: 26). Accelerating economic development in Turkey requires dealing with the three traps of low confidence, weak governance and a large informal sector, which have in the past prevented stronger sustained growth. All of them are 'institutional'. A third example is the growth of efficient consumer-oriented food supply chains, which requires an overhaul of the institutions currently operating in the agrifood sector, an issue dealt with in Chapter 6 of this book. A fourth example is the difference between potential and actual agricultural production. We deal briefly with this question in Chapter 4, but more relevant is the observation that many countries produce far below their potential or even below 'normal' agricultural production because of poorly functioning institutions. A last example concerns the effectiveness of the education system in maximising the potential of Turkey's labour force, and the effectiveness of the research and extension services. Although serious under-performance in these areas is partly an issue of resources, poorly functioning institutions are also heavily implicated.

Chapter 2 gives a general description of the Turkish economy and the regional income distribution in Turkey. Chapter 3 deals with the institutional context of Turkish agriculture, giving a brief account of key historical developments with implications for Turkey's institutions. How existing institutions function in specific areas, and with what results, is studied in Chapters 5 to 10. In Chapter 4 agricultural production is considered from an economic perspective. Chapter 5 examines the quality and performance of the agricultural labour force, and the employment and living conditions of the rural population. Chapter 6 provides an overview of the agrifood sector from a chain perspective. In Chapter 7, agricultural, food, rural and structural policies are discussed. Chapter 8 deals with the trade position of Turkey, its trade relations with its main trading partners and its conduct within the WTO. Chapter 9 focuses on the environment in relation to agriculture and Chapter 10 deals with plant and animal health.

Chapters 11 and 12 offer syntheses of the previous chapters, focusing on the consequences of Turkish accession first for the European Union (Chapter 11) and then for Turkey (Chapter 12). Agriculture, food, rural development and structural policy are still the main focus, but now from a perspective that is less descriptive, and more analytical and speculative. The central question is: what will happen if Turkey becomes a member of the EU in 2015? Here it is necessary to make a number of assumptions. Our quantitative analysis depends, of course, on the precise numbers we have assumed in order to describe changes between now and 2015. However, our qualitative analysis and conclusions are more robust in that they depend on assumptions about general trends and developments that are less controversial.

The book concludes with a short chapter discussing the opportunities, threats and challenges of the Turkish accession for agriculture, food and rural development. A summary of the conclusions from each chapter is given in an Appendix.

Limits of the study

It is not our intention to describe in detail the state Turkey has currently reached in implementing the *acquis communautaire.* Rather, we concentrate on how Turkey's economy and rural society are performing in the agrifood sectors and the rural economy. Our study is complementary to the work done by the European Commission (European Commission, 2003a, 2003b; 2004a, 2004b). Where useful, we comment on these studies and incorporate a small number of tables and figures from these sources, often in updated form, in order to give the reader a coherent picture. We have excluded fisheries from our analysis, since this sector was outside the expertise of the research team. Given that agriculture and fisheries have virtually no overlap in terms of policies and issues, the lack of attention to fisheries does not affect the conclusions we can draw[3].

The political history of Turkey is outside the scope of this book[4]. The same holds for the discussion of whether and in what respects Turkey is part of 'Europe'. It is undeniable that most of Turkey is geographically in Asia. Because of this, however, Turkey is – and historically has been - at the intersection of Europe, Asia and (to an extent) Mediterranean Africa. This unique position means that Turkey has played, and will continue to play, a special role in its relationships with Europe and with countries further to the east and south.

References

European Commission (2003a) 2003 Regular report on Turkey's progress towards accession. SEC (2003) 1212, Brussels.

European Commission (2003b) Agricultural situation in the candidate countries. Country report: Turkey. DG-AGRI, Brussels, November 2003.

European Commission (2004a) 2004 Regular report on Turkey's progress towards accession. SEC (2004) 1201, Brussels.

European Commission (2004b) Issues arising from Turkey's membership perspective. SEC (2004) 1202, Brussels.

European Council (2003) Presidency conclusions: Copenhagen, 12 and 13 December 2002. Cover Note, 15917/02, Brussels.

European Council (2004) Presidency Conclusions. Brussels European Council 16/17 December 2004. Cover note from the Presidency to Delegations. 16238/04. CONCL 4. Brussels, 17 December, 2004.

Flam, H. (2003) Turkey and the EU: Politics and economics of accession. Seminar Paper No. 718, Institute for International Economic Studies, Stockholm University.

Hughes, K. (2004) Turkey and the European Union: Just another enlargement? A Friends of Europe Working Paper, June 2004, Brussels.

Kalshoven, C. and Küçükakin, S. (2004) Opportunities beyond the Bosphorus: Exploring growth, FDI and trade flows in view of Turkey's EU-accession. Study commissioned by the Dutch ministry of Economic Affairs. ABN-AMRO, Netherlands, May 2004.

Lejour, A.M., de Mooij, R.A. and Capel, C.H. (2004) Assessing the economic implications of Turkish accession to the EU. CPB document No 56. CPB Netherlands Bureau for Economic Policy Analysis, The Hague, March 2004.

OECD (2004) *Economic Survey – Turkey* 2004. OECD, Paris.

Quaisser, W. and Reppegather, A. (2004) EU-beitrittsreife der Türkei und konsequenzen einer EU-mitgliedschaft. OstEuropa-Institut Working Paper 252. OstEuropa-Institut, München.

Togan, S. (2004) Turkey: Toward EU accession. *World Economy* 27 (7), 1013-1045.

Togan, S. and Ersel, H. (2005) Macroeconomic policies for EU accession. In: S. Togan and B. Hoekman (eds.) *Turkey: Economic Reform and Accession to the European Union*. World Bank.

---

[3] Fisheries were also excluded from the Issues study of the European Commission (2004b:35-36) where only some very general remarks are made on fishery policy and the increasing responsibility the EU would acquire for the Mediterranean and Black Sea.

[4] See for example Flam, 2003; Zürcher, 2004.

World Bank (2004) Country at a glance tables. http://www.worldbank.org/data/countrydata/countrydata.html (accessed 25 November 2004)
Zürcher, E.J. (2004) *Turkey: A Modern History,* 3rd edn. Tauris, London.

2

Turkey's Economy and Regional Income Distribution

Arie Oskam, Natasha Longworth and Ayselin Yildiz

2.1 Introduction

This chapter provides the macroeconomic context within which other chapters analyse agrifood issues and the consequences of Turkey's accession to the EU. After a brief overview of the economy in Section 2.2, Section 2.3 gives information on economic growth, inflation, employment, exchange rates, trade and the balance of payments on current account. The Turkish economy has gone through a period of instability and high inflation, which makes it difficult to measure the underlying growth trend. The growth of Turkey's economy will, however, strongly influence its future position relative to the European Union, and will, together with the exchange rate at and after accession, determine the size of budgetary flows. Section 2.4 gives a brief discussion of this issue. Section 2.5 analyses inter-household and regional income distribution in some detail, since large regional differences in income in an acceding country have implications for targeting EU structural and rural policy.

2.2 Turkey at a glance

Table 2.1. Turkey's land and population resources

Land area		Population	
Area (million hectares), of which:	77.9	Population (million, 2003)	70.7
Agricultural land (excl. pastures)	26.4	Average annual % growth (1995-2003)	1.7
Forests	20.2		

Source: OECD, 2004a

Table 2.2. Key indicators of Turkey's economy

National accounts			
	2003[1]	Average annual growth 1993-2003[2]	%
GDP EUR billion	212.3	GDP EUR billion	2.7
GDP per capita (EUR)	3,000	GDP per capita (EUR)	0.9
GDP per capita (PPS)	5,930		

Source: [1]European Commission, 2004a; [2] World Bank, 2004

Table 2.2 contd. Key indicators of Turkey's economy

Structure of the economy			
Sectoral shares of GDP, 2003[2]	%	Average annual growth 1993-2003[3]	%
Agriculture	11.5	Agriculture	1.0
Industry	24.2	Industry	2.2
Services	60.9	Services	3.0
Construction	3.4	Construction	n.a
Employment (2003)			
Employment by sector[2]	%		%
Agriculture and forestry	33.9	Labour force participation rate[4]	48.3
Industry	18.2	Unemployment rate [4]	10.5
Construction	4.6	Labour force growth p.a., 1997-2003[3]	2.3
Services	43.3		
Merchandise trade			
	2003[2]		2003[3]
Exports (EUR billion)	41.7	Imports (EUR billion)	61.2
Exports (% of GDP)	27.7	Imports (% of GDP)	31.0
Exports to EU-15 (% of total)	51.8	Imports from EU-15 (% of total)	45.6
Sectoral share of exports[1]	%	Sectoral share of imports[1]	%
Agricultural products	11.2	Mineral fuels and oil	16.6
Food	10.0	Machinery and equipment	22.8
Agricultural raw material	1.1	Vehicles	7.8
Manufactured products	83.8	Iron and steel	8.0
Mining products	4.2	Other products	44.8
Other products	0.8		

Source: [1]*OECD, 2004a;* [2]*European Commission, 2004a;* [3]*World Bank, 2004;* [4]*SIS, 2003a (see chapter 5 for definitions)*

Figure 2.1. Map of Turkey
Source: CIA, 2004

2.3 Key features of the Turkish economy

Turkish GDP in 2003 was about 2.3 per cent of EU-15 GDP and 49 per cent of GDP in the New Member States (NMS). Table 2.3 shows economic indicators for Turkey, the AC-2, the NMS and the EU-15. For purposes of comparison, GDP is also given in purchasing power standards (PPS).[1] Turkey's average per capita GDP in PPS is lower than the NMS average, but similar to that of Bulgaria and Romania.[2]

Table 2.3. Economic indicators for Turkey, Bulgaria, Romania, NMS and EU-15 in 2003

Indicator	Turkey	Bulgaria	Romania	NMS	EU-15
GDP (EUR billion)[1]	212.3	15.2	50.4	437.0	9300.8
Population (million people)[1, 2]	70.7	7.8	21.7	*74.7	*377.8
GDP per capita (EUR, market rates)[1, 2]	3000	2260	2320	7854	24320
GDP per capita (EU-15=100) [1, 2]	12.3	9.3	9.5	32.3	100
GDP per capita (PPS) [1, 2, 3]	5930	6520	6340	*11300	23270
GDP per capita (PPS) (EU-15=100) [1, 2, 3]	25	27	27	48.6	100
Unemployment rate (%) [2, 4]	10.5	13.6	6.6	14.3	8.0

* 2002 figures

Source: [1] *Eurostat, 2004a;* [2] *European Commission, 2004a, b and c;* [3] *Eurostat, 2004c;* [4] *Eurostat, 2004b*

For years, the Turkish economy has experienced large fluctuations in national income and a very high level of inflation. Both elements make it difficult to observe long-term economic growth rates. Moreover, economic growth measured in a stable foreign currency depends very much on the exchange rate, which is influenced by political and fiscal uncertainties and resulting international money and capital flows (OECD, 2004a: 11). The following sections discuss these issues.

Instability of economic growth

Recent economic crises caused severe recessions in 1994, 1999 and 2001. An exchange market crisis in 1994 led to a decline in real GDP, triggered by a loss of confidence in economic policy and concerns over Turkey's ability to service external and internal debt (OECD, 1995:1). In 1999, a fragile global economy, political uncertainty and severe earthquakes led to a decrease in real GDP (OECD, 2001:21). The most severe recession occurred as a result of the banking and currency crisis in late 2000 and early 2001, which resulted in a growth rate

[1] The purchasing power standard (PPS) is an artificial currency that equalises the purchasing power between different currencies. Converting from a foreign currency to euros at the market exchange rate does not take into account that the market rate may not fully reflect the relative purchasing power of the two currencies. The PPP (purchasing power parity) rate, from a foreign currency into euros, is the ratio of the cost of a basket of selected goods and services in the foreign currency at local prices relative to the cost of the same basket in euros at representative EU prices. The PPP rate converts from a foreign currency into euro PPS.

[2] As Derviş *et al.* (2004) point out, the NMS figures show their position one year before accession, and in the case of AC-2 several years from accession, whereas Turkey's accession is much further away.

of –7.5 per cent in 2001 (OECD, 2002: 9). Growth of 7.8 per cent in 2002 brought the economy nearly back to its 2000 level.

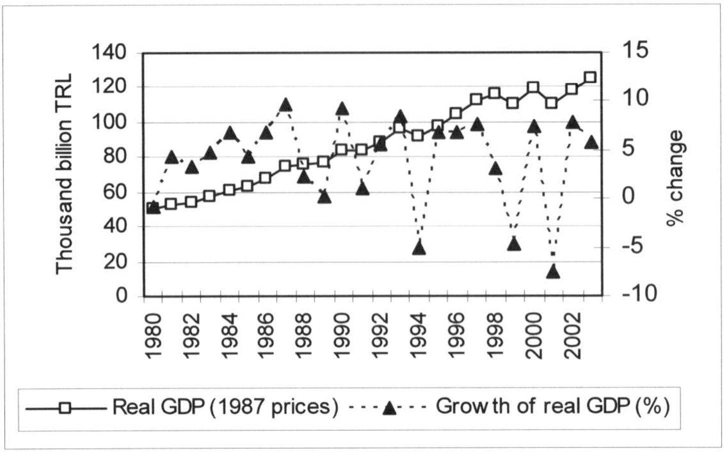

Figure 2.2. Real GDP and real GDP growth rate in Turkey, 1980-2003
Source: IMF, 2004

Figure 2.2 shows that, while real GDP more than doubled between 1980 and 2000, growth rates have been very erratic in the last decade. Following the 2001 crisis, real GDP was lower than its 1997 level, but returned to its pre-crisis level in 2002.

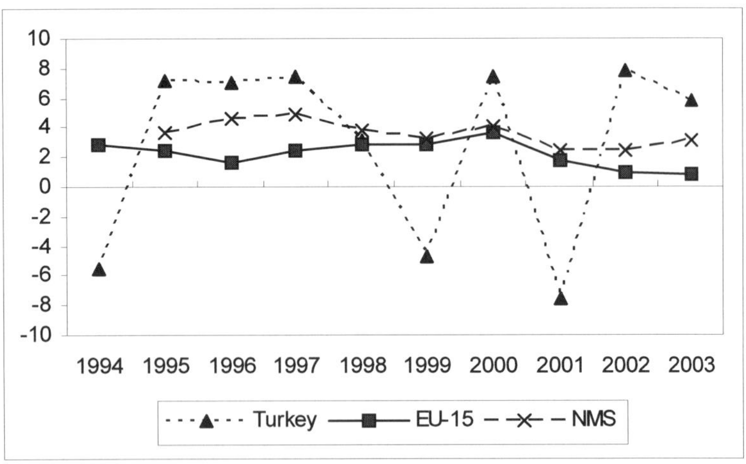

Figure 2.3. Real GDP growth rate, Turkey, EU-15 and NMS, percentage change from previous year
Source: Eurostat, 2004b; OECD, 2004b

Figure 2.3 shows the recent economic growth path of Turkey compared to NMS and EU-15 averages. Although Turkey frequently records much higher growth rates than in the NMS, this is offset by severe contractions of a magnitude not seen in the NMS or EU-15.

Table 2.4 compares average long-term growth rates for Turkey with the World Bank (WB) rates for low- and middle-income countries. Turkey's long-term growth of real GDP per capita is much lower than the growth of real GDP because of high population growth. The average growth rate of both total and per capita GDP was higher in the 1980s than the 1990s, due mainly to the two economic crises in the latter decade. Average growth in the current decade is lower than both previous decades. The period is, however, very short and influenced by the 2001 economic crisis.

Table 2.4. Growth rates of real GDP and real GDP per capita for Turkey, 1980 – 2003

	1980-89	1990-99	2000-03	1980-03
Real GDP per capita, Turkey	2.5	2.1	1.5	2.1
Real GDP, Turkey	4.2	3.9	3.4	3.9
Real GDP (average, low-income countries)	4.7	4.3		
Real GDP (average, middle-income countries)	2.9	3.2		

Source: IMF, 2004; World Bank, 2003

Table 2.4 shows that Turkey's growth performance was better than the average for middle-income countries (WB classifications), but worse than low-income countries. Furthermore, middle-income countries showed improved growth in the 1990s, while Turkey's average growth deteriorated.

Inflation

Price inflation in Turkey has been decreasing since 1994 but until 2003 it was consistently above levels imaginable within the European Union. Since the 1970s, many empirical studies have suggested that inflation has a negative impact on medium- and long-term growth. Inflation hinders the efficient allocation of resources by obscuring relative price changes, the most important signal for efficient economic decisions (Fischer, 1993)[3]. In empirical work on the effect of inflation on Turkey's economic growth, OECD (2001: 189) estimated that a 10 per cent increase in inflation leads to a 0.25 percentage point decline in the growth rate of per capita GDP. The results suggested that causality runs strongly, but not exclusively, from inflation to growth. OECD (2001) found little evidence that variation in inflation has affected economic growth in the case of Turkey.

Figure 2.4 shows the Consumer Price Index (CPI), the Wholesale Price Index (WPI) and the GDP deflator. For comparison, average inflation in the EU-15 for 2000-2003 was 2.1 per cent per year (Eurostat, 2004d). For short-term changes

[3] This literature is summarised in OECD (2001). There appear to be complex effects of inflation on output growth. Khan and Senhadji (2001) suggest thresholds of 1-3 per cent for industrialised countries and 11-12 per cent for developing countries. Below these thresholds, inflation has a weak positive effect on economic growth, while higher levels of inflation have a negative impact on economic growth.

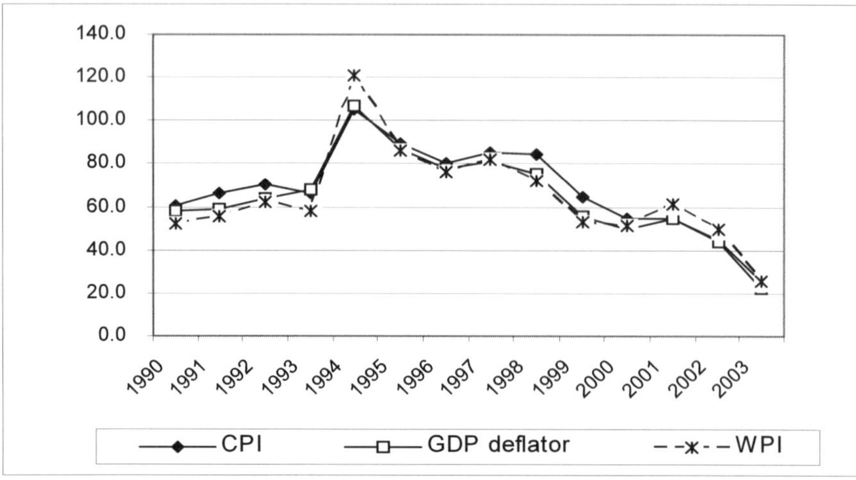

Figure 2.4. Inflation in Turkey, percentage change in the CPI, WPI and GNP Deflator
Source: OECD, 2004b; SPO, 2004a

during this period, the differences between the various inflation measures for Turkey are sometimes much larger than the annual economic growth rate, which indicates the sensitivity of the measured real economic growth rate to the price index selected. As of October 2004, annual WPI inflation was 15.5 per cent and CPI inflation 9.9 per cent. The expected annual CPI inflation for 2004 is 9.5 per cent (CBRT, 2004a).

In January 2005, a currency reform replaced Turkish lira with New Turkish lira, at the rate of 1 New Turkish lira (YTL) = 1 million Turkish lira (TRL).

Employment

Turkey's official unemployment level in 2003 was not very different from that of EU-15, and was lower than in both Bulgaria and the NMS (see Table 2.3). However, comparisons are probably deceptive because of the way unemployment is measured in Turkey. About 50 per cent of Turkey's economic activity is in the 'informal economy' (i.e. unregistered for tax and social security purposes) (OECD, 2004a: 8, 146), but this feature is thought to create only minimal bias in the unemployment statistics. Employment in rural areas and agriculture is explored in Chapter 5.

With almost no labour regulation in the informal economy, job insecurity is pervasive amongst unregistered workers. However, as wages in the informal sector are determined by market supply and demand, the informal labour market is flexible. Labour regulations are applied in the formal sector, and taxes and social security charges are paid, which makes the official labour market less flexible. The size of the informal sector has increased markedly in recent years, but this has also contributed to avoiding very high unemployment (Togan and Ersel, 2005).

The exchange rate

In the presence of rapid price inflation, the exchange rate may adjust gradually or via shocks. Both these phenomena have occurred in Turkey's recent past. From 1999 to 2001, the Turkish lira underwent a number of gradual adjustments. It was then floated on 22 February 2001 following the banking and currency crisis. It immediately fell by almost one-third, and ultimately almost two-thirds, against both the euro and US dollar (OECD, 2002: 9).

Figure 2.5. TRL/EUR exchange rate
Source: European Central Bank, 2004

Figure 2.5 illustrates the volatility of the Turkish lira/euro exchange rate. In 2003 the lira gained over 16 per cent in nominal terms against the US dollar, but remained steady against the euro. Atasoy and Saxena (2003) estimated the equilibrium real exchange rate for Turkey over the period 1980-2003. They found that the lira was overvalued prior to the 1994 and 2001 crises, but was close to its equilibrium level in 2003 despite claims of overvaluation. According to Togan and Ersel (2005), Turkey's exchange rate is still in disequilibrium because of a substantial current account deficit that is not sustainable.

Trade

In 2003 Turkey's visible trade deficit was EUR 18.6 billion, with exports valued at EUR 41.5 billion and imports at EUR 60.1 billion. Exports have been steadily increasing in real terms (Figure 2.6), while imports have followed a less regular pattern. Intermediate goods account for 70 per cent of imports, followed by capital goods (16 per cent) and consumption goods (11 per cent). Most intermediate goods are processed industrial inputs, and processed and unprocessed fuels and oils. Manufactured goods account for about 90 per cent of total merchandise exports (SPO, 2004a).

The EU-15 member states have a relatively constant share of Turkish exports (an average of 52 per cent over the last 5 years) and a decreasing share in Turkish imports over the same period (46 per cent in 2003). Trade flows between the NMS and AC-2 countries and Turkey have increased over the last few years, but remain relatively small. The US share of both exports and imports has been declining in recent years.

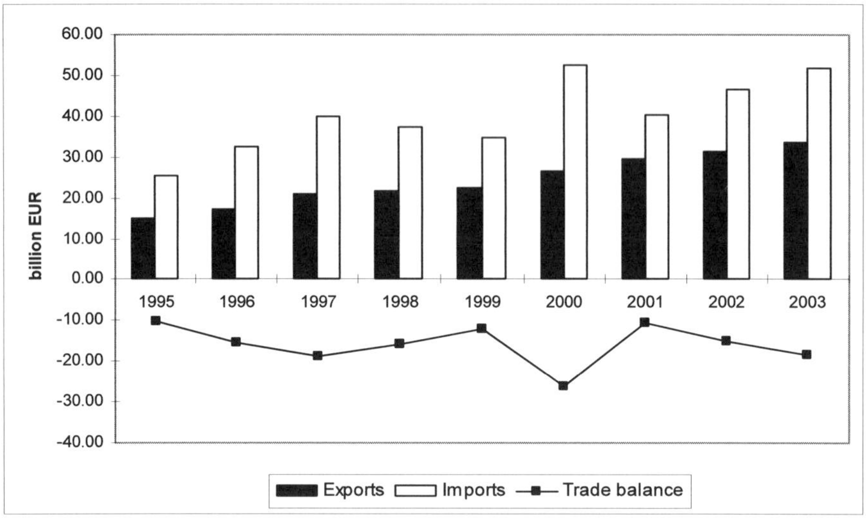

Figure 2.6. Total real exports, imports and trade balance for Turkey, EUR bn (1995 values)
Source: Eurostat, 2003 and 2004e

Current account and debt service

Table 2.5 provides an overview of the current account balance[4] in Turkey for selected years. Negative entries in the current account balance are net outflows. A current account deficit indicates that a country is using net capital inflows and/or foreign currency reserves to finance domestic consumption and investment. Turkey typically has a negative balance on goods and net income from abroad, and a positive balance on services and current transfers. The current account was in surplus in 2001 due mainly to lower imports of goods.

The current account deficit is shown as a percentage of GDP in Table 2.5[5]. The deficit was very high in 2000 preceding the 2001 crisis. Togan and Ersel (2005) suggest that the probability of a current account crisis in Turkey increases as the current account deficit relative to GDP increases above the 3.5 per cent level. Table 2.5 shows that the current account deficit was 4.5 per cent

[4] Calculated as the sum of individual balances on goods (exports minus imports), services (including tourism), income and current transfers.

[5] Current account data are provided in US dollars and GDP in Turkish lira. Both have been converted into euros before calculating the current account deficit as a percentage of GDP.

of GDP in 2000 and 3.7 per cent in 2003. It was still increasing at the beginning of 2004.

Table 2.5. Current account balance in Turkey, selected years, EUR million

	1990	1995	1999	2000	2001	2002	2003
Current Account	-2061	-1837	-1055	-7711	2662	-1195	-6208
A-Goods	-7419	-10328	-7998	-17244	-2931	-5719	-11002
B-Services	3900	7554	5879	8927	7170	6187	8249
C-Income	-1969	-2517	-2778	-3143	-3926	-3576	-4262
C32-Interest Expenditure	-2563	-2670	-3560	-3789	-4317	-3469	-3601
C321-Long Term	-2158	-2284	-2694	-2972	-3354	-3183	-3359
C322-Short Term	-405	-386	-865	-817	-963	-286	-242
D-Current Transfers	3428	3454	3841	3749	2350	1913	806
D1-Worker Remittances	2549	2613	3557	3581	2188	1520	572
Current account as % of GDP	-2.2	-1.8	-0.8	-4.5	2.1	-0.8	-3.7

Source: CBRT, 2004b; SPO, 2004a and b; own calculations

Not only the size but also the sources of financing are important for the sustainability of the current account deficit. The most sustainable source is foreign direct investment, but this has remained weak in Turkey (less than EUR 1 billion in 2003). To compensate, the current account deficit has been almost entirely financed by debt-creating capital inflows. Short-term funds such as portfolio investments and trade credits have played an important role (OECD, 2004a). Interest payments are a significant item in the Turkish current account. Table 2.5 shows that interest expenditure on short-term debt was particularly high over the period 1999-2001, but has since fallen. The importance of worker remittances in the current account has been declining.

Total foreign debt in Turkey reached nearly 62 per cent of GDP at the end of 2003, which reflects a significantly higher level of indebtedness than in most other emerging countries (OECD, 2004a).

2.4 Measuring the size of the economy and the standard of living

The size of the Turkish economy may be measured in terms of real or nominal GDP, denominated in Turkish lira or another currency. When expressing GDP in a foreign currency, amounts may be converted at market exchange rates or at rates reflecting standardised purchasing power (PPP rates). These choices depend on how the GDP measure is to be used, and they imply different interpretations. When comparing the size of two economies at a given point in time, the choice between nominal and real units is unimportant, whereas the comparison may be quite sensitive to the choice of exchange rate. When the objective is to compare average living standards between countries, and if market exchange rates are influenced by factors such as currency risk, lack of convertibility, political instability etc, the use of PPS is recommended. By contrast, when comparing levels of a country's GDP in its own currency at different points in time, or measuring the growth of its economy in GDP terms, the choice of deflator to remove the effect of price inflation may be crucial (see Section 2.3).

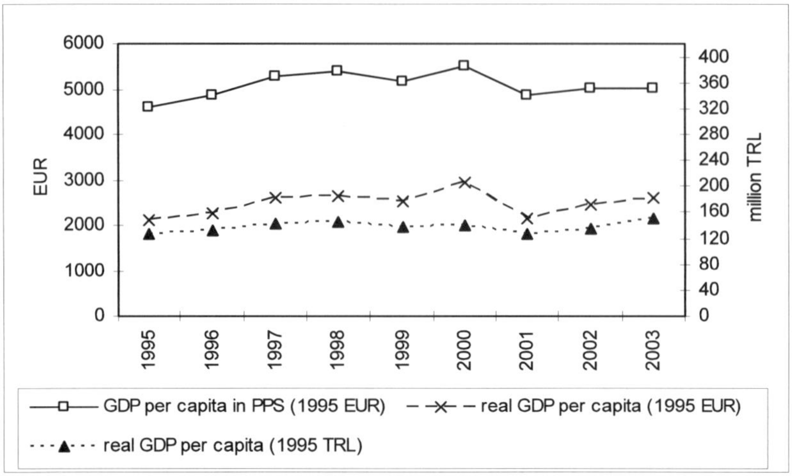

Figure 2.7. Trends in per capita GDP in Turkey, 1995-2003
Source: Eurostat, 2004a

Figure 2.7 shows GDP per capita denominated in real terms in euros, in PPS (euro inflation removed), and in real Turkish lira (1995 prices). The gap between the two series in euros has remained fairly stable since 1995. It widened slightly during 2001 as financial markets discounted the lira more heavily due to the economic crisis. The gap between the two series has narrowed since then.

Eurostat has recently improved its PPS series in order to remove a number of anomalies present in intertemporal comparisons (Eurostat, 2004f). The new series for Turkey shows a decline in its position relative to EU-25 over the last few years. In 1995, Turkey's GDP per capita in PPS was 30 per cent of the EU-25 level, while in 2003 this had decreased to 27 per cent. This is despite economic growth rates in domestic currencies that were on average higher than EU-25 rates over this period. This indicates that the gap in average living standards between Turkey and EU-25 has increased.

Turkey's future contribution to the EU budget would depend on its GDP measured in euros. Conversion to euros for this purpose would depend on market exchange rates. Clearly, if these rates are strongly affected by an unwillingness to hold Turkish lira (and not just by the lira's purchasing power in markets for goods and services), Turkey's budget contribution would be lower. The gap between per capita GDP in Turkey's regions and average EU per capita, measured in PPS, is relevant for eligibility for structural funds.

2.5 Income distribution in Turkey

This section provides an overview of the income distribution in Turkey. It first examines the distribution of income over households, and compares household income distribution over time, and between rural and urban areas. It then considers regional income distribution at both the NUTS I and NUTS II levels. Turkey is also compared to selected EU-15, NMS and AC-2 countries in terms of regional income disparities. The evaluation of the regional income disparities is based on the results of the latest Household Income Surveys carried out by Turkish State Institute of Statistics (SIS).

Developments in the household income distribution

Table 2.6 provides data on the household income distribution in Turkey over the last three decades. The surveys were conducted by different state departments, but the sample design did not change much between the surveys. Turkey's income distribution was slightly less unequal during the 1970s and 1980s.

Table 2.6. Household income distribution in Turkey

Percentage of households	SPO (1963)	SPO (1973)	SIS (1987)[6]	SIS (1994)[6]
1st 20%	4.5	3.5	5.2	4.9
2nd 20%	8.5	8.0	9.6	8.6
3rd 20%	11.5	12.5	14.1	12.6
4th 20%	18.5	19.5	21.1	19.0
5th 20%	57.0	56.5	49.9	54.9
Gini Coefficient[7]	0.55	0.51	0.43	0.49

Source: Yeldan, 2000: p.5

The total number of households has been divided into 5 equal groups of 20 per cent according to their incomes from the lowest to the highest. The first 20 per cent represents the poorest households. Over the past three decades, the richest 20 per cent of households received 50 per cent or more of total disposable income. Compared to other OECD countries in 1987 and 1994, Turkey had the highest Gini coefficient and the lowest GDP per capita (SPO, 2001b: 9).

Table 2.7 compares results from the 1987, 1994, 2002 and 2003 surveys and focuses on urban and rural income distribution. Income inequality increased significantly between 1987 and 1994 in Turkey in general and urban areas in particular. Successive economic crises, and migration flows from rural to urban areas, have helped to increase the inequality of income distribution particularly

[6] The Household Income and Consumption Expenditure Survey covers 26,400 households in 1987 and 26,256 households (18,264 households in urban and 7,992 households in rural areas) in 1994.

[7] The Gini coefficient is an inequality measure comparing the actual distribution with a hypothetical situation where everybody gets an equal share. The Gini coefficient varies from zero to one, zero representing perfect equality and one representing the maximum possible degree of inequality (one household receives everything).

in urban areas. As indicated by the Gini coefficients, income distribution and inequality in rural areas remained nearly constant, whereas income inequality significantly increased in urban areas. This reflects both the emergence of high-income groups and the higher rate of unemployment (low income groups) in urban areas. Average rural income per household was 24 per cent lower than urban in 1987. This difference increased to 42 per cent in 1994 (Cakmak, 1998:12).

Compared with the mid-1990s, the income distribution in urban areas, and in Turkey as a whole, has become less unequal, and closer to that in rural areas. Nonetheless, considerable income inequalities still remain.

Table 2.7. Comparison of income distribution surveys for 1987, 1994, 2002 and 2003

Percentage of Households	Turkey				Urban				Rural			
	1987	1994	2002	2003	1987	1994	2002	2003	1987	1994	2002	2003
1st 20%	5.2	4.9	5.3	6.0	5.5	4.8	5.5	6.1	5.2	5.6	5.2	6.4
2nd 20%	9.6	8.6	9.8	10.3	9.3	8.2	9.7	10.3	10.0	10.1	10.3	11.0
3rd 20%	14.1	12.6	14.0	14.5	13.6	11.9	13.9	14.5	15.0	14.8	14.7	15.0
4th 20%	21.2	19.0	20.8	20.9	20.8	17.9	20.5	20.8	22.0	21.8	21.7	21.2
5th 20%	49.9	54.9	50.1	48.3	50.9	57.2	50.4	48.3	47.9	47.7	48.0	46.3
Gini coefficient	0.43	0.49	0.44	0.42	0.44	0.51	0.44	0.42	0.42	0.41	0.42	0.39

Source: Yeldan, 2000: p.6; SIS, 2004; SPO, 2001b

According to Gini coefficient comparisons, Turkey's income inequality is high relative to EU countries and this inequality has remained relatively stable over the years (World Bank, 2000). Inequality in Turkey stems from differences in endowments (land, labour and capital), opportunities faced in the labour market, education and employment status (World Bank, 2000:4).

Regional income distribution

More than 10 per cent of total household income inequality in Turkey is explained by rural-urban differences. Regional factors account for another 11 per cent (World Bank, 2000:5). According to the UNDP report, the share of overall inequality explained by differences in regional means grew by 10 per cent over the period 1975-1995 (Akder, 2001: 25).

Turkey demonstrates high regional income disparities when compared to selected large EU-15, NMS and AC-2 countries. The average ratio of GDP per capita between the richest and the poorest NUTS II regions averages 2.4 in these countries whereas it is 5.62 in Turkey, as shown in Table 2.8.

Regional income inequalities arise mainly because of differences in types of economic activities pursued, together with differences in productivity between sectors. For instance, poorer regions generally have a bigger share of their resources employed in agriculture, where productivity is usually lower (World Bank, 2000:12). It is clear that improving the productivity of agriculture would improve the situation of underdeveloped regions.

Table 2.8. Regional disparities in terms of GDP per capita ratio at NUTS II level, 2001

Countries	Region with highest GDP per capita (EUR)	Region with lowest GDP per capita (EUR)	Disparity Ratio
Germany	42,800	15,072	2.83
Italy	30,131	13,047	2.31
Poland	8,332	3,746	2.22
Romania	4,232	1,590	2.66
Spain	21,649	10,303	2.10
Turkey	4,588	815	5.62

Source: Eurostat, 2004a; SIS, 2003b; own calculations

In Turkey, productivity differences between provinces are growing over time (World Bank, 2000:11), and regional inequality at province level has been increasing. As a result, Turkish provinces are diverging: richer provinces (mainly those from Marmara or historically important port cities around the Aegean and the Mediterranean coast) are converging towards each other and getting richer, while poor provinces are falling further behind. The main driving forces are fast structural change as a result of migration from rural to urban areas, and the increasing capital use mainly in western Turkey (SPO, 2000: 23-24).

Nearly all Turkey's development plans have given priority to reducing regional development disparities and implementing effective regional programmes.[8] Underdeveloped regions have always received priority in public investment distribution plans and programmes. In this regard, the main targeted aims have been to increase the welfare level of priority regions for development to the level of the national average, to reduce urban-rural development differences generally, to promote sustainable development, and to exploit more effectively the local resources (SPO, 2000: 25). These aims have not been fulfilled (SPO, 2001a: 73). Except for some areas near main cities and some areas in the GAP region, the aim of creating development centres within the priority regions for development was not realised. Similarly, coordination of investment plans has not been achieved (SPO, 2000:31). The main reasons are the large number of rural settlements and their scattered and fragmented structure, difficulties in terms of geographical structure, climatic conditions, inefficient policy-making mechanisms and insufficient delivery of physical and social infrastructure.[9]

Among the 12 regions defined at NUTS I level in Turkey, the East Marmara, Istanbul, Aegean and West Marmara regions generally have the highest average levels of per capita GDP whereas the lowest levels are found in Northeast Anatolia, Central Anatolia and the Southeast Anatolia regions. Moreover, the trends seem to be unchanged over time.

---

[8] Turkey has been carrying out its social, economic and cultural development through Five-Year Development Plans since the beginning of the 1960s (SPO, 2003:51).

[9] As of 2002, there are 78,625 settlements in total in rural areas, 36,527 of which are villages and 42,098 of which are sub-village settlements.

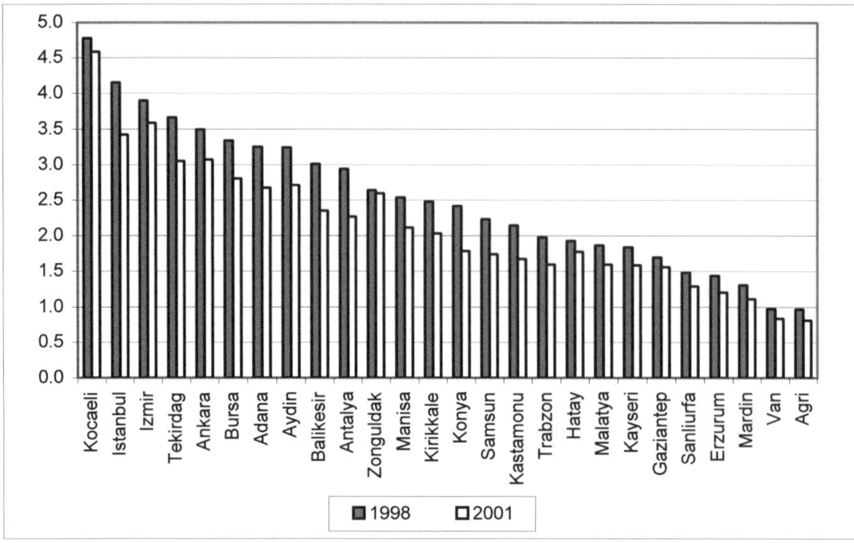

Figure 2.8. GDP per capita in NUTS II regions (nominal thousand EUR), 1998 and 2001
Source: SIS, 1999; SIS, 2003b

Income distribution *within* each region at NUTS level I is also very unequal. As already mentioned, mainly urban regions have higher Gini coefficients (the highest is Istanbul, with 0.43) and mainly low-income rural areas have lower inequality, the lowest being the East Black Sea, with 0.35 (SIS, 2004). Increasing population in urban areas also reduces access to public services and employment opportunities and worsens living conditions (SPO, 2001b: 65).

Regional income disparities are more visible at NUTS II level[10]. While GDP per capita decreased in 2001 due to the economic crises that hit Turkey in the same year, the GDP per capita ratio between the richest region (Kocaeli) and the poorest region (Agri) increased from 4.9 in 1998 to 5.6 in 2001. In 1998, the average GDP per capita was almost 80 per cent lower in Agri than in Kocaeli. This difference increased to 81 per cent in 2000 and 82 per cent in 2001 (see Figure 2.8).

GDP per capita in the 12 priority development NUTS II regions is almost half the country average. Similarly, according to the 2003 survey of rankings of provinces and regions by socio-economic development levels, these 12 NUTS II regions (except Gaziantep, Kayseri, Konya) ranked below the averages for

[10] 12 of the 26 NUTS II regions have been given priority for development (Kastamonu, Samsun, Trabzon, Erzurum, Agri, Malatya, Van, Gaziantep, Sanliurfa, Mardin, Konya and Kayseri). Except for Gaziantep, Konya and Kayseri provinces, these NUTS II regions are also within the scope of Priority Regions for Development. 'First Degree Priority Regions for Development' were first identified in the Third Five-Year Development Plan. First Degree Priority Regions for Development cover 49 provinces and 2 administrative districts. The population of these regions represents 36 per cent of the total population (SPO, 2003:38).

Turkey[11]. The results of this survey further support the conclusion that the five most developed NUTS II regions of the country are Istanbul, Ankara, Izmir, Bursa and Kocaeli (SPO, 2003:42).

Among the 26 NUTS II regions, Ankara has the highest level of employment in the service sector and Istanbul has the highest rate of employment in the industrial sector (SPO, 2003:43). In terms of industrial development, Istanbul, Kocaeli, Tekirdağ, and Bursa regions are the most developed regions of the country.

By contrast, the 12 priority development NUTS II regions rank highest in terms of agricultural employment. Among these 12 NUTS II regions, Konya, Kayseri and Gaziantep are more developed since they also have established industry and service sectors. These 12 NUTS II regions also have low literacy rates. The three regions with the lowest literacy rates (Sanliurfa, Van, Mardin) register below 70 per cent. Moreover, in these regions the literacy rate for the female population is also below 60 per cent (SPO, 2003:43).

2.6 Summary

The Turkish economy is around half the size of the combined NMS, but the Turkish population is only slightly smaller. Therefore, GDP per capita is much lower in Turkey than in the NMS, and at a similar level to the AC-2. Turkey has had a relatively low average growth rate in recent years, with wide fluctuations and a number of severe recessions. Inflation is still much higher than in both EU-15 and the NMS, although it has fallen in the last few years. Official unemployment rates are similar or lower than rates in the NMS and AC-2 countries.

Average long-term growth rates in Turkey have been heavily influenced by recent economic crises. This resulted in a lower average per capita GDP growth in the 1990s compared to the 1980s. The current account deficit reached similar levels in 2003 to those seen prior to the 2001 crisis, at around 3.7 per cent of GDP. Most of the current account deficit is financed by debt-creating capital inflows, particularly since foreign direct investment inflows are low. Interest expenditure on external debt is a significant item in the current account balance. Worker remittances have decreased in importance. Turkey has a high level of foreign debt, around 62 per cent of GDP by the end of 2003.

Turkey's per capita GDP measured in euros doubles when exchange rates that are standardised for purchasing power differences are used. However, using these same exchange rates, Turkey's average per capita GDP appears to have worsened in recent years relative to that of the EU-15, falling from 30 per cent of the EU-15 level in 1995 to 27 per cent in 2003. Therefore, the gap in average living standards has increased.

Turkey has a relatively unequal income distribution and the inequality persists over time. Inter-household disparities due to regional location, sector of

---

[11] The survey indicates the development performances of the provinces by using the composite index of variables, which are classified as demography, employment, education, healthcare, industry, agriculture, construction, financial and other welfare indicators.

employment and productivity level, educational attainment and employment status largely explain the unequal income distribution. Inequality is high between urban and rural areas, and is more pronounced within the urban population than in the rural population, although the average income level of rural groups is lower. Regional income distribution is also very unequal. It is clear that the regional income differences in Turkey are largely determined by urban-rural and west-east differences, which are due to the relative importance of agriculture in the region as well as the relative productivity of each region's agriculture. More generally, inequality in economic development is increasing over time when considering indicators such as literacy rate, employment and GDP per capita. In a number of regions, agricultural productivity, industrial potential and investment are lagging far behind due to a lack of physical and social infrastructure. Comparison of GDP per capita at NUTS II levels with several current and future EU members shows that the inequality of GDP per capita across regions is much higher in Turkey. This has significant implications for structural policy in an enlarged EU.

References

Akder H. (2001) 2001 national human development report for Turkey. United Nations Development Programme, New York.

Atasoy, D. and Saxena, S.C. (2003) Misaligned? Overvalued? The untold story of the Turkish liras. University of Pittsburgh Working Paper. University of Pittsburgh, Pittsburgh.

Cakmak, E.H. (1998) Agricultural policy reforms and rural development in Turkey. Economic Development and Poverty Reduction Workshop, *Mediterranean Development Forum*, 3-6 September, 1998.

CBRT (2004a) October inflation and outlook. Central Bank of the Republic of Turkey press release, 08-11-04. CBRT, Ankara.

CBRT (2004b) Electronic Data Delivery System. CBRT, Ankara. Electronic.

CIA (2004) *The World Factbook*. Central Intelligence Agency, Washington DC.

Derviş, K., Gros, D., Öztrak, F., Bayer, F. and Işik, Y. (2004) Relative income growth and convergence. *EU-Turkey Working Papers*, No. 8, September 2004. Centre for European Policy Studies.

European Central Bank (2004) Euro foreign exchange reference rates. http://www.ecb.int/stats/exchange/eurofxref/html/index.en.html (accessed 20 August, 2004).

European Commission (2004a) 2004 Regular report on Turkey's progress towards accession. SEC (2004) 1201, Brussels.

European Commission (2004b) 2004 Regular report on Romania's progress towards accession. SEC(2004) 1200, Brussels.

European Commission (2004c) 2004 Regular report on Bulgaria's progress towards accession. SEC(2004) 1199, Brussels.

Eurostat (2003) *External and Intra-European Union Trade – Statistical Yearbook*. Office for Official Publications of the European Communities, Luxembourg.

Eurostat (2004a) Eurostat Database. Eurostat, Luxembourg.

Eurostat (2004b) An overview of the economies of the new member states. *Statistics in Focus*, Theme 2 – 17/2004. Eurostat, Luxembourg.

Eurostat (2004c) Quarterly accounts: Second quarter 2003, the GDP of the candidate countries. *Statistics in Focus*, Theme 2 – 2/2004. Eurostat, Luxembourg.

Eurostat (2004d) *EC Economic Data Pocket Book, Quarterly*. Number 1/2004. Office for Official Publications of the European Communities, Luxembourg.

Eurostat (2004e) *External and intra-European Union trade, monthly statistics*. Number 7, 2004. Office for Official Publications of the European Communities, Luxembourg.

Eurostat (2004f) Purchasing power parities and related economic indicators for EU, candidate countries and EFTA. *Statistics in Focus*, 37/2004.

Fischer, S. (1993) The role of macroeconomic factors in growth. *Journal of Monetary Economics* 32, 485-512.

IMF (2004) World Economic Outlook Database. IMF, Washington DC.

Khan, M.S. and Senhadji, A.S. (2001) Threshold effects in the relationship between inflation and growth. *IMF Staff Papers*, 48(1). IMF, Washington DC.

OECD (1995) *OECD Economic surveys: Turkey 1994-1995*. OECD, Paris.

OECD (2001) *OECD Economic surveys: Turkey 2000-2001*. OECD, Paris.

OECD (2002) *OECD Economic surveys: Turkey 2001-2002*. OECD, Paris.

OECD (2004a) *OECD Economic surveys: Turkey 2004*. OECD, Paris.

OECD (2004b) *OECD Economic Outlook*, No 75, statistical annex tables. OECD, Paris.

SIS (1999) Data for GDP per capita in 1998 at NUTS II. Gross Domestic Product Bulletins, 14 December 1999. SIS, Ankara.

SIS (2003a) Sayilarla Turkiye. SIS, Ankara.

SIS (2003b) Data for GDP per capita in 2001 at NUTS II. Gross Domestic Product Bulletins, 16 May 2003. SIS, Ankara.

SIS (2004) Results of the 2003 Household Budget Survey, 06-10-2004. SIS, Ankara (*in Turkish*).

SPO (2000) The eighth five-year development plan: Regional development expert committee report. State Planning Organisation, Ankara (*in Turkish*).

SPO (2001a) Long term strategy eighth five-year development plan 2001-2005. State Planning Organisation, Ankara.

SPO (2001b) Ameliorating the income distribution and struggle with poverty. Expert Commission Report, Ankara (*in Turkish*).

SPO (2003) Preliminary national development plan (2004-2006). State Planning Organisation, Ankara.

SPO (2004a) Economic and social indicators 1950-2003. State Planning Organisation, Ankara.

SPO (2004b) Main economic indicators. State Planning Organisation, Ankara.

Togan, S. and Ersel, H. (2005) Macroeconomic policies for EU accession. In: Togan, S. and Hoekman, B. (eds.) *Turkey: Economic Reform and Accession to the European Union*. World Bank and Oxford University Press (in press).

Undersecretariat of Foreign Trade (undated) Turkey's Foreign Trade 1990-2003. http://www.byegm.gov.tr/references/turkeys-foreign-trade.htm (accessed 15 November 2004).

World Bank (2000) Turkey: Economic reforms, living standards and social welfare study. Report no. 20029-TU. World Bank, Washington DC.

World Bank (2003) *World Development Indicators 2003*. CD-ROM. World Bank, Washington DC. http://www.worldbank.org/data/countrydata/countrydata.html (accessed 15 October 2004).

World Bank (2004) Country at a glance tables.

Yeldan, E. (2000) The impact of financial liberalisation and the rise of financial rents on income inequality, the case of Turkey. United Nations University WIDER Working papers, No. 206, Finland.

3

The Institutional Framework of Turkey and Turkish Agriculture

Tugrul Temel

3.1 Introduction

This chapter describes the current institutional framework of Turkish agriculture, beginning with a brief account of the key historical developments that have influenced this framework. Specifically, it covers institutional developments that have implications for the alignment of Turkish agricultural institutions with those of the European Union (EU), with the aim of providing insights relevant to Turkey's possible accession to the EU and to expectations of future developments.

Institutions are humanly devised constraints that shape human interaction, and that provide the structure of formal and informal economic, political and social behavioural rules (North, 1990). Three basic concepts are used in describing institutional developments: institutional environment, institutional arrangements and organisations. Institutional environment refers to the formal and informal values and basic ground rules of a society, such as traditions, norms and religion. From an economic perspective, the institutional environment includes political, social and legal rules that form the basis for production and market exchange, such as the rule of law, private ownership, the enforceability of contracts and so on (Davis and North, 1971). Taking shape within the given institutional environment, institutional arrangements refer to formal and informal rules of conduct for specific types of interaction. Lastly, organisations are defined as formal and informal entities that rule the governance process.

Four general remarks provide insight into the dynamics of institutional change, and explain the relevance of this assessment of Turkey's institutional setting. First, formal institutions are easier to create or modify than informal ones, for the same reason that it is easier to change specific institutional arrangements than to modify the institutional environment: the latter requires changes in the social fabric that usually take a long time. Second, changes in formal institutions are easier to implement if they conform with the general informal rules of the society. Third, there are no guidelines or explicit procedures for changing or modifying informal rules. The fourth remark is that the same institutions, formal or informal, might yield different economic, social or political outcomes in different countries. These remarks suggest, first, that one of the greatest challenges Turkey will face if it becomes an EU member concerns the adaptation and re-orientation of its institutions, and second, that it may be unrealistic to expect formal institutions in Turkey to perform as effectively and efficiently as they do in the EU for some years after accession.

This chapter is organised in four sections. After this introduction, Section 3.2 provides a brief history of the development of the institutional environment of modern Turkey. It describes formal and informal rules relating to political, social and economic development, competition, property rights, and education and science. Section 3.3 presents the institutional arrangements in three areas of Turkish agriculture: agricultural resources; agricultural research, technology and innovation; and agricultural production, markets and trade. Section 3.4 concludes the chapter.

3.2 Institutional Environment

Political, social and economic ground rules

History and ideology of Atatürk's reforms

The Ottoman Empire was founded in northwestern Turkey at the end of the thirteenth century. From 1517 onwards, the Ottoman Sultan was also the Caliph of Islam, and the Ottoman Empire was synonymous with the *Khilafa* (the Islamic State). The Turkish elite overthrew Ottoman rule, following victory in the Liberation War (1919-1922) against British, French, Italian and Greek forces. The Republic of Turkey was declared in 1923, and adopted parliamentary democracy in which religion was separated from all state, educational and legal affairs. The ultimate goal was to establish a nation state organised around the Kemalist principles[1] of republicanism, nationalism, populism, reformism, etatism and secularism.

The Kemalist principles were regarded as 'fundamental and unchanging', and were written into the Constitution of the Republic of Turkey in 1937. The principle of republicanism was contained in the constitutional declaration that sovereignty is vested in the nation. Populism included the notion that all Turkish citizens are equal. Accordingly, the *millet* system[2], which had provided communal autonomy to other ethnic and religious groups, was abolished. Reformism legitimised the radical means by which changes in Turkish political and social life were implemented. Etatism emphasised the central role of the state in directing economic activities. This concept was cited particularly to

---

[1] Named after Mustafa Kemal Atatürk (1881-1938), the founder of the Turkish Republic and its first President (1923-1938).

[2] The Ottoman Empire divided subject peoples into the domain of the faithful, the Muslims, and the domain of war, the non-Muslims. An individual's obligations and rights were determined by his membership of one of these groups. The non-Muslim community was divided into *millets*, administrative units organised on the basis of religious affiliation rather than ethnic origin. The four non-Muslim *millets* in the Ottoman state were Armenian, Catholic and Orthodox Christians, and Jews.

justify state planning and large-scale investment in state-owned enterprises. Atatürk's economic policy aimed to prevent foreign interests from exercising undue influence on the economy. Secularism was also included in the constitutional declaration as an unchanging principle of the Republic. The Islamic *Sufi* orders were suppressed, religious schools were closed and public education was secularised. The reforms summarised in Table 3.1 were initiated in 1923-1935 to support the process of change towards a modern westernised state.

Table 3.1. Major reforms during 1923-1935

Year	Reform
1923	Republic of Turkey with capital at Ankara proclaimed (October 29).
1924	Caliphate abolished. Traditional religious schools closed, *seriat*[3] abolished. Constitution adopted.
1925	Fez outlawed by the Hat Law. Veiling of women discouraged; Western clothing for men and women encouraged. Western calendar adopted.
1926	New civil, commercial and penal codes based on European models adopted. New civil code ended Islamic polygamy and divorce by renunciation and introduced civil marriage. *Millet* system ended.
1928	New Turkish alphabet (modified Latin) adopted. State declared secular. Constitutional provision establishing Islam as official religion deleted.
1933	Islamic call to worship and public readings of the Kuran required to be in Turkish.
1934	Women given the vote and the right to hold office.
1935	State role in managing the economy written into the constitution.

Source: Sansal, undated

Impacts of Atatürk's reforms: political, social and economic

Turkish society has evolved both as a consequence of and a response to the major socioeconomic changes guided by the Kemalist ideology. In 1923, Atatürk founded the Republican People's Party as a vanguard party for the new regime. The single-party period effectively ended in 1946 with the participation of the newly established Democrat Party in the national elections. Immediately after its election victory in 1950, the Democrat Party launched policies to promote private enterprise and foreign investment. However, state investment and involvement remained and even increased.

Traditionally, the military leadership has believed that the army should stay out of politics, with a condition that a major role of the army was to act as guardian of the constitution. Observant of the government's impotence against the rise and increasing popularity of anti-Kemalist and anti-secular ideas by the late 1950s, the military command concluded that the government had departed from Kemalist principles and that the Republic was in imminent danger of disintegration. With the military intervention of May 27, 1960, the Kemalist principles and the role of the state in economic affairs were reinstated. The basic principles were once more confirmed in the newly prepared constitution, which further extended civic

[3] *seriat*: legal system founded on the Islamic religion.

rights and democratic mechanisms for improving social dialogue.

In a memorandum issued on 12 March, 1971, the army demanded a strong government in order to curb violence and implement economic and social reforms, including the land reform stipulated by the 1961 constitution. The Prime Minister resigned and was replaced immediately after the army's warning. Almost 10 years later, continuously worsening economic conditions, rising religious movements and armed clashes between left and right movements led to another military intervention on 12 September, 1980. Whereas the 1960 and 1971 interventions targeted institutional reforms, the 1980 intervention aimed simply to restore the order created by the earlier interventions.

The key lesson that Turkish society has drawn from these successive military interventions is that "the army saves the Republic when things go wrong". This expectation, still held by the large majority of the population, reflects strong trust in the army. In the 1990s, the army capitalised on this trust by repeated public warnings to incumbent governments, thereby mobilising civil society organisations around the Kemalist principle of secularism. Anti-secular groups were curbed and a new government was installed shortly after the military-backed political campaign started. This indirect intervention served as a socially and economically less costly means for reinstating the order, while at the same time informing civil society of the change in the rules of conduct. It was the first time that the military relied on the support of civil society organisations.

On the social front, changes have been rather slow despite radical reforms launched in the early years of the Republic. The sympathy for ruling civilian and military elites that impose policies and use force to meet popular resistance has not changed much. Although the reforms aimed to create the pattern of social relations based on secular and economic values, the vast majority of the rural population still subscribes to traditional patriarchal family values and observance of the religious ethic. Again, although women have received unprecedented legal rights, their status still remains a complex issue, largely due to the traditional value system that works against the schooling and labour participation of women.

On the economic front, the Izmir Economics Congress of 1923 suggested a protectionist development strategy, with cautious promotion of the private sector, and recognised the critical role of agriculture in economic development. In 1925, a heavy tax-in-kind on peasants (the tithe) was removed. The Law on Industrial Promotion (1927) led to the growth of the sugar, cotton, flour milling, coal, iron and gasoline industries. Sumerbank and Etibank were established in the 1930s to meet the credit demands of the private sector.

By 1923, landownership was concentrated within a small group with large holdings. The Law on Land Reform was adopted in 1945. The opening of new areas to cultivation also made land available to farmers without holdings. Although land reform was enacted relatively early in the Republican period, implementation was delayed for several reasons. First, large landowners effectively blocked most action and the government often lacked the will to implement the reform (US Library of Congress, 1995). Second, mainly in the Kurdish-populated areas of the east and south-east, the reform was not compatible with local informal relationships. The underlying structure of these relationships was that feudal-style landlords, who owned the land, also controlled everything in the villages concerned, thereby providing a form of social safety net (Pamuk, 1982, 1985).

In the process of extending land ownership, peasants had to pay for the land they obtained over a 20-year period (Ataturk ve Turkiye Cumhuriyeti, undated). The state-owned Agricultural Bank of Turkey (Ziraat Bank) supported small peasants by providing low-cost credit, although merchants rather than peasants were the main beneficiaries of the Bank's lending until the 1950s. Accompanying this process, peasant cooperatives, agricultural credit cooperatives, agricultural colleges and agricultural institutes were organised. Experimental agricultural stations and agricultural parks were established to introduce new seeds and new agricultural products to farmers.

Evolution of interest groups and formal institutions

Interest group activities in Turkey gained momentum in the 1950s. By that time, the military had internalised a view of itself as the only national body responsible for safeguarding the Kemalist principles, which defined a broad area for internal military intervention. Even today, with various law-and-order issues still viewed as falling within the realm of national security, the army exerts pressure on elected governments.

Labour unions emerged with the Trade Union Law of 1947. In 1963 a further law legalised strikes, lockouts, and collective bargaining. But all labour unions (except for Turk-is, which had remained outside the political domain) were banned immediately after the 1980 coup. The Progressive Labour Union (DISK), the second largest union at the time, started its activities again only in 1991.

In 1952, merchants' and industrialists' associations started to grow, including the Turkish Exporters' Association (TTA), the Union of Chambers of Commerce and Industry (TOBB), the Turkish Confederation of Employers' Unions (TISK) and the Turkish Industrialists' and Businessmen's Association (TÜSİAD). Among these, only the TÜSİAD has been concerned with the widening economic inequalities between regions and social classes, which it perceived as jeopardising Turkey's chances of entering the EU.

Religious movements also grew steadily, particularly after 1970. With increasing economic problems, religious sentiments gained strength especially in rural areas and lower-class urban neighbourhoods. The 1980 coup only temporarily interrupted the trend toward increased religious observance but without interfering with the informally organised *sufi* religious groups (*tarikat*) and other voluntary religious associations. In fact, with an article in the 1982 constitution, the army introduced compulsory religious instruction in all schools. After the 1980 coup, the military government regained state control over another interest group, the universities, by establishing the Council of Higher Education. During the 1980s, universities were strictly controlled by the Council, but this control gradually eroded in the 1990s.

The ideological and social forces summarised above underlie the complex political balance in Turkey. They shape the background against which Atatürk's legacy has been incorporated and transformed into political, economic and social rules, which have been developing throughout the rest of the twentieth century. The main developments over that period are summarised in Table 3.2.

Table 3.2. A chronology of policy and institutional developments in Turkey: 1923-present

Period/ motivation	Policies	Institutional environment
1923-1945 Catching up with the West	▪ Closed economy ▪ Import substitution policy ▪ Agriculture for food self-sufficiency	Strong nationalism, etatism and secularism; single party politics; strong military; advanced women's rights and gender equality; limited social dialogue; strong property rights and law enforcement (tangible); moderate state aid to initiate the private sector; strong industrial research-education and supporting institutions; strong agricultural education-extension and supporting institutions; moderate land tenure institutions but weak enforcement.
1946-1960s Collaborating with the West, adopting Western institutions (participation & democratisation)	▪ Partially open economy ▪ Import substitution policy ▪ Agriculture supports industry	Moderate nationalism; intermittent etatism; moderate secularism; strong military control; multi-party politics; strong property rights and law enforcement (tangible); moderate state aid to the private sector; enhanced institutions for industrial and agricultural development; strong social dialogue; improved research and education; weak land institutions.
1970-1980 Establishing social stability by modifying the institutions adopted earlier	▪ Partially open economy ▪ Import substitution policy	Weak nationalism and etatism; moderate secularism; strong military control; multi-party politics; strong property rights and enforcement; moderate state aid to the private sector; poor institutions for industrial and agricultural development; weak social dialogue; weak land institutions and enforcement; preparations for a competition framework started.
1980-present Integrating with and competing in international markets	▪ Open economy (except agriculture) ▪ Export promotion policy ▪ Customs Union completed, process of EU application speeded up	Moderate privatisation, competition and secularism; strong military control; multi-party politics; strong property rights and law enforcement (tangible); moderate state aid to the private sector; improved institutions for industry; subsidised agriculture with old institutions; limited social dialogue; moderate national innovation policy and institutions; establishment of Scientific and Technical Research Council; more religious content in primary education; IPR institutions in place but weak enforcement; Act on the Protection of Competition adopted in 1994.

Competition rules

In the early 1980s, Turkey embarked on structural adjustment policies at the instigation of the International Monetary Fund (IMF). With these policies, the economic role of the state was substantially restricted and markets were recognised as the means for determining prices. Over the past 25 years, various laws and regulations have been adopted to support and promote competition.

However, success in this area depends not only on appropriate legislation and administrative capacity, but also on how effectively the laws are enforced. The following paragraphs describe the extent to which enforcement has followed.

Competition law and implementing agencies

The alignment of Turkish legislation with that of the EU started in 1994 with the adoption of Law 4054 on the Protection of Competition. The Law applies equally to all economic activities and all firms (domestic and foreign) that influence market competition. At present, the Competition Board and Competition Authority are responsible for implementing the Law. Appointed in early 1997, the Board exercises the Competition Act's decision-making powers. Its members are appointed by the Council of Ministers from a pool of candidates nominated by several ministries and stakeholders (Turkish Competition Authority, 2003). The Competition Authority, a legally separate entity from the government, became operative in late 1997.

Law 4054 is compatible with EU competition rules in some areas, while in others it lags behind. It does not yet include a clause like Article 86 of the EC Treaty, which explicitly brings public undertakings within the scope of the Law. This gap is especially important as state-owned or state-controlled enterprises still play a role in certain sectors of the Turkish economy, and they will continue to do so until privatisation and liberalisation policies are effectively implemented. On the positive side, Article 6 of the Law prohibits the abuse of dominant power in markets, which is almost identical to Article 82 of the EC Treaty. Article 7 prohibits mergers and acquisitions aimed at creating a dominant position and inhibiting competition in the markets for goods and services (Mumcu and Zenginobuz, 2001).

In the area of state aid, 2003 saw some new action. The Competition Authority prepared a draft law on the monitoring and supervision of state aids, and a new autonomous institution has been foreseen for the monitoring and supervision task (Turkish Competition Authority, 2003). It is important to note that for such aids to be compatible with the Community *acquis*, they need to follow the regional aid guidelines of the EU (European Commission, 2004).

Enforcement

The Board began concrete enforcement of Law 4054 in 2002. Six fertiliser production and distribution firms were fined on the basis of Article 4 of the Law, as they engaged in price-fixing agreements and created difficulties for other importers via a concerted practice (Turkish Competition Authority, 2003). In general, courts in

Turkey function slowly due to lengthy processes. The parties involved usually try first to resolve conflicts by themselves without the courts' involvement. With respect to enforcement of the Competition Authority's decisions, the situation is unlikely to be any different, especially as the Authority has been in place for only seven years and the courts require judges specialised in competition rules.

The domestic markets for cigarettes, tobacco products and alcoholic beverages were opened to competition first from imported brands over 10 years ago. Then, joint ventures between Turkish and foreign companies (such as Phillip Morris–Sabanci) started producing a few brands domestically. In addition, arrangements are under way for the privatisation of the Directorate General for Tobacco and Tobacco Products, Salt and Alcohol Industry (TEKEL). A new law on the sugar market has begun reform in that market. Starting from the 2002-03 season, the government no longer sets the sugar price. But the market will still be regulated by the Competition Board (Turkish Competition Authority, 2003).

Cultural attitudes to competition

Competitive behaviour is not yet embedded in social life, and cultural events and artisan groups initiated in the early years of the Republic still support non-competitive behaviour to some degree. Competitive attitudes are not actively encouraged in public schools. Traditional culture is supportive of solidarity-based community businesses especially in rural areas. Some traditional entities such as Ahi Ervan (a form of artisans' association active mostly in rural areas) are still in support of collaborative rather than competitive business (Karaman, undated).

Property rights

The 1982 Constitution gives everyone the right to own and inherit property (Article 35), and freedom to work and contract (Article 748). The Property Law recognises private ownership as an essential principle (Article 618 of the Civil Code). In the Land Title Law (2644) of 1934, private ownership of real estate was extended to foreigners on the basis of reciprocity (Article 35) (National Assembly of the Republic of Turkey, undated). The 'reciprocity principle' states that a foreigner can become the owner of real estate in Turkey if a Turkish citizen has the same right in the foreigner's country of origin. However, Article 87 of the Village Act denies foreign legal and natural persons the right to ownership of property outside a village centre. Another restriction relates to the Act regarding Military Prohibited Areas and the Security Areas.

In 1995, Turkey started to build an intellectual and industrial property rights system. The steps taken by Turkey in the previous decade in the field of intellectual and industrial property rights are embodied in various laws, decree laws and regulations. Effective enforcement of these laws and regulations, however, is weak at present. Moreover, Turkey is a signatory to various international treaties on intellectual and industrial property rights. Following

Article 90 of the Turkish Constitution, international treaties duly ratified are treated the same as internal law (PricewaterhouseCoopers, 2003).

The Turkish Patent Institute, a special governmental authority with administrative and financial autonomy and connected to the Ministry of Industry and Commerce, was established in 1994 under the Decree Law 544 for the Establishment and Functions of the Turkish Patent. The Institute carries out the administration of intellectual and industrial property rights. Its scope and duties, as enumerated in Decree Law 544, are to organise, monitor and perform all actions necessary for the protection of rights (Turkish Patent Institute, undated).

There is no legislative arrangement in Turkey to protect the property rights of owners of traditional knowledge of indigenous natural resources. Traditional knowledge indirectly used in the development of a new process or product is not subject to any kind of protection and is not covered by Turkish Decree Law 551 on patenting.[4]

The Breeders' Rights Law, enacted in 2003, aims to protect varieties and plant breeders' rights. Under this law, organisations or persons are able to apply for the protection of their varieties, and plant breeders will be able to demand royalties for the amount of the seeds of the varieties marketed.

Informal rules hindering formal property rights

Intellectual and industrial property rights embody specialised rules, supported by particular values of Western societies, which have evolved over centuries. In Turkey, rudimentary forms of these values are gradually emerging, especially in economic affairs with respect to industrial property rights. However, the very same values in relation to intellectual property rights encounter resistance from social networks where relationships are essentially motivated by trust and sharing rules. In these networks, knowledge is regarded as a public good and shared freely. Knowledge sharing is regarded as a means of inter-generational knowledge transfer, especially practised in closed rural communities and/or artisans' associations.

Education, science and innovation

Education

Education in Turkey has always been at the centre of political dialogue. With the adoption of the Latin alphabet, Atatürk intended to distance young generations from the influence of religious movements and to pave the way for westernisation. However, since the 1950s, the curriculum of primary, secondary and high schools included either compulsory or optional religion courses.

[4] This issue may be incorporated in the WTO's Agreement on Trade-Related Intellectual Property (TRIPs), currently under review in the Doha Development Round. However, for a WTO member to benefit from any international agreement on this matter, rights to indigenous knowledge must first be recognised in its own legislation.

In 1997, the duration of compulsory schooling was extended from five to eight years. This is expected to increase enrolment rates to 100 per cent in primary school, 75 per cent in secondary school and 37 per cent in higher education by 2005. In 2003, the enrolment ratios in primary school were 93.6 and 90.2 per cent for boys and girls respectively, whereas these ratios for secondary school enrolment were 50.2 and 42.4 per cent (OECD, 2004a). As well as improving the educational level, a key motivation behind this reform was to protect the secular foundation of the country by keeping pupils longer under the secular education system.

Currently, Turkey has 53 public and 24 private universities, and 2 vocational colleges. Although the ratio of university graduates to the total labour force rose from 5.7 to 7.3 per cent between 1995 and 1999 (OECD, 2004a), the greatest challenge is still to meet the increasing demand for skilled labour in key sectors. In particular, there is an urgent need to meet high demand for qualified labour in the information and communication technologies (ICT), machinery manufacturing and chemistry sectors. The traditional sectors of civil engineering and agriculture record a significant excess of graduates compared to potential demand.

Private sector investment in education increased significantly during the 1990s. A number of private primary schools, high schools and universities have been established, but they are concentrated in a few big cities (Istanbul, Ankara and Izmir). Private universities are polarised with respect to subject range and geographical location. Typically, a private university specialises in business, economics, engineering, informatics and natural sciences, and is located in the west. Public universities are scattered all over the country and education quality varies with region. The Council of Higher Education is responsible for the coordination of all higher education institutions.

Science and innovation

Science and innovation policies were introduced in the 1960s with the establishment of the Scientific and Technical Research Council of Turkey (TÜBITAK), which is the only organisation responsible for formulating and coordinating national science and technology policies. The State Planning Organisation (SPO), also established in the 1960s, is responsible for preparing national economic targets and allocating funds for projects. The Undersecretariat of the Treasury and the Undersecretariat of Foreign Trade are responsible for implementing innovation policies. Most recently, in conjunction with the preparation of the 8th Five-Year Development Plan (2002-2005), the specialised science and technology committee formulated a major programme for implementing science and innovation policies over the period 2003-23. The role of ICT in the implementation is noted as a critical element, but currently there is no well-defined policy or agency responsible for activities in this area (Sayan, 2004; Cebeci and Gul, 2003). The General Directorate of Agricultural Research of the Ministry of Agricultural and Rural Affairs (MARA) is carrying out a research master plan partly supported by the World

Bank. Within the National Agricultural Research System, this plan prioritises research areas and provides funds for research projects.

TÜBITAK has indicated the need for new legislation to restructure public research organisations within the innovation system framework. Especially stressed is the need for a framework linking university, public research and industrial research. Efforts have been made since 1993 to create an innovation-facilitating environment. To this end, various Decree Laws were adopted during 1995-1998, covering R&D assistance to industry, establishment of a Turkish Patent Institute, protection of trademarks, protection of intellectual property rights in software and the setting up of university-industry cooperation centres (TÜBITAK, undated). Two agencies, the Technology Monitoring and Evaluation Board and the Technology Development Foundation, are responsible for the distribution, respectively, of technology-targeted subsidies and of the World Bank loans for industrial R&D projects (OECD, 2004a; World Bank, 2004).

The World Bank (2004) notes significant progress in the development of Turkish technology infrastructure. The progress includes harmonisation of the technology infrastructure with European standards, improvements in the Standardisation and Testing Systems through support to the Turkish Standards Institute, a strengthened regime for Intellectual Property Rights through support to the Turkish Patent Institute, more use by industry of metrology services as a result of upgrading the National Metrology Institute, and the setting up of the Technology Development Foundation of Turkey (TTGV) to support a number of technology financing programmes. At present, innovation infrastructure in Turkey faces the problem that Turkish Telecom, a publicly owned company, is still the key actor in the telecommunications sector. Although legal problems with respect to its privatisation have been solved, the actual transfer of ownership to private hands has not occurred (Sayan, 2004).

Informal education and innovation systems

The informal education and innovation systems of Turkey have not yet internalised the underlying meanings of such concepts as competition and property rights. In addition, the social value system does not encourage the application of these concepts in social and economic affairs. For a large majority of people, especially in rural areas, basic teachings at home are supportive of 'collaboration' rather than 'competition', and of 'common use' instead of 'private property rights'.

Institutional environment: summary

Table 3.3 summarises the information given in this section on the institutional environment prevailing in three 'stylised' periods: the late Ottoman period, the early decades of modern Turkey and the period from the 1950s to the present. These institutional environments are compared with that of the EU.

Table 3.3. Characteristics of the Ottoman Empire, Republic of Turkey and EU

The Ottomans	Kemalist principles (1923-1950)	Republic of Turkey (1950-present)	European Union
Ruled by Caliphate, religion and tradition; temporary constitutions in 1876 and 1908	Parliamentary democracy; Kemalist principles in the constitution	Parliamentary democracy; the Kemalist principles in the constitution	Multi-level government; EU-level Commission, Council, Parliament, Court
Ottoman Empire	Nationalism – strong state sovereignty	Somewhat weakening state sovereignty	Basic philosophy: mixture of supranational and intergovernmental
Ottomanism - Multi-national Empire; Rule of Caliphate	Republicanism – sovereignty vested in the nation	Sovereignty interrupted by military interventions	National identities respected
Millet system – separation of non-Muslim from Muslim, & religious freedom to non-Muslims	Populism - all people in Turkey are equal and all of them are Turkish citizens	Some rights of minorities recognised only recently	Respect for human rights and fundamental freedoms of minorities
Elite's rule – mixture of bureaucrats from minorities in all spheres of life	Etatism – state to regulate economic activity in areas where private enterprise is inadequate	Weakening etatism – strengthening market economy, privatisation of state monopolies after 1980	Multi-level government, limited role for state
Minimal scope for change due to traditional and religious institutions	Reformism – radical means to replace traditional with modern institutions	Significant economic reforms after 1980 but in some aspects a stagnant society	Economic integration as a driver of integration in other policy areas
Religion an essential part of governance of the Empire	Secularism - separation of state and religion; of religion from cultural, educational and legal affairs; independence of institutions from religion & religious institutions	Weakening secularism – state often used to support religious thought and institutions for political purposes	Separation of state and 'church', though differences exist between member states

3.3 Resource, technology and market institutions

Resource institutions

This section describes institutions that relate to the use and ownership of land, water, labour and the environment. Specific land and water institutions discussed in the following paragraphs include property rights, formal rules and regulations on water use and ownership rights, and land markets. In relation to labour institutions, the following areas are addressed: employment insurance and benefits, minimum wage setting, employment protection, early retirement and pensions, bargaining and labour unions, severance payments and labour contracts. Finally, formal regulations and informal rules concerning environment quality are described. For presentational simplicity, these institutions are explained, together with the roles of the relevant administering and implementing organisations (Table 3.4).

Land institutions and implementing organisations

A major problem in Turkish agriculture nowadays is declining productivity. Contributing factors are land fragmentation, land inheritance law, large-scale water resource development unsuitable to the needs of small farms and underdeveloped water use rights (Cakmak and Kasnakoglu, 2002).

Modern cadastral work started with the application of Land Registry and Cadastral Law 2613 in 1934 (updated several times since). The Directorate of Land Registry and Cadastre is responsible for all tasks related to land registration and land surveying. It employs 14 thousand people, of whom 40 per cent work in land registry activities and 60 per cent in cadastral activities (Yomralioglu, 2003). The 1055 Land Registry Offices throughout the country are tasked to perform contract and registration transactions concerning all types of real estate. Progress has been slow and the register is still incomplete. It was estimated that, in November 2003, 75 per cent of all parcels in rural areas were registered and legally occupied, and that of the remaining unregistered 25 per cent, four-fifths were legally occupied and one-fifth illegally occupied. The Ministry of Justice has responsibility for legal issues such as the identification of owners. Difficulty in establishing the details of land ownership has been one of the factors delaying completion of the register. The land register also serves as the basis for land taxation.

The absence of adequate cadastral maps, clearly defined land titles and rights for transferring land continue to hamper land market developments and transactions in some rural areas (Cakmak and Kasnakoglu, 2002). These unfavourable conditions can also make it difficult for farmers to access agricultural credit, which is already limited due to the recent restructuring of the operations of the Agricultural Bank of Turkey and reorganisation of Agricultural Sales Cooperatives (TZOB, 2004). On the positive side, each Land Registry Office has undertaken to digitalise its information (Yomralioglu, 2003), and MARA is in the process of building a digital database to store information on the availability of different types of land, types of land tenure and the procedures for buying, selling and leasing land (Demir and Duman, 2003). This database has been used to implement direct income support payments and will also be used for all other agricultural support payments.

Land fragmentation has been the reality of Turkish agriculture for a long time. The SPO, the Ministry of Agriculture and Rural Affairs (MARA), and the Food and Agriculture Organisation (FAO) have identified the absence or poor functioning of land institutions as factors behind this fragmentation. Among the

Table 3.4. Institutions and governing organisations

Institutional environment	Intended functions of institutional arrangements	Administering and implementing organisations
Political, social and economic rules ▪ Competition rules ▪ Property rights ▪ Education, science and innovation	To organise agricultural resources ▪ Land (ownership, renting, inheritance) ▪ Labour (contracts, social security) ▪ Water (use and distribution) ▪ Environment (soil, water, air)	Land, labour, water, environment Directorate of State Hydraulic Works, Water Users' Org., Ministries of Agriculture & Rural Affairs (MARA) and Environment-Forest; Social Security Inst.; Public Retirement Pension, Turkish Emp. Org., labour and employer unions, Bağ-Kur (union for self-employed in agriculture), Public Minimum Wage-setting Commission, High Accreditation Council
	To organise agricultural research, technology and innovation ▪ NARS, Agricultural Knowledge and Information System, R&D partnership ▪ Science and innovation activities ▪ Technology generation and transfer	Research, technology, and innovation Scientific and Tech. Research Council, Undersecretariats of Treasury and Foreign Trade, State Planning Org., Council of Higher Education, Ministries of Education, Health, Environment & Forestry; universities; MARA; Agr., research organisations, Provincial Ext. Adm., Chambers of Agr., farmers' organisations, NGOs, development organisations
	To organise agricultural production, markets and trade ▪ Farming ▪ Marketing contracts ▪ Agricultural product markets and trade ▪ Seeds, agricultural chemicals, livestock, credit and technology markets	Production, markets and trade Turkish Competition Authority, Chambers of International Trade, High Accreditation Council, Ministry of Foreign Trade, Ministry of Trade and Industry, State Planning Org., the Court of Appeals, the Council of State, the Inter-University Board; Agricultural Banks, Agricultural Sales Cooperatives, Agricultural Credit Cooperatives

institutions blamed are inheritance and property laws, commercial laws in the rural context, the arrangements for leasing and distributing state land to farmers, land use policies, and the complex structure of agricultural infrastructure organisations.[5] Inheritance Law, amended in 2001, prevents the division of land holdings below a specified level that would provide enough income for a family of four. Farmers' emotional attachment to their land is also a factor contributing to fragmentation; such attachment is likely to make farmers reluctant to accept new tenure arrangements even if these arrangements are economically beneficial. Land consolidation by the authorities started on a very small scale in the 1960s. The Land Consolidation Statute, decision 7/18231 (1979), and Law 3083 (1984) now form the basis for land consolidation programmes that are under the control of the GDRS, and for large-scale irrigation projects, under the GDAR (beginning in 1989) (Gür and Demirel, 2002). Completion of a cadastral survey in the area concerned is a pre-condition for land consolidation. From 1961 to 2000, 255 thousand hectares involving 152 thousand landowners in 26 areas were consolidated by GDRS, whereby the number of plots was reduced from 263 to 138 thousand. From 1989 to 2002, GDAR consolidated 129 thousand. Although 4.5 million hectares have been brought under irrigation, just 8 per cent of this land has been consolidated. Gür and Demirel (2002) concluded that continuing fragmentation has prevented many smallholdings from benefiting from irrigation.

Labour institutions and implementing organisations[6]

The Ministry of Labour and Social Security (ÇSGB) is responsible for employment and workers. In 2000, three State agencies were brought together under the Social Security Organisation: the Social Insurance Organisation (SSK) for wage and salary earners, Bağ-Kur (BK) for the self-employed, including those in agriculture, and the Turkish Employment Organisation, which is in charge of providing public employment services. They all are administratively and financially independent. Approximately 60 per cent of the insured population are under SSK, 25 per cent under BK, and 15 per cent under the Retirement Fund (which is available only to civil servants).

The government sets the minimum wage, the retirement benefits of civil servants and others, unemployment insurance, and social insurance. Minimum wage legislation has been in force nation-wide since 1974, but its effective enforcement is weak especially in rural areas. The Work Law 1475 stipulates that minimum wages have to be adjusted every year by a Committee reporting to the ÇSGB and remain in

[5] For further information see Devlet Planlama Teskilati: http://ekutup.dpt.gov.tr/tarim/oik534.pdf; Ministry of Agriculture and Rural Affairs: http://www.tarim.gov.tr/arayuz/6/menu.asp; also http://www.fao.org/

[6] This section draws on the views of Taymaz and Ozler (2003) and Tunali (2003), and the most recent information on labour market institutions in Turkey. The reader is referred to www.ssk.gov.tr, www.bagkur.gov.tr, www.emeklisandigi.gov.tr, www.iskur.gov.tr, and www.turkis.org.tr for more information. It should be noted that in the reviewed documents agricultural labour market institutions are not studied in depth.

effect throughout the year. As of 2004, the monthly minimum wage for age 16 and above is USD 293 and that for below age 16 is 85 per cent of USD 293. Until mid-1989, a lower minimum wage was set for agriculture and forestry.

To reduce the State's dominance in the provision of retirement benefits, a new personal pension scheme for workers and the self-employed was legislated in April 2001. Based on voluntary contributions, the scheme is designed to supplement public pensions. Employers can also contribute to the system on behalf of their employees. The fund is managed by private pension companies and is overseen by the Treasury. As farmers do not receive the state pension, this scheme represents a new opportunity for them to secure their retirement. However, to date very few farmers take part in this scheme (E. Cakmak, Istanbul, October 2004, personal communication). The new retirement system that started in 2002 sets the minimum age for retirement for a new entrant at 58 for women and 60 for men. The legal framework for unemployment insurance (Law 4447 of 1999) has been implemented from 2002. It covers workers registered with the SSK and does not cover civil servants or the self-employed.

Union Law 2821 on collective bargaining (1983) regulates employee/employer relations. White-collar unions in the public sector re-emerged in 1990. Most civil servants and the regular employees of state enterprises are represented by unions, but are legally prevented from striking. The Supreme Arbitration Board resolves conflicts involving these groups.

Arrangements regarding part-time, flexible work, workplace safety and children's work conditions were made to align Turkey's labour market legislation to that of International Labour Organisation and EU standards. But Turkish Labour Law still does not adequately protect part-time and fixed-term employees, who are especially numerous in agriculture (Tunali, 2003). The changes introduced mainly address the short-term concerns of employers by facilitating labour market flexibility. According to OECD (2004b), about 50 per cent of Turkey's workforce operates in the unregistered economy with minimal legislative protection.

Water institutions and implementing organisations[7]

Since the 1926 Civil Code on water use, only one special law has been enacted on surface water use for hydropower production and thermal waters. Various customary rules and regulations developed locally are applied when conflicts arise among water users. The court settlement is final but the court does not have a reference regulation to solve the problems.

A draft Law concerning water resource use was prepared in 1968 but has not been enacted yet. The 1982 Constitution states that water resources are 'natural' wealth and under the authority of the state. However, legislation on water rights and ownership is complex. Water resource development is the responsibility of the state, except for some privately owned small springs and other water resources. Again, a

[7] For detailed information on water developments in Turkey, see Unver and Gupta (2003) and http://www.fao.org/ag/agl/swlwpnr/reports/y_nr/z_tr/tr.htm#waterr.

special law arranges groundwater use, and licenses that cover only user rights are issued by the General Directorate of State Hydraulic Works (DSI) upon users' request. Water use rights can be neither transferred nor sold.

Water User Associations have various problems with respect to finance, administration and machinery. Sub-surface drainage and other on-farm land improvements are carried out by the General Directorate of Rural Services (GDRS), but are poorly maintained by farmers as ownership is not assigned. There is no monitoring and evaluation of the small-scale irrigation schemes realised by GDRS. Research findings on improved irrigation techniques are not effectively transferred to public extension service and farmers. Water management advice is not clearly included in the existing extension programmes for improved water management.

Governmental and non-governmental organisations are involved in water resource development in Turkey. The DSI under the Ministry of Public Works and Settlements deals with major irrigation projects, hydropower development and water supply to cities. The GDRS under MARA deals with the construction of village water supplies, small-scale irrigation projects and research on soil-water-plant relationships. The Electrical Power Resources Administration under the Ministry of Energy and Natural Resources conducts surveys of hydroelectricity generation. The MARA assists in the development of water resources through research and planning. Non-governmental organisations include user organisations such as irrigation cooperatives and water user associations, and they operate and maintain irrigation schemes transferred to them.

The 1954 Law concerning the organisation and duties of the DSI allows the transfer of management responsibility for publicly constructed irrigation schemes to local authorities and water unions. In 1997, the DSI initiated a process of drafting new and generic legislation governing the status and operation of water users' unions. In recent years, the process of transferring water management to water unions and farmers' groups has begun. These unions have to register all irrigators in their areas as members, though it is unclear what the term 'irrigators' means. Farmers pay only an annual crop- and area-based fee towards the cost of operation and maintenance of water resources (Unver and Gupta, 2003).

Environment institutions and implementing organisations

The 1983 Environmental Law adopted the Polluter Pays Principle. A large number of regulations endorsed since then, in support of the Law, specify emission and discharge standards, and require polluting industries to obtain discharge permits (OECD, 1999). National environmental administration is the remit of the Ministry of Environment and Forestry (MoEF)[8], in affiliation with the authority for the Management of Specially Protected areas. A 1992 regulation introduced mandatory environmental impact assessment (EIA).

Future environmental institutions are likely to grow around three issues. The

---

[8] The Ministry of Environment (MoE), created in 1991, was merged with the Ministry of Forestry (MoF) in 2003 to become the Ministry of Environment and Forestry (MoEF).

first issue is that all appropriate sectors of economic activity and all relevant governmental policies should take environmental concerns into account. To this end, Turkey adopted the EIA regulation for all relevant sectors. However, it is hardly applied; for example, there is *no* irrigation project using the established guidelines. The absence of skilled human resources seems to be the main reason for this. The second issue relates to the enforcement of environmental laws and regulations, which has implications for behavioural changes at the individual and community levels. Enforcement is especially demanding, as environmental concerns require the collaboration of multiple organisations under different jurisdictions. The third issue concerns stakeholders' participation in priority-setting processes, which requires changes in the dominant environmental policy and management perspective (Okumus, 2002).

The SPO has prepared the National Environmental Action Plan (NEAP) in cooperation with the MoEF. Furthermore, environmental protection foundations are established in every city with a MoEF provincial branch. They generate income from paid services such as measuring industrial emissions, vehicle exhaust emissions and noise levels to fund local environmental projects. A large number of NGOs specialise in environmental issues in Turkey. They are mostly involved in public awareness and public participation activities. Universities are also beginning to establish themselves in this field. Today, 22 environmental research centres and 11 environmental engineering departments are in place. The Ministry of National Education is aware of the need for environmental education in biology, geography and philosophy programmes, but has taken no action. Private sector investment in the environment and compliance with regulations are woefully short (Markandya, 2003).

Partnership approaches involving government and the private sector are increasing. Examples include voluntary agreements between the cement industry and the government to reduce particular emissions, as well as between the automobile industry and the government (OECD, 1999). However, similar partnerships between the government and grassroots associations do not exist, because grassroots activities and their collaboration with international agencies are tightly controlled and limited by law (Okumus, 2002).

Technology institutions

Agricultural research policy, priorities and implementing agencies

The organisation of agricultural research policy is spread over a large domain. TÜBITAK, MARA, SPO and the Ministries that control the state-enterprise research organisations all address aspects of national agricultural research policy. TÜBITAK and SPO identify national research priorities in all domains, including agriculture, forestry and veterinary science, and provide financial support to individual and collaborative research projects. The General Directorate of Agricultural Research (GDAR) under MARA and the General Directorate of Rural Services (GDRS) organise and implement agricultural research programmes through their respective research organisations. The GDAR has identified high priority research areas such as oilseeds and food legumes, dairy and beef, industrial crops, cereals, fruit and vegetables. The GDRS, on the other hand, carries out research on soil, water, irrigation, land and mechanisation. Research priorities identified recently include catchment areas, soil, water and investment management. There is a strong need for much greater coordination between the bodies in charge of agricultural research policy design and implementation.

Agricultural research priorities are identified in the context of conventional priority-setting workshops with the participation of public research administrators, staff from various national and regional research institutes, and stakeholders from the private sector (Yalvac, 1999). Research programmes are designed in line with the priorities identified, but research impact assessment is not a common practice yet. The effective implementation of the priorities identified seems to suffer from poor human resources and limited research funds. Furthermore, the allocation of human resources and public research funds shows regional imbalances (Uzunlu *et al.*, 1999). Recently, with increased regionalisation of research, interaction with farmers and agricultural industries has been strengthened, but available resources have been fragmented and overall coordination has weakened.

National Agricultural Research System (NARS)

The NARS (for an overview up to 1999, see Uzunlu *et al.*, 1999), comprises four groups of research and development organisations.

- Group 1 includes 82 public research organisations - 56 are associated with GDAR, 12 with GDRS, 11 with the Research Directorate of the MoE, and three with sugarbeet, tea and tobacco state enterprises. The Sugarbeet Research Organisation of the Sugar Factories Enterprise is under the Ministry of Industry and Trade; the General Directorate of Tea Establishments (ÇAYKUR) and TEKEL are both affiliated with the Ministry of Finance and Customs.
- Group 2 includes universities governed by the Council of Higher Education. Most of these universities have, apart from their agricultural faculties, various units or departments specialised in agricultural sciences. Currently, there are 23 agricultural and 19 veterinary faculties. Agriculture-related training and vocational schools are all under the Ministry of Education.
- Group 3 includes independent research organisations established by public-private collaborations: the Marmara Research Centre and the Nuclear Agricultural and Animal Research and Training Centre.
- Group 4 includes various national and international development organisations, including NGOs such as Farmers' Unions and Agricultural Chambers of Commerce. The Southeastern Anatolia Project (GAP) under the auspices of the Prime Ministry manages socioeconomic development in the less developed south-eastern region. The GAP administration collaborates with ICARDA in joint projects and proposes to develop a regional agricultural research and training centre in the south-eastern region.

National extension system

For many years, considerable national and international effort has gone into building up the national extension system[9]. Yet there is dissatisfaction with its performance at all levels. MARA organises the national extension services. MARA's effectiveness, however, in coordinating the contributions of the universities, its own directorates and research units, and their links with the extension services has been strongly criticised (see, for example, Tekinel and Yazar, 2000; Ozcatalbas *et al.*, 2004). Critics consider that poor communication between the research and extension services has hampered the dissemination of new technologies to farmers for several decades. Feedback from farmers to extension agencies and then to research organisations is also poor.

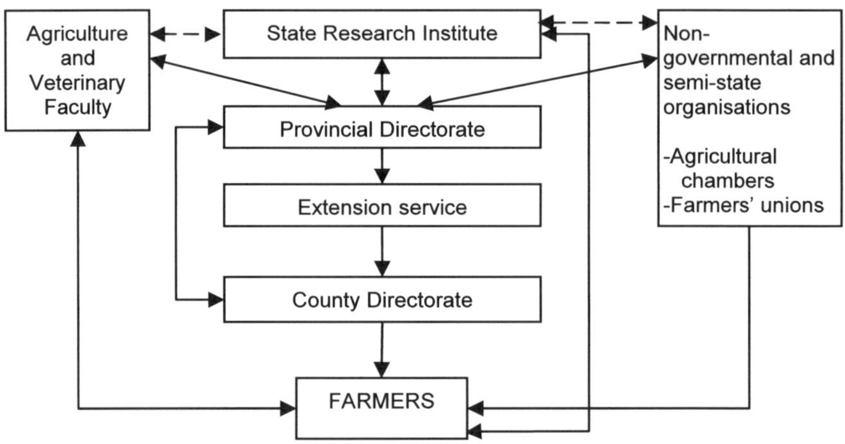

Figure 3.1. The structure of the Turkish extension system
Source: Güney et al., 1999: 66

Figure 3.1 shows the current structure of Turkey's extension system. The structure is rather formal and uses indirect channels to reach farmers. Village group technicians, the closest agents to farmers, are placed in a group of 4-5 villages and live in the village. They address farmers' problems through the Farmers' Training Division that provides training to the lead farmer, who then disseminates agricultural technology to farmers in the village (Güney *et al.*, 1999; Kumuk and Van Crowder, 1996). Television and radio broadcasts are also used in extension work (Kumuk and Van Crowder, 1996). Extension activities follow a rigid schedule with little input from farmers, with the result that programmes tend to lack relevance to local farm problems. In many cases,

[9] For example, two large World Bank projects, Agricultural Extension and Applied Research Project (AEARP) I (1984-91) and II (1990-97), concentrated on strengthening links between farmers, extension and research.

village group technicians, who are supposed to live in their work areas but often do not, have become essentially little more than purveyors of centrally prepared messages. As well as structural and organisational weaknesses, the extension system has faced farmer reluctance to engage in extension activities because of limited access to credit. More positively, NGOs have recently started to provide extension services, and independent producers' associations and cooperatives have started participating in extension activities.

Technology and new institutions

Biotechnology, and information and communication technologies (ICT), offer new opportunities for agricultural development.[10] However, exploiting such opportunities is conditional on the extent to which countries can adapt their existing legal, regulatory, policy and institutional frameworks. For example, a biotechnology information system is needed in order to set up a biotechnology regulatory framework. Presently, Turkey does not have such a system. The application of intellectual property rights in biotechnology research requires appropriate legislation to prepare the ground for the private sector to invest in biotechnology. ICTs are vital in establishing an agricultural knowledge and information system (AKIS). Only recently did the Turkish agricultural authorities initiate projects aimed to create AKIS (Demir and Duman, 2003).

Conventional research and emerging formal research institutions

The current research and extension systems operate in a very isolated manner. Many public research organisations have become detached from farmers, mainly due to the weak extension system. Traditionally, investment in agricultural research was the sole responsibility of the public sector. This convention still dominates not only among researchers but also among farmers and their organisations. Despite recent examples of public-private collaboration, especially in seed development, many researchers and research managers are still conditioned by the view that agricultural research is the responsibility of the public sector.

Agricultural research systems in EU countries emphasise interactions and feedback between research and development (Kern, 2000). The dominant model of public national agricultural research organisations has given way to that of innovation systems that include a broader range of interacting partners, such as universities, agricultural research institutes, NGOs and the private sector. Currently, rudimentary examples of such interactions, although unplanned, are also taking place in Turkish agriculture in the context of the GAP.

To date, the general tendency among agricultural and research policy-making bodies in Turkey has been to design and implement policies without consulting stakeholders. Through the implementation of internationally funded

[10] Sayan (2004) discusses the compatibility of Turkey's culturally and ethnically diverse population with the development of information technology infrastructure.

projects, however, this traditional approach is gradually being replaced by one that emphasises stakeholder participation in agricultural research priority setting. The growing emphasis on agricultural innovation systems assigns a critical role to non-governmental and grass-root organisations in uncovering indigenous technologies and disseminating them in a wider knowledge system.

Social networks and technology transfer

Social networks are instrumental in the development and adoption of new or improved technology. Typically, technology is diffused among the members of the network through social relations. Actual application, however, involves imitation of those who have already used it, and depends on the level of trust in the social relations involved. With the setting up of village institutes in 1940, Turkey adopted an agricultural development strategy based on social networks. These institutes used specially trained teachers who lived and worked in their own communities to improve farming methods and promote Kemalist reforms in villages. This model viewed technology transfer as a trust issue. The institutes were closed in 1953 during the period of the Democratic Party. Intermediary agencies, including NGOs, farmers' organisations, cooperatives, international development organisations and marketing associations, form another network that bridges the gap between social and economic networks. Within the innovation system framework, the intermediary network plays the role of knowledge facilitator. In Turkey, economic and social networks are isolated, while the intermediary network is gradually expanding through the participation of NGOs, farmers' organisations, cooperatives and community representatives.

Market institutions

Competition and agricultural markets

A major goal of agricultural policy is to ensure stable income and living standards for rural populations. This is often in conflict with the aim of competition policies to ensure maximum welfare for society as a whole. Therefore, in many countries competition rules and agricultural market institutions do not go hand-in-hand.

A large number of farms in Turkey are small[11], with less than 5 hectares of land, and lack the capital for expansion. Farmers purchase inputs in oligopolistic markets, and often suffer from the presence of an oligopsonistic or monopoly wholesale marketing sector (see Chapter 6). General competition rules should be tailored to promote competition in these highly concentrated markets. In the area of agricultural credit, Agricultural Bank and a few private banks started in 2002 to extend low-interest credit to those farms large enough to use their farms as collateral. The agricultural machinery market similarly tends to exclude small farms, which lack the finance and incentives to buy expensive machines. The

[11] A large majority of agricultural enterprises are small, the number of small enterprises is increasing and the size of an average enterprise is declining (see Table 6.4).

leasing market and cooperatives are therefore better options for them (Turkish Competition Authority, 2003). In agricultural product markets, however, competitiveness is gradually developing.

Agricultural insurance markets are also underdeveloped because of the lack of farm credit. The government has assumed responsibility for preparing a regulatory framework to facilitate the growth of the insurance market. A draft law was prepared in 2002 to promote the development of an agricultural insurance system, but still has to be approved by Parliament (Turkish Competition Authority, 2003; TZOB, 2004).

In Turkey, agricultural reforms have often been triggered by large budget deficits, not as part of a market reform strategy. Hence, as soon as budget deficits are reduced, the reforms are relaxed and the old rules re-emerge. The situation is different now since permanent change in Turkish institutions is required for EU accession. In this respect, non-distorting policy interventions and a complete withdrawal of the public sector from agricultural markets are both critical. Under the first of these headings, reduction in market price support and the adoption of a direct income support (DIS) scheme are already under way (see Section 7.2). With respect to public sector withdrawal, there is still a long way to go. Agricultural credit is one such area. Currently, only a few private banks extend credit to wealthy farmers and two-thirds of small farms use credit from (illegal) brokers (Turkish Competition Authority, 2003). Operational restructuring of the state-owned Agricultural Bank in 2001[12], whereby farm assets and productivity play a greater role in credit assessment, together with the lack of real competition in this sector, could well force small farms to turn more to informal credit markets.

Currently, the asymmetry of competition rules impedes their application in Turkish agriculture. Public agencies, responsible for ensuring the proper organisation of competition rules, also represent public enterprises operating as players in agricultural markets. That is, some players in particular markets have the privilege of designing and applying competition rules, whereas rules that govern these privileged actors' activities are not yet well developed. On the other hand, rules that govern farmers' associations and cooperatives are relatively well organised and monitored (Turkish Competition Authority, 2003).

The Competition Authority is a young organisation. At this stage of its development, one can only comment on its general approach to agricultural markets rather than analyse the implementation of competition rules. Whether the Authority is fully independent at this early stage of its development is questionable. For example, the Authority does not intervene to correct anti-competitive behaviour of tomato producer cooperatives when they try to overcome the oligopsonistic power of agricultural processors. The Authority shows a similar attitude towards the growing practice of contract farming, which

---

[12] Strictly speaking, since 2000 (when the Agricultural Bank became a joint stock company) it is no longer a State Economic Enterprise. It describes itself as "a joint-stock company whose structural and operational characteristics are those of a private sector concern but whose capital happens to be state-owned" (Ziraat Bank, 2003).

works against competition policy. The Authority's independent activities are hindered by the fact that public enterprises are also competition policy makers. For example, complaints have been made that the publicly-owned sugar enterprise sets sugar prices too low, which forces private sugar companies out of the market. But the Authority has acknowledged that sugar prices are determined by special laws approved by the council of Ministers. This indicates that the Authority lacks full power in issues relating to the implementation of competition rules (Turkish Competition Authority, 2003).

Adapting institutions to the EU's Common Market Organisations

The EU's common agricultural policy (CAP) places each agricultural product under a common set of market rules throughout the Union. Institutional prices are being gradually reduced towards world market levels, and being replaced by direct aids as the basic support mechanism for Community farming. The effective implementation of the *acquis* requires that intervention agencies are capable of performing tasks such as regular market and price monitoring, public storage, and sales and stock control. The *acquis* further specifies precise rules for producer organisations (Togan *et al.*, 2003).

Within the framework of the Agriculture Reform Implementation Project (ARIP), launched in late 1999, Turkey has been reforming its price support and input subsidy policies. In this context, the Agricultural Sales Cooperatives Unions (ASCUs) were restructured and the Turkish Grain Board (TMO) acquired new functions. The farmer registration system, which is a necessary tool for the payment of direct income support under the DIS scheme, was launched. It utilises the existing land registry records and MARA's Farm Registry System (providing data on the number of farmers, the demographic characteristics of the farmers, assets, number and the size of parcels, and land use). In 2003, nearly 2.8 million farmers and 16.65 million hectares were registered for DIS payments (see section 7.2). The registration system developed under the DIS is a step towards an Integrated Administration and Control System (IACS) (Togan *et al.*, 2003).

A second key element of the ARIP framework has been the privatisation of agricultural State Economic Enterprises (SEE). The new sugar law, which was adopted in April 2001, aims to open up the sweetener market (defined to include isoglucose) to competition and reduce state interference, whilst maintaining a system of sugar production quotas as in the EU. In the tobacco sector, Parliament adopted a new law restructuring the Directorate General for Tobacco and Tobacco Products, Salt and Alcohol Industry (TEKEL), converting it into a commercial enterprise that will operate under free-market conditions. The processing facilities of TEKEL are to be privatised. A tobacco law was adopted in January 2002, which intends to end state subsidised tobacco purchases and to introduce auction sales, with individual purchasing contracts between producers and buyers. Finally, privatisation of the tea factories of ÇAYKUR began in 2001. Some other firms were also liquidated, such as the Turkish Agricultural

Supply Corporation, the state firm responsible for input supply (Togan *et al.*, 2003; Turkish Competition Authority, 2003).

Food safety and food quality standards

Regarding food safety legislation and control systems, Turkey has incorporated some of the *acquis* into its legislation, mainly in the fields of packaging materials and food. The Establishment and Duties of Province Control Laboratories were revised in early 2001. Accreditation has been initiated for some of the laboratories involved in tests organised by the Food Analyses Performance Assessment Scheme (FAPAS) and TÜBITAK. Finally, food legislation in Turkey has been continuously updated since 1985. All stages of food production are targeted for inspection, with legislation aimed to protect public health against all possible food-related diseases. Turkey has accepted the Codex Alimentarius, embodying international food safety standards, in 1997, and subscribes to other international standard-setting bodies (including the OIE and IPPC).

The Turkish Standards Organisation (TSE) has been using internationally recognised certification (ISO) guidelines since 1994. Ninety-three per cent of Turkish standards are now based on European and international standards, while over 90 per cent of EU standards have been adopted as Turkish standards. Products that bear EU certificate marks are now directly granted a conformity certificate by the TSE and the Ministry of Health. With the enactment in 1999 of the Law on the Organisation and Functions of the Turkish Accreditation Council, Turkey reformed the institutional framework for accreditation. The Accreditation Council, which began operation in 2001, assesses and audits laboratories, certification and inspection bodies, and certifies the competence of conformity assessment bodies.

Agricultural policy implementation agencies[13]

Four key agencies are responsible for implementing agricultural policies, including MARA, the Ministry of Industry and Trade, the Agricultural Bank of Turkey and the Treasury. MARA's main task is to assist in the elaboration and implementation of agricultural policies. MARA also performs commercial functions through an affiliated state economic enterprise, the Turkish Grain Board (TMO). The TMO functions as a buffer stock agency to stabilise producer and consumer grain prices. The TMO provides signals to merchants about the future directions of the market by announcing purchasing prices, which are later revised based on market conditions. Under the World Bank funded reform programme (ARIP), it is intended that TMO prices will be increasingly linked to world prices so that ultimately state procurement becomes a 'buyer of last resort', as is the case in the EU.

---

[13] This section draws on information from Togan *et al.* (2003), DPT (2000), TZOB (2004), Turkish Competition Authority (2003).

The Agricultural Bank is the main provider of agricultural credit for crop and livestock production. The credit is distributed to farmers through the Agricultural Credit Cooperatives (ACCs) and the Agricultural Sales Cooperatives Unions (ASCUs). The farm credit subsidy has been eliminated. Since these changes, credit has been limited to those farmers with enough collateral. A few private banks have also started to provide credit but under very restrictive conditions.

In the past, the ASCUs operated under the control of the Ministry of Industry and Trade. They were authorised to set prices for the members' commodities, and to implement support purchases from producers on behalf of the state. They were also authorised to set up facilities such as warehouses, and processing and packing plants, and to market commodities in accordance with wholesale and retail market practices. Within the ARIP framework, financial aid is granted to assist the restructuring and transformation of ASCUs into genuine cooperative organisations, i.e. independent, financially autonomous and self-managed cooperatives that sell and process members' production.

Trade policy

The Undersecretariat of the Prime Ministry for Foreign Trade (DTM) acts to formulate, administer and co-ordinate Turkey's foreign trade policies. Private sector concerns are customarily taken into account. Turkey has enacted new legislation on, *inter alia*, customs, anti-dumping and countervailing measures, standards and other technical regulations, banking, energy and government procurement to comply with its obligations under WTO Agreements (WTO, 2003).

3.4 Conclusions

This chapter describes the current institutional situation of Turkish agriculture within the wider economy. Following the description of the economy-wide institutional environment, including a brief summary of historical developments, institutional arrangements for agricultural resource use and technology, and for production, markets and trade are described.

Although Turkey's institutional development process shows several different phases, the process has always targeted western institutions and values. The first phase (1923-1950) started with the establishment of modern Turkey, based on rules that were supported by nationalist aspirations and rather new to Turkish society. The influence of these rules was pervasive throughout society. The second phase (1950-1984) began with the multi-party system and views that openly opposed some of the basic principles of Kemalism. Institutions were shaped by the interplay of internal and external economic developments. This is the period in which market institutions were first introduced to Turkish society; but accompanied by the old formal and informal institutions. Finally, in the last phase (1984-2004), the government has formally announced that the goal is the market system, and that organisational and institutional requirements of this system must be fulfilled.

The economic bottlenecks repeatedly experienced by Turkey and the rapidly changing international economic and political system have made accession to the EU a vital goal for Turkey. Nonetheless, especially since the start of the Customs Union with the EU in 1996, Turkish governments have been busy with the adoption of purely administrative arrangements, but have made little effort to develop administrative and social capacities. In fact, current weak enforcement should be seen as a sign of the lack of these capacities.

Several conclusions emerge from the analysis in this chapter. In view of EU membership, Turkey may encounter new challenges. The military's task, as spelled out in the constitution, of safeguarding the basic Kemalist principles is at a turning point. Significant changes have recently been observed regarding the conflict between the state and the Kurdish minority and between secular and anti-secular forces. Any proposal for a change in the constitution is most likely to be perceived as a threat to the general character of the Republic.

The state planning approach has a long history in Turkey. Low trust by the state towards the public, a legacy from the Ottomans, has been translated into a large number of laws and regulations with prohibitions, sanctions and fines. Current competition rules and property rights are no exception to this growing bureaucracy. It is ironic that there are no credible sanctions to apply in the case of bureaucratic mismanagement and failures. The bureaucracy has not yet internalised principles such as accountability and transparency. Until now, public officials have generally not been sanctioned for their mistakes. On the economic front, the remaining elements of the command-control approach, such as state planning exercises, create constraints on the growth of competitive markets and their institutions. On the social front, the dominant perspective in social and family relations is still patriarchal.

Critical institutional arrangements that shape the use of factors of production are still not in place or are implemented ineffectively. Land property rights and rules for land transactions, for instance, need to be better defined if they are to support the growth of land markets. Water use, water ownership rights and pricing rules, especially in the context of the Southeastern Anatolia Project, need to be improved considerably. Labour and farming contracts, and their enforcement, should be improved, and labour force regulations should be further modified to protect part-time and fixed-term employees, who are widespread in agriculture. Environmental safeguards need to be integrated into economic activities in all sectors, and the Environmental Impact Assessment regulation should be enforced in practice.

Turkey's agricultural technology system is under-performing. Success depends on much better coordination than at present between bodies responsible for the design and implementation of agricultural research and technology policy. The public agricultural research system should be rationalised in line with current agricultural priorities. At the same time, a major effort is needed to improve the performance of the national extension services, and farmers and their organisations should be empowered to create demand for public extension services. Again, better coordination between the various bodies on the supply side, and with users, is of paramount importance.

References

Ataturk ve Turkiye Cumhuriyeti (undated) Reformlar ve Atilimlar: Izmir Iktisat Kongresi. http://www.ataturk.net/cumh/izmir.html.

Cakmak, E.H. (2004) Structural change and market opening in agriculture. *CEPS EU-Turkey Working Papers,* No. 10, September 2004. Centre for European Policy Studies, Brussels.

Cakmak, E.H. and Kasnakoglu, H. (2002) Assessment of agricultural policies and the impact of EU membership on agriculture in Turkey. Unpublished manuscript.

Cebeci, Z. and Gul, A. (2003) Information and communication technology in Turkish agriculture: The problems, expectations and suggestions. *International Congress on Information Technology in Agriculture, Food and Environment*, 7-10 October, 2003, Ege University, İzmir, Turkey.

Davis, L.E. and North, D.C. (1971) *Institutional Change and American Economic Growth*. Cambridge University Press, Cambridge.

Demir, K. and Duman, A. (2003) AB ve Turkiye acisindan tarimda girdi kullanimi. Turktarim, 151.

DPT (2000) Sekizinci 5 yillik kalkinma plani - Tarimsal politikalar ve yapisal duzenlemeler ozel ihtisasi komisyonu raporu. Devlet Planlama Teskilati, Ankara.

European Commission (2004) Twenty-third report on competition policy, 2003. SEC(2004) 658, Brussels.

Güney, O., Oztürk, D. and Biçer, O. (1999) The present situation of extension service organisation in Turkey. In: Rubino, R. and Mohrand-Fehr, P. (eds.) *Systems of sheep and goat production: Organization of husbandry and role of extension services.* CIHEAM-IAMZ, Zaragoza, pp. 63-67.

Gür, M. and Demirel, Z. (2002) Development in rural land arrangements and legal regulations in Turkey. Paper presented at an International Congress on Land Consolidation from the Field, 19-26 April 2002, Washington DC.

Karaman, M. (undated) Tarih Defteri. http://tarihdefteri.8m.com/ahilik.htm.

Kern, S. (2000) Evolution of Dutch innovation policies from a system's perspective: Towards a new paradigm. Paper presented at the *5th Annual Conference of the European Network of Industrial Policies*, November 2001, Vienna.

Kumuk, T. and Van Crowder, L. (1996) Harmonizing T&V extension: Some experiences from Turkey. Agricultural Extension and Education Service (SDRE), FAO Research, Extension and Training Division, Rome.

Markandya, A. (2003) Turkey – towards EU accession: Environmental acquis. Paper presented at *Turkey : Towards EU accession*, Bilkent University, 10-11 May, 2003, Ankara.

Mumcu, A. and Zenginobuz, U. (2001) Competition policy in Turkey. Paper prepared for the *8th ERF Annual Meeting*, 15-17 January 2002, Cairo.

National Assembly of the Republic of Turkey (undated). The Constitution. http://www.tbmm.gov.tr/anayasa/constitution.htm.

North, D.C. (1990) *Institutions, Institutional Change, and Economic Performance.* Cambridge University Press, Cambridge.

OECD (1999) *Environmental Performance Reviews: Turkey*. OECD, Paris.

OECD (2004a) Small and medium-sized enterprises in Turkey: Issues and policies. OECD, Paris.

OECD (2004b) *OECD Economic Surveys: Turkey 2004*. OECD, Paris.

Okumus, K. (ed.) (2002) *Turkey's environment: A review and evaluation of Turkey's environment and its stakeholders*. The Regional Environment Center for Central and Eastern Europe, Szentendre, Hungary.

Ozcatalbas, O., Brumfield, R.G. and Ozkan, B. (2004) The agricultural information system for farmers in Turkey. *Information Development* 20, 97-104.

Pamuk, S. (1982) Osmanli Tariminda Üretim Iliskileri. Toplum ve Bilim17, 3-50.

Pamuk, S. (1985) Anadolu'da Küçük Köylülük Üzerine Tezler 13, 102-110.

PricewaterhouseCoopers (2003) Acquisition of intellectual and industrial property rights in Turkey. PricewaterhouseCoopers, November 2003.

Sansal, B. (undated) Attaturk's Reforms. http://www.allaboutturkey.com/reform.htm.

Sayan, S. (2004) Factors and impacts in the information society: An evaluation for Turkey. Bilkent University, Tender No J04/04/2002.

Taymaz, E. and Özler, S. (2003) Labor market policies and EU accession: Problems and prospects for Turkey. Paper presented at *Turkey: Towards EU accession*, Bilkent University, 10-11 May 2003, Ankara.

Tekinel, O. and Yazar, A. (2000) For an efficient management of water demands: Farmer training service in the LSP in Turkey. Fen ve Mühendislik Dergisi 3(1), 1-8.

Togan, S., Bayaner, A. and Nash, J. (2003) Analysis of the impact on agricultural markets and incomes of EU enlargement to Turkey. Paper presented at *Turkey: Towards EU accession*, Bilkent University, 10-11 May 2003, Ankara.

TÜBITAK (undated). Tubitak website: Policy implementation. http://www.tubitak.gov.tr / (accessed 2 August 2004).

Tunali, I. (2003) Background study on labour market and employment in Turkey. Report prepared for the European Training Foundation, Italy.

Turkish Competition Authority (Turk Rekabet Kurumu) (2003) Faaliyet Raporu. Turk Rekabet Kurumu, Ankara.

Turkish Patent Institute (undated) Turkish Patent Institute website. http://www.turkpatent.gov.tr/ (*in Turkish*).

TÜSİAD (2002) Avrupa Birligine cevre mevzuatina uyum sureci. Yayin no. TUSIAD-T/2002-9/331. TUSIAD, Istanbul.

TZOB (2004) Tarimda gelismeler ve sorunlar. Turkiye Ziraat Odalari Birligi, Ankara.

Unver, O. and Gupta, R.K. (2003) Water pricing: Issues and options in Turkey. *Water Resources Development* 19(2), 311-330.

US Library of Congress (1995). *Turkey: A Country Study*. http://countrystudies.us/turkey/60.htm

Uzunlu, V., Düsünceli, F., Talug, C., Uyanik, M., Yalvac, K. and Istanbulluoglu, E. (1999) NARS of Turkey. In Casas, J., Sohl, M. and Hafaz, H. (eds.) *The National Agricultural Research Systems in the West Asia and North Africa Region*. Icarda, Aleppo, Syria.

World Bank (2004) Draft agenda. Innovation and competitiveness, practitioners workshop, 19-21 April 2004, Istanbul. WTO (2003) Trade policy review: Turkey. WT/TPR/S/125.

WTO (2003) Trade policy review: Turkey. Report by the Secretariat. WT/TPR/S/125, WTO, Geneva, 19 November 2003.

Yalvac, K. (1999) Advanced agronomic training needs and priority agricultural research in Turkey . In: Dupuy, B. (ed.) *Advanced Training for Agricultural and Food Managers in the Countries of the Mediterranean Area*. CIHEAM-IAMM, Montpellier, pp. 35-53.

Yomralioglu, T. (2003) Cadastral Template, Country Report: Turkey. November 2003. http://www.swisstopo.ch/cadastraltemplate2003/countryreport/Turkey.pdf (accessed 31 July 2004).

Ziraat Bank (2003) 2003 Annual Report. Ziraat Bank, Ankara.

4

Agricultural Production, Prices and Trade

Natasha Longworth

4.1 Introduction

Agriculture plays an important role in the Turkish economy, one that differs markedly from its role in the European Union. This chapter provides an overview of Turkey's agricultural sector, covering production, prices and trade. The agricultural sector in Turkey is compared with that of the European Union and other acceding countries. For purposes of comparison, the EU-25 is shown as the EU-15 (member states of the European Union prior to the May 2004 enlargement) plus the NMS (the ten member states that acceded to the EU in May 2004). Where relevant, the AC-2 (Bulgaria and Romania) are also mentioned. This chapter is complementary to the situation report published by the European Commission (2003). Where possible, it uses updated information; it also provides more detail, particularly regarding fruit and vegetable production, and agricultural prices and incomes.

Statistical definitions and methodology differ between the EU-25 and Turkey, and data used in this chapter are therefore often not fully comparable. Figures provided should be interpreted as indicative of the pattern of differences that exist between the European Union and Turkey and their orders of magnitude. Furthermore, the high rate of inflation in Turkey complicates the interpretation of time series in money terms.

The agricultural economy in Turkey is characterised by a large informal or unregistered sector, and significant subsistence or semi-subsistence agriculture. This makes it difficult for statisticians to provide a complete picture. Data used in the chapter are sourced where possible from relevant organisations in Turkey, mainly the State Institute of Statistics. Data in monetary terms have been converted from Turkish lira to euros using the average of daily exchange rates from the European Central Bank (European Central Bank, 2004).

The chapter is structured as follows. Section 2 outlines the role of agriculture in the general economy, showing its shares in GDP, employment and foreign trade. Section 3 attempts to assess the importance of land and labour for Turkish agriculture. Section 4 covers agricultural production: its size, diversity and economic value, with more detail for specific sectors including crops, fruit and vegetables and livestock, and the agronomic production potential of Turkey. Section 5 looks at output price indices for agriculture, as well as trends in real prices for specific agricultural products. Turkish output prices are compared with those of the EU-15. Section 6 addresses agricultural incomes in Turkey, compared to agricultural incomes in the EU-15. Section 7 presents self-sufficiency levels and outlines foreign trade in agricultural products. The last section summarises the chapter.

4.2 Agriculture in the economy

Agriculture has an important role in the Turkish economy, with high shares in GDP and employment. Although they remain much higher than in the EU-15, these shares have been declining over time.

Table 4.1. Indicators of the agricultural economy – Turkey, NMS and EU-15

Agricultural economic indicators	Turkey		EU-15		NMS	
	2000	2003	2000	2002	2000	2002
Share of agriculture in GDP (%)[1, 2]	13.6	11.5	2.0	1.9	3.5	3.1
Employment in agriculture (million)[2, 3]	7.8	7.2	7.1	6.9	3.9	3.9
Share of ag. in civilian employment (%)[2, 3]	36.0	33.9	4.2	4.0	n.a.	13.4
Agricultural imports/total imports (%)[2, 4]	3.6	4.2	5.7	5.7	5.6	4.5
Agricultural exports/total exports (%)[2, 4]	13.9	11.2	6.2	3.9	5.5	4.4

Sources: [1] *European Commission, 2004b;* [2] *European Commission, 2004a and 2002;* [3] *SIS, 2004f;* [4] *Undersecretariat of Foreign Trade, 2004 and SIS, various issues.*

Table 4.1 shows that the share of agriculture in GDP, employment and exports remains high in Turkey relative to the EU-15 and, to a lesser extent, to the NMS. The share of agriculture in GDP has been steadily decreasing over the last two decades, from 25 per cent in 1980 and 17 per cent in 1990 (SPO, 2004) to 13.6 per cent in 2000. The share of agriculture in total employment has also been declining, from 44 per cent in 1996/1997 (Cakmak, 2004) to 34 per cent in 2003. Around seven million people are employed in Turkish agriculture, about the same number as in the entire EU-15.

The share of agricultural imports in total imports is quite similar in Turkey, the NMS and EU-15, but agricultural exports have a much larger share of total exports in Turkey. The share of agricultural exports in total exports declined between 2000 and 2003 in all country groups. Turkey maintains a trade surplus in agricultural and food products.

Year-on-year changes in real agricultural output in Turkey fluctuate widely (see Figure 4.1). Since 1998 the growth rate of the agricultural sector has largely matched that of GDP in Turkey, with the exception of 2003, when the agricultural sector declined and the total economy grew. Growth in the agricultural sector also depends on weather conditions, although this effect is difficult to isolate as some areas and products may experience adverse conditions while others may experience ideal conditions.

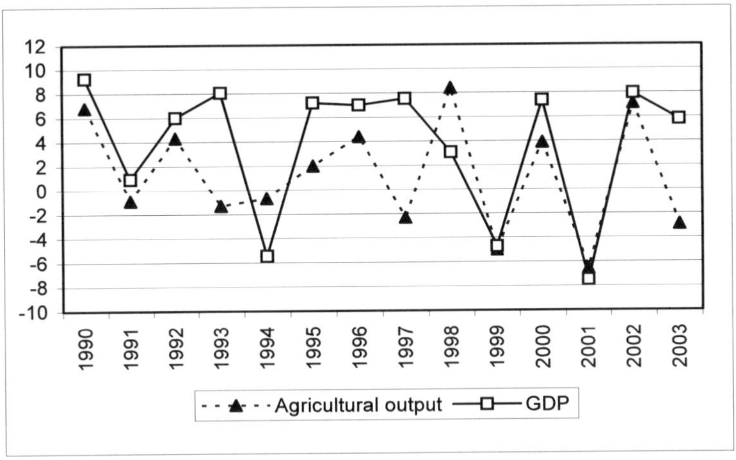

Figure 4.1. Growth rates in real agricultural output compared to GDP, Turkey, percentage change
Source: SPO, 2004; OECD, 2004a

4.3 Labour and land in agriculture

Labour

The Household Labour Force Survey of 2003 estimates that 7.17 million workers were employed in agriculture[1]. This represents 33.9 per cent of the total employment and 30.3 per cent of the labour force. For comparison, in 1990 8.7 million workers (47 per cent of the labour force) were employed in agriculture.

The share of rural[2] employment in total employment (47 per cent) is much greater than the share of rural population in total population (39 per cent), indicating both that unemployment is lower in the rural than the urban population (6.5 per cent as against 13.8 per cent), and that the labour force participation rate is higher in rural than in urban areas (56 per cent as opposed to 44 per cent) (SIS, 2004f). The overall labour force participation rate has been declining since the 1950s, associated with rural to urban migration. In rural areas, most household members participate in household-based production activities, while in urban areas there is more specialisation in either market or non-market production (Tunali, 2003). Over 63 per cent of the labour force in rural areas was employed in agriculture in 2003.

Agricultural employment is characterised by high rates of unpaid family labour, particularly amongst women. In 2003, 51 per cent of agricultural employment was classified as unpaid family labour (Cakmak, 2004) while 44 per cent were either employers or self-employed. Only five per cent were wage earners. Unpaid family labour accounted for 80 per cent of female agricultural labour, suggesting that the proximity of house and work allows these females to assist in agricultural production

[1] Employment in agriculture also include hunting, forestry and fishing.

[2] The rural population covers those in settlements of less than 20 thousand persons. The urban population lives in settlements with more than 20 thousand persons.

while maintaining child raising and household activities. The fact that half of agricultural labour consists of unpaid family labour is key in explaining the high labour force participation rates and low unemployment rates in the rural population.

Turkey's rural labour force and agricultural work force differ from those of EU countries because of high labour force participation rates, low unemployment and relatively high rates of illiteracy. In general, however, the category of 'unpaid family labour' makes cross-country comparisons difficult, since this category is treated differently in the statistics of different countries.

Land

Estimates of land used in agriculture vary. Table 4.2 gives an overview of statistics on agricultural land area, gathered from a number of sources. Even allowing for differences in definitions, there remain large discrepancies in agricultural area estimates. Furthermore, the trends in agricultural area differ according to the statistical source. The reasons for these discrepancies are discussed in Box 4.1.

Table 4.2. Definitions and estimates of agricultural land in Turkey 1980-2001, thousand ha

Element	Source	Definition	1980	1990	2000	2001
Total surface area	FAO	Total area of Turkey, including area under inland water bodies	77,482	77,482	77,482	77,482
Agricultural area	FAO	Sum of arable land and permanent crops and permanent pastures	38,579	39,677	39,050	40,888
Arable land	FAO	Area temporary crops, temporary meadows, market and kitchen gardens, temporary fallow (< 5 years)	25,354	24,647	24,138	25,938
Permanent crops	FAO	Area of crops that remain for long periods and need not be replanted after each harvest, including fruit and nut trees and vines	3,125	3,030	2,534	2,585
Permanent pasture	FAO	Land used permanently (>5 years) for grassy forage crops, either cultivated or wild	10,100	12,000	12,378	13,167
UAA	Eurostat[1]	Arable land including temporary grass, fallow, green manure, permanent grassland, area permanent crops, crops under glass and other agricultural areas			41,264	38,883

[1] The original source for the Eurostat data was not clear

Table 4.2 contd. Definitions and estimates of agricultural land in Turkey 1980-2001, thousand ha

Element	Source	Definition	1980	1990	2000	2001
Cultivated area	SIS	Area sown (land used for cereals, pulses, industrial crops, forage crops and other field crops with a growing cycle of under one year) and fallow land	24,920	24,192	23,588	23,002
Fallow land	SIS	Land at rest for a period before cultivation and no yield can be gained	8,188	5,324	4,826	4,914
Other agricultural land	SIS	Vegetable gardens, vineyards, fruit trees and olive trees. From 1995 only the enclosed area of fruit and olive trees is given	3,615	3,664	3,346	3,349
Total agricultural land	SIS	Total area covered with crops, vegetables, fruits, flowers and other permanent crops and fallow land	28,535	27,856	26,934	26,351
Cultivated area	Agricul-tural Census	Estimate derived from Village Head Census. Data from 1991 and 2001 census		21,103		22,156
Cultivated area	Agricul-tural Census	Estimate derived from Household Survey. Data from 1991 and 2001 census		21,449		17,164

Source: FAO, 2004b; Eurostat, 2004; SIS, 2004a, b and c

Given the large discrepancies in estimates, it is difficult to identify concrete trends in agricultural area in Turkey. However, the following tentative conclusions can be drawn regarding changes in agricultural area from 1980-2001 based on the data in Table 4.2, SIS publications and the discussion above: cultivated land has declined slightly and fallow land more so; the area of vegetable gardens has increased, the area of olive trees and vineyards has decreased and the area of fruit and nut trees has remained fairly constant. In 2001 it would appear that between 22 and 26 million hectares of land were cultivated (field crops, fallow land, vegetable gardens, orchards and vineyards). Native rangeland pastures have decreased and cultivated pasture has probably increased.

It is difficult to reach a conclusion regarding total agricultural area: FAO estimates that agricultural area has increased by almost 2 million hectares, while Eurostat figures show a decline in agricultural area of around the same amount. In 2001 the true agricultural area was somewhere between 35 and 41 million hectares. Once farm and land registers are completed, Turkey will have more accurate information regarding the extent and use of agricultural land.

Box 4.1 Measurement of agricultural land in Turkey

FAO estimates are based on data obtained from SIS, which are then reclassified according to FAO definitions. However, SIS definitions are not always consistent with FAO definitions, which leads to potential distortions in the data. Moreover, SIS uses various collection methods. The figures published in the annual SIS publication 'Summary of Agricultural Statistics' are collected by the Agricultural Directorates of MARA and are based solely on the information provided by agricultural technicians according to their regional observations (SIS, 2004c: IX). Data from the agricultural census (conducted every 10 years) are collected by two methods: village information survey and agricultural holdings survey. The village information survey is a full census method and covers all the provinces and districts with fewer than 25 thousand inhabitants. Questionnaires are issued to all *muhtars* (village heads) of all settlements with village status. The agricultural holdings survey is sample-based and in 2001 was applied to 10 agricultural holdings (chosen by sampling methods) in each of 4998 sample villages. The agricultural census is designed to obtain data that are not collected in the annual statistics, such as the size and types of agricultural holdings and land use. Figures from the agricultural census provide more detailed information on land use.

One element appears crucial in explaining the different estimates of total agricultural land, namely permanent pasture. The FAO defines permanent pasture as land used, for five or more years, for herbaceous forage crops, either cultivated or growing wild. According to FAO estimates, permanent pasture was around 10 million hectares in 1980 and had increased to 13 million hectares by 2001. However, other sources indicate that the pasture resource has been steadily declining. Karagöz (2003) reports estimates of around 21.7 hectares of pastures in 1980, and about 10 million hectares (in line with FAO) for 2000. The estimates reported by Karagöz (2003) appear to be based on the 1982-1984 land survey conducted by the GDRS.

The difficulty appears to centre on the issue of rangelands and common pasture areas. Turkey traditionally had a large area of natural rangeland that was, and still is, the main source of grazing for ruminant livestock. These rangeland areas are common property and are grazed free of charge. Until recently, boundaries of pasture areas were not clearly defined or assigned to village communities. It is therefore possible that estimates of agricultural area largely ignored or under-estimated these resources. Furthermore, since native rangeland vegetation usually included large shrubby areas, these may not have been included in some pasture estimates. In 1998, the Pasture Law (Law 4342) attempted to address these issues and assigned grazing areas to villages after defining boundaries (Karagöz, 2003). This should improve the accuracy of statistics relating to pasture.

In 1991, estimates of cultivated area from the survey of village heads and agricultural households carried out for the agricultural census were quite similar, but substantially lower than the annual figures provided by SIS. However, in 2001 the village head census yielded an estimate of cultivated area that is circa 5 million hectares more than the agricultural household survey. This may result from the sampling procedures of the household survey. It is also possible that the *muhtars* act strategically when answering the village information survey, thereby misrepresenting agricultural area.

4.4 Agricultural Production

Size and regional diversity of agricultural production

Agricultural production in Turkey is highly diversified due to the range of climatic and topographical conditions.

Using world prices to calculate the value of output quantities, FAO (2004a) reports that in 2003, Turkey's top commodity was wheat, followed by cows' milk, tomatoes, grapes and cotton lint. Other important commodities are barley, indigenous cattle and chicken meat, olives and apples. By international standards, Turkey is a major agricultural producer. Turkey ranks in the top five world producers for chickpeas, chillies and peppers, cotton, cucumber, eggplants, green beans, lentils, nuts (hazelnuts, pistachios, chestnuts, walnuts) onion, sugarbeet, tomatoes, watermelons and melons, stone fruit, figs, olives, and sheep's milk. Turkey is the world's largest producer of apricots, hazelnuts and figs (FAO, 2004a).

Table 4.3. Regional agricultural production diversity in Turkey

Region	Avg. temp °C	Days with humidity	Days with snow	Principal Products
Central-North	11	60	22	Rainfed cereals, food and forage legumes. Extensive small ruminants, intensive dairy cattle.
Aegean	16	65	0	Rainfed cereals, olives, citrus, grapes, tomatoes, vegetables, figs, irrigated cotton, tobacco. Extensive small ruminant and beef cattle, intensive dairy cattle. Proximity to main Turkish towns and export markets.
Marmara	14	70	10	Cereals, sunflower, olives, tomatoes, fruit and vegetables, poultry. Important cattle region.
Mediterranean	18	62	0	Rainfed and irrigated cereals, olives, cotton, citrus, maize, fruit and vegetables. Significant goat meat production, other livestock less important. Proximity to main Turkish towns and export markets.
North-East	7	60	100	Mainly extensive livestock production and subsistence cereal production.
South-East	8	50	1-80	Rainfed cereals and food legumes, irrigated cotton, nuts, grapes, fruits. Extensive sheep and goats. Increasing use of irrigation (GAP project).
Black Sea	14	75	10	Rainfed hazelnuts, vegetables, maize, sugarbeet, tea. Cattle production and extensive sheep raising.
Central-East	12	55	30	Rainfed cereal, food legume production. Extensive small ruminant production, local and cross-bred cattle for milk
Central-South	11	60	22	Rainfed cereals, food legumes, sugarbeet, vegetables and forage production.

Source: OECD, 1994; Karagöz, 2003; European Commission, 2003a

Agricultural production in Turkey has a distinctive regional distribution based on geographic and climatic factors. Most agricultural production originates from the coastal regions; the highest production is in the Mediterranean and Aegean regions (European Commission, 2003a) which are well suited to fruit and vegetable production. Table 4.3 shows climatic conditions and main agricultural products in the nine agricultural regions of Turkey. Over 55 per cent of Turkey is high altitude (> 1000 m), while only 10 per cent can be classified as low altitude (0-250 m) (Karagöz, 2003).

Value of agricultural production

In 2002, the value of agricultural production in Turkey was around EUR 29 billion. This was similar to the production value in the NMS but only 11 per cent of total production in the EU-15. The main agricultural product categories, key products and production value are shown in Table 4.4. Where data were available in a similar format, estimates for the EU-15 and the NMS are included for comparison.

Agricultural production in Turkey is mainly crop production, with 77 per cent of total production. This is a much higher share than in the EU-15. However, the official statistics may miss a significant amount of livestock production. As an example, meat statistics are based on registered data from the municipal slaughterhouses and the Moslem festival of sacrifice. The USDA (Sarigedik, 2003a) estimates that this accounts for only 70 per cent of total production. This may also be true for milk.

Moreover, the calculation of grazing livestock products is based on the application of fixed yield coefficients to the number of animals slaughtered or milked. Since 1984, the same coefficients, obtained from the 1984 General Census of Livestock, have been used. It is highly likely that yields have increased in the last twenty years and therefore official statistics will underestimate the true size of animal production. By contrast, the statistics should be more representative for poultry as production generally occurs in larger commercial operations. Taking all these factors into account, livestock production would still be small in comparison with the EU-15, probably in the vicinity of 30-35 per cent of total output.

Field crops (mainly cereals and industrial crops) have the largest share in Turkey's total output value. Fruit and vegetables together account for more than 40 per cent of production value. By comparison, fruit and vegetable production is far less important in the EU-15 and the NMS. For EU countries, the separate categories of fruit and vegetables include only those products covered by common market organisations (they exclude grapes, bananas, sweet corn, peas, fodder beans and olives). However, for EU-15, the *total* share of fruit and vegetables in production is 19.9 per cent (European Commission, 2004a), which is still a much smaller share than in Turkey. According to the official statistics, livestock products provide 23 per cent of Turkey's production value. The official share of sheep and goat meat in production value is just 1.3 per cent, although sheep are important in extensive agriculture in some low-income areas (Karagöz, 2003).

Table 4.4. Value of agricultural output in Turkey (EUR million), 2002[3,4]

	Turkey		EU-15		NMS	
	EUR mn	%	EUR mn	%	EUR mn	%
Total	28,940	100.0	270,911	100.0	26,085	100.0
% of EU-15 output		10.7		100.0		9.6
Crops	22,411	77.4	154,310	55.0		
Field crops	10,118	35.0				
Cereals	4,969	17.2	35,336	13.0	4,765	18.3
Wheat	*3,396*	*11.7*	*17,668*	*6.5*	*2,207*	*8.5*
Barley	*1,042*	*3.6*	*7,852*	*2.9*	*808*	*3.1*
Pulses	929	3.2				
Industrial crops	2,280	7.9				
Tobacco	*380*	*1.3*	*1,122*	*0.4*	*54*	*0.2*
Sugarbeet	*894*	*3.1*	*4,768*	*1.8*	*700*	*2.7*
Raw cotton	*973*	*3.4*				
Oilseeds	521	1.8	4,487	1.7	835	3.2
Tuber crops	1,418	4.9				
Potatoes	*959*	*3.3*	*6,170*	*2.3*	*1,319*	*5.1*
Vegetables	5,318	18.4	22,716	8.4	1,750	6.7
Fruits	6,975	24.1	17,668	6.5	1,265	4.9
Livestock products	6,529	22.6	126,090	45.0		
Milk	2,448	8.5	39,262	14.5	4,038	15.5
Cows' milk	*2,130*	*7.4*				
Meat	1,643	5.7				
Beef	*1,268*	*4.4*				
Poultry meat	1,036	3.6				
Eggs	946	3.3	5,328	2.0	1,023	3.9

Source: SIS, 2004a; European Commission, 2004a; own calculations

Turkey's output of cotton and nuts exceeds that of EU-25, whilst its production of textile fibres, tomatoes and onions is half or more of the EU-25 total (Table 4.5). Even for cereals, sugarbeet and oilseeds, where Turkey does not have a particular comparative advantage, accession of Turkey would increase the production capacity of the EU by over 10 per cent.

Comparing the AC-2 countries with Turkey, the agricultural land area in Bulgaria and Romania is only 13 and 36 per cent, respectively, as large as Turkey's agricultural area (Eurostat, 2004). Agricultural output in Bulgaria and Romania is also roughly one-seventh and one-third, respectively, the size of Turkey's output.

---

[3] Output data for the EU-15 and the NMS were available as shares of total agricultural output, broken down into products with common market organisations and those without. Therefore, not all sub-categories can be matched with those used for the Turkish data.

[4] Data on production value are published annually by SIS in the publication Agricultural Structure. For livestock products, it provides two tables of value, one for animal products (meat, milk, hides etc.) and one for 'livestock'. The livestock table estimates the value of production of the entire livestock herd, and may be an estimate of the capital value. In European Commission (2003a) it appears that these livestock 'production' figures have been aggregated with animal products, which significantly increases the estimated share of the livestock sector in production value. We have omitted livestock figures, and include only the value of animal products, on the assumption that the production value of almost all livestock is in the meat, hides and milk that it produces in one year. Including an estimate of the value of the livestock herd on top of this figure is double counting.

Table 4.5. Comparison of Turkey's output quantities (thousand tons) with EU-15 and EU-25

Product	Year	Turkey	% EU-15	% EU-25
Cereals	2002	30667	14.3	11.5
Wheat	2002	19500	18.7	15.6
Barley	2002	8300	17.3	14.7
Textile fibres	2002	989	73.7	72.2
Cotton lint	2002	988	152.8	152.8
Raw tobacco	2001	145	43.8	39.6
Sugarbeet	2002	16500	13.9	11.7
Oilseeds	2000	2140	14.8	12.2
Potatoes	2002	5200	11.2	7.8
Apples and pears	2001	2540	22.9	n.a.
Stonefruit	2000	1557	24.9	22.6
Citrus	2002	2493	23.4	23.1
Olives[1]	Avg 01-02	1200	11.0	11.0
Nuts	2002	843	114.4	111.5
Tomatoes	2001	8425	56.4	54.1
Watermelons	2002	4575	250.6	n.a.
Onions	2001	2375	61.9	49.3

[1] An average or production from 2001 and 2002 is used due to the typical uneven yield pattern of olives.
n.a. Data unavailable
Source: SIS, 2004a; European Commission,2004a; Eurostat (2004)

The production shares of the crop and livestock sectors in the AC-2 countries are similar to those of the EU-15 rather than to those of Turkey. Within the crop sector, there are some complementarities between AC-2 and Turkey, with AC-2 more specialised in cereals and oilseeds, and weaker in fruit, vegetables, pulses and treenuts. Bulgaria's tobacco production in 2003 was nearly 40 per cent as high as production in Turkey, and Romania's potato output was 75 per cent as high as Turkey's. Meat production is relatively much more important in AC-2 than in Turkey: Bulgaria's production is one-third that of Turkey, and Romania produces three-quarters as much meat as Turkey. Not only do the AC-2 countries produce more (relative to total size) of those meats that Turkey produces, but the meat category for AC-2 also includes pigmeat production[5].

Crop production

Cereals contributed 17 per cent of Turkey's total agricultural output in 2002. The share of cereals in agricultural production was similar in Turkey and the NMS, but much lower in the EU-15 (see Table 4.4). Table 4.6 shows the area of production for field crops in Turkey over the period 1980-2003. In 2003, 60 per cent of this

[5] Output comparisons in this paragraph are based on FAO production statistics for 2001, 2002, 2003 (FAO, 2004b).

area was used for cereal production. Cereals area peaked in 2000 and has been declining since then. Total cereal production in 2003 was 30.7 million tons, of which wheat and barley were 19 and 8.1 million tons. Official statistics do not identify durum wheat separately; the Turkish Field Crops Central Research Institute and the USDA estimate that durum wheat production is around 10 per cent of total wheat production (European Commission, 2003a). Other cereals grown are rye, oats, spelt, maize, millet, rice, canary grass and mixed grain (SIS, 2004a).

Table 4.6. Area in field crop production, thousand ha, 1980-2003

	Cereals	Pulses	Industrial crops	Oil seeds	Tuber crops	Fallow land	Total
1980	13292	732	1226	1362	268	8188	25067
1985	13845	1433	1258	1490	304	6025	24354
1990	13711	2283	1392	1557	286	5324	24553
1995	13817	1867	1401	1537	355	5124	24101
2000	13963	1539	1388	1319	319	4826	23354
2002	13786	1595	1426	1430	300	5040	23577
2003	13414	1514	1285	1377	292	4991	22873

Source: SIS, 2004b; SIS, 2001

Figure 4.2 shows wheat and barley yields for Turkey over the period 1980 to 2002. Both wheat and barley yields exhibit a small growth trend over this period, growing by an average of 10 kg per hectare per year for wheat and 14 kg per hectare per year for barley, although most of this growth occurred between 1980 and 1990. Cereal yields in Turkey are far below those achieved in the EU-15, and range from 31 per cent of EU-15 yield for wheat to 57 per cent for oats. However, the EU-15 average is heavily influenced by yields obtained in the United Kingdom, Germany, Denmark and France, which are exceptionally high by world standards[6]. The average wheat yield in Turkey is only marginally lower than those achieved in Argentina, Canada and Pakistan and slightly higher than that achieved in Australia (FAO, 2003a). The agro-ecological conditions in the latter group of countries are more comparable with those in Turkey.

The area planted to pulses has varied over the period 1980-2000 but has remained around 1.5 million hectares since 2000 (see Table 4.6). The two main pulse crops are chickpeas and lentils, with 41 and 26 per cent of the pulse production area in 2002. Although pulses have a relatively low share in production value (just 3 per cent in 2002), chickpeas and lentils are significant exports. Other pulses include broad beans, peas, dry beans, kidney beans and vetches.

The main industrial crops are sugarbeet, tobacco and cotton, which contributed respectively 3.1, 1.3 and 3.4 per cent to total agricultural output in 2002. The share

[6] Based on average cereal yields for the period 1999-2002, using SIS (2004b) data for Turkey and data from European Commission (2004a) for EU-15.

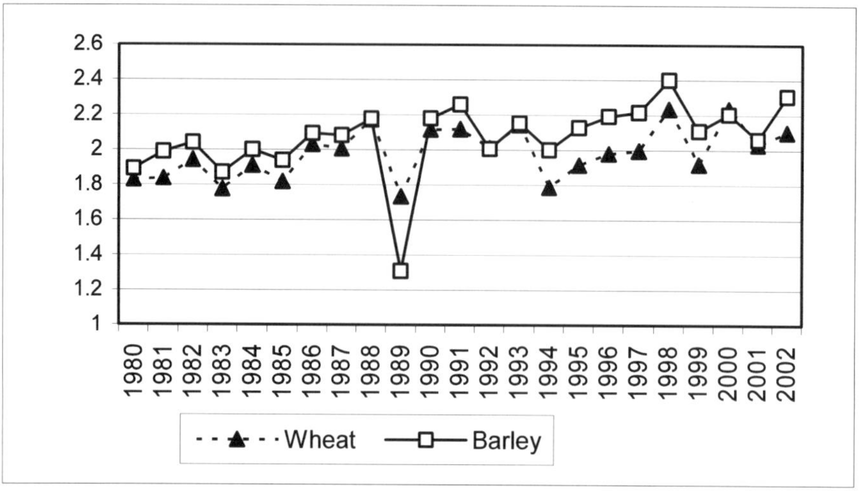

Figure 4.2. Wheat and barley yields in Turkey, ton/ha, 1980-2002
Source: SIS, 2004b and 2001

of sugarbeet in total production in Turkey is greater than in the EU-15 (and similar in the NMS). Turkey's average sugarbeet yield (2000-2003) was 41.4 tons per hectare, compared with 55 in the EU-25 and 59.1 in EU-15. Tobacco is also more important in Turkey compared to both EU-15 and the NMS. The largest production area is cotton (49 per cent of area in industrial crop production) followed by sugarbeet (25 per cent) and tobacco (14 per cent). Tobacco and sugarbeet areas declined by 32 and 28 per cent respectively over the period 1999 to 2002 in response to decreasing real prices, and stricter application of sugar quotas. Other industrial crop products include hemp and flax fibre, poppy capsules, aniseed, hops, dry pepper, cumin and lupin.

Fruit and vegetable production

Fruit and vegetables have high shares in total agricultural production in Turkey and also play an important role in agricultural exports. A number of products are important both in domestic production and for export. These products include tomatoes, hazelnuts, grapes, olives and citrus. Table 4.7 gives a breakdown of the value of fruit and vegetable production, separately identifying the major components. Grapes and tomatoes together account for about 30 per cent of total value.

Table 4.8 shows the trends in vegetable production over the period 1980 to 2003. Total vegetable production has more than doubled over this period. In particular, large increases have occurred in the production of fruit-bearing and tuberous vegetables. This production increase has been mainly due to tomatoes, green peppers and cucumbers for fruit-bearing vegetables, and to carrots, green onions and to a lesser extent red radish and turnips for tuberous vegetables.

Table 4.7. Fruit, nut and vegetable output value, 2002

	EUR mn	%		EUR mn	%
Total fruit and vegetables	12293	100.0	Total fruit and vegetables	12293	100.0
Vegetables	5318	43.3	Fruits (continued)		
Leafy or stem	492	4.0	Stone fruits	2124	17.3
Leguminous	332	2.7	*Apricots*	*165*	*1.3*
Fruit-bearing	4265	34.7	*Olives*	*1410*	*11.5*
Watermelon	*721*	*5.9*	Citrus	602	4.9
Tomatoes	*1903*	*15.5*	*Lemons*	*199*	*1.6*
Other vegetables	229	1.9	*Oranges and mandarins*	*380*	*3.1*
Fruits	6975	56.7	Nuts	1070	8.7
Green tea	176	1.4	*Hazelnuts*	*646*	*5.3*
Pome fruits	1004	8.2	Other fruits	2000	16.3
Apples	*791*	*6.4*	*Grapes*	*1599*	*13.0*

Source: SIS, 2004a

Tomatoes are the most important vegetable product in Turkey. They are grown throughout the country but production is centred in the Marmara and Aegean regions, where climatic conditions are ideal (Sirtioglu, 2004). There is a recent trend to increased greenhouse production in southern Turkey to provide for fresh consumption in urban areas during winter (Sirtioglu, 2004: 3). The bulk of production occurs on small family farms, and almost all planting and all harvesting is still done by hand. About 25 per cent of tomato production is processed, the rest is destined for fresh consumption. Of this 25 per cent, about 35 per cent is processed as tomato paste, 10 per cent for tinned tomatoes and the rest for tomato sauce, tomato juice, dried tomatoes and other products (Sirtioglu, 2004: 4). Tomato production was 9.8 million tons in 2002, compared to 6 million in 1990 and 3.5 million in 1980 (SIS, 2004b and 2001).

Table 4.8. Vegetable production in Turkey, thousand tons, 1980-2003

	Leafy[1]	Fruit-bearing[2]	Leguminous[3]	Tuberous[4]	Other[5]	Total
1980	1094	10111	443	277	65	11990
1985	1289	12989	542	380	58	15258
1990	1420	13958	560	451	68	16457
1995	1492	16101	602	670	80	18945
2000	1671	19284	660	653	90	22358
2002	1684	20596	686	643	90	23699
2003	1697	20679	709	827	108	24020

[1] Cabbage, leek, spinach, head lettuce, leaf lettuce, black cabbage, parsley, artichokes, celery, garden orach, mint, purslane, dill, cress, rocket. [2] Tomato, watermelon, melon, cucumber, green pepper, eggplant, pepper, squash, pumpkin, okra. [3] Bean, pea, broad bean, calavence, kidney bean. [4] Carrot, onion (green), red radish, garlic (green), horseradish, turnip, Jerusalem artichokes. [5] Cauliflower, asparagus.

Source: SIS, 2004b and 2001

Production of fruit has increased by 55 per cent since 1980 (Table 4.9). In particular citrus output almost tripled over the period. Within the citrus category, production of lemons and oranges increased more than grapefruit. By contrast, grape and soft fruit production stagnated. Dominating this category is grape production, which has remained very stable over the period.

Table 4.9. Production of fruit and nuts in Turkey, thousand tons, 1980-2003

	Nuts[1]	Pome fruit[2]	Stone fruits[3]	Citrus[4]	Grape and soft fruits[5]	Total
1980	470	1819	738	875	4010	7912
1985	422	2352	806	983	3851	8414
1990	630	2407	1094	1474	4042	9647
1995	715	2599	1106	1782	4107	10309
2000	758	2901	1557	2222	4182	11620
2002	843	2666	1333	2493	4138	11473
2003	789	3097	1606	2488	4312	12292

[1] Hazelnuts, walnuts, chestnuts, almonds, pistachios. [2] Apples, pears, quinces, loquats and medlar. [3] Olive, peaches, apricots, cherries, plums, sour cherries, wild apricots, cornel, oleaster. [4] Oranges, mandarins, lemons, grapefruit, sour oranges. [5] Grapes, figs, strawberries, bananas, pomegranates, mulberry, persimmons, carobs, kiwi, raspberry and avocado.
Source: SIS, 2004b and 2001

Hazelnuts are a key export product and Turkey dominates the world market. Figure 4.3 shows the trend in the number of hazelnut trees and production over the period 1980-2003. Hazelnut production is concentrated along Turkey's Black Sea coast, extending about 25 kilometres inland. Hazelnuts require little effort to cultivate, and inputs and labour costs are therefore low. The bulk of work occurs during the harvesting period in August (Sarigedik, 2003b). Best estimates suggest that the current area in hazelnuts is around 650 thousand hectares. Most eastern

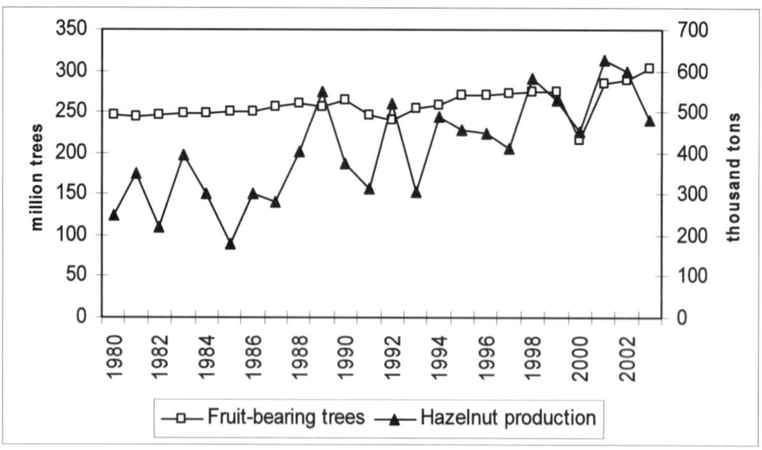

Figure 4.3. Number of trees and production of hazelnuts in Turkey, 1980-2003
Source: SIS (2004b and 2001).

producers have small orchards of around one to 2.5 hectares, and use hazelnut production for supplementary income, while western producers have larger orchards (10-15 hectares).

Hazelnut yield in Turkey is lower than in other major exporting countries. Yield is just above one ton per hectare which is 40 per cent of the average US yield and 50 per cent of the yield in Italy (Lundell *et al.*, 2004). Figure 4.3 shows that despite attempts to limit hazelnut production over the last few years, the number of fruit-bearing trees was the highest ever in 2003. However, production decreased from the 2001 peak. As the historic production pattern shows peaks and troughs, this decrease may or may not be permanent.

Grape production has remained remarkably stable since 1980 at around 3.5 million tons, despite a decrease in the area of production. In 1980, 820 thousand hectares were used for grape production, but this had decreased to 580 thousand hectares by 1990 and to 530 in 2003. Sarigedik (2003c) indicates that more adaptable varieties and improved cultivation practices have contributed to the increase in yield. About 45 per cent of Turkey's total grape production is consumed fresh, 35 per cent is dried and 15 per cent is processed, mainly for wine. A large proportion of the dried grapes are exported as raisins, an important export product for Turkey. It is estimated that seedless sultana varieties represent around 35 per cent of total grape production and are grown mainly in the Aegean provinces of Izmir, Manisa and Denizli. Sultana production has increased its share of total production due to high export demand and attractive domestic prices (Sarigedik, 2003c).

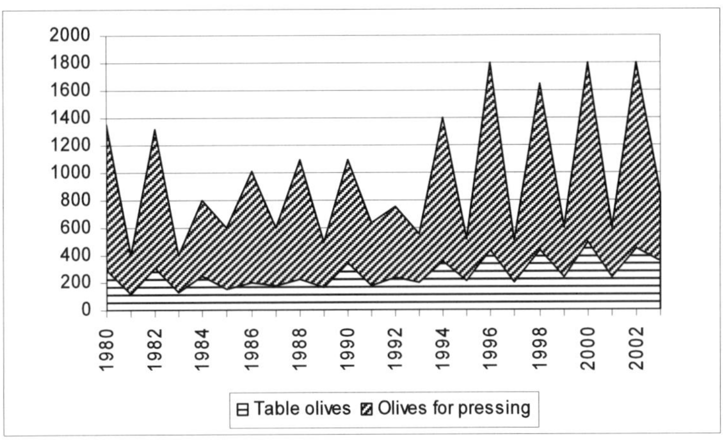

Figure 4.4. Olive production in Turkey, thousand tons, 1980-2003,
Source: SIS (2004b and 2001)

Olive production has been following an increasing trend. The number of fruit-bearing trees has increased from around 74 million in 1980 to 87 million in 2003. In 2003, 625 thousand hectares were planted to olive trees, compared with 556 thousand in 1995. Data from earlier years are not comparable due to a change in definition (SIS, 2004b). Despite steady trends in the number of fruit-bearing trees, however, output shows typical large variability from year to year, which appears to have increased in recent years. Olive production is sold for table consumption and for pressing.

Figure 4.4 shows that in low-yield years about half of the olive production is marketed as table olives, while in high-yield years the majority of the harvest is pressed. In exceptional yield years, there appears to be about one million tons of olives destined for pressing.

Citrus production nearly tripled between 1980 and 2003. Recent growth in grapefruit, tangerines and mandarins has been at the expense of oranges and sour oranges. Oranges have a share of 50 per cent in total products, followed by lemons and mandarins (22 per cent each) and grapefruits. Citrus is a traditional crop grown mainly in the southern Mediterranean and Aegean coastal plains. Sarigedik (2003d) estimates that about half of production takes place in larger orchards with modern technology. About 25 per cent of citrus fruit is exported (primarily oranges) and attractive export markets have led to the recent growth in grapefruit and mandarins. Figure 4.5 shows a sharp increase in total citrus production since 1997, while the increase for oranges and lemons was smaller.

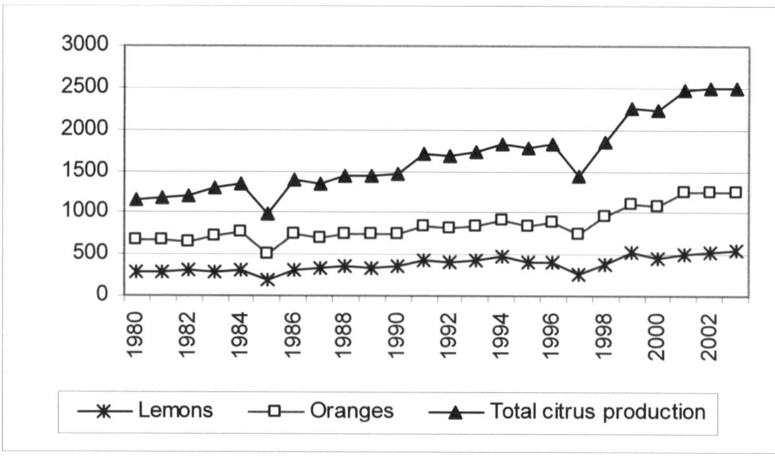

Figure 4.5. Citrus production in Turkey, thousand tons, 1980-2003
Source: SIS, 2004b and 2001

Livestock production

Total livestock numbers in Turkey have been steadily declining (Table 4.10). However, the figures for grazing livestock should be interpreted with caution since the reliability of official statistics is controversial (Sarigedik, 2003a). The changes between 2002 and 2003 suggest that numbers may be stabilising. However, it is too soon to establish any reversal in trend.

Table 4.10. Livestock numbers in Turkey (million head)

	1980	1985	1990	1995	2000	2002	2003	Total % change 1980-2000	Total % change 2000-2003
Cattle	15.89	12.47	11.38	11.79	10.76	9.80	9.79	-41.2	-9.0
Sheep	48.63	42.50	40.55	33.79	28.49	25.17	25.43	-47.4	-10.7
Goats	19.04	13.34	10.98	9.11	7.20	6.78	6.77	-88.8	-6.0
Poultry	58.58	61.05	96.67	129.02	258.17	245.78	277.53	326.9	7.5

Source: SIS, 2004b and 2001

Traditionally, lamb has been the preferred meat. However, tastes in urban areas are shifting to beef and poultry (Sarigedik, 2003a), which may have contributed to the decrease in sheep and goat numbers. Perhaps of greater importance is the prolonged social and political unrest in rural regions of the east and south-east of Turkey (Sarigedik, 2004), which may have impacted on sheep and cattle numbers. Poultry livestock numbers have more than doubled since the early 1990s and the sector has benefited from food safety concerns relating to beef.

Official meat production statistics are based on registered slaughterings from large commercial processing facilities, and also include animals slaughtered during the Celebration of Sacrifice. The USDA Foreign Agricultural Service estimates that this accounts for about 70 per cent of total production (Sarigedik, 2004). Prior to 1990, SIS added 10 per cent to the number of slaughtered animals to account for unofficial slaughter, but this was discontinued in 1990. Meat production is calculated by applying yield coefficients to the number of slaughtered animals. The series since 1984 use yield coefficients obtained from the 1984 General Census of Livestock (SIS, 2004a).

In value terms, the most important meat product is beef, followed by poultry. Meat production (including poultry meat) was 9.3 per cent of total agricultural output in 2002. Figure 4.6 shows beef, sheep and goat production in Turkey for the period 1980-2003. The strong increase in beef production in 1984 coincides with the change in yield coefficients used by SIS. Beef production has declined steadily since 1998; a similar trend is seen for sheep and goat meat. The production of poultry meat has tripled since 1995, despite declining real prices over this period. Figures for poultry production prior to 1995 are unavailable.

Milk, most of which is cows' milk, has a share of 8.5 per cent in total production. The production of cows' milk was circa 9.5 million tons in 2003. Cows' milk production decreased from 1995 to 2002 in line with declining livestock numbers and despite relatively stable real prices for cows' milk. Only about 20 per cent of milk production has a registered status, reaching the modern processing industry (G. Terberg, Ankara, November 2004, personal communication). Dairy farmers with milk cooling units receive a substantial price premium. Sheep and goat milk production occurs on a much smaller scale, although Turkey is a major world producer for sheep milk (770 thousand tons in 2003). Sheep and goat milk production have declined steadily in recent years, with a total decline of 30 and 25 per cent respectively over the period 1995-2002. Total milk production increased by 2 million tons from 2002 to 2003, reflecting increased production of cows', sheep and goat milk.

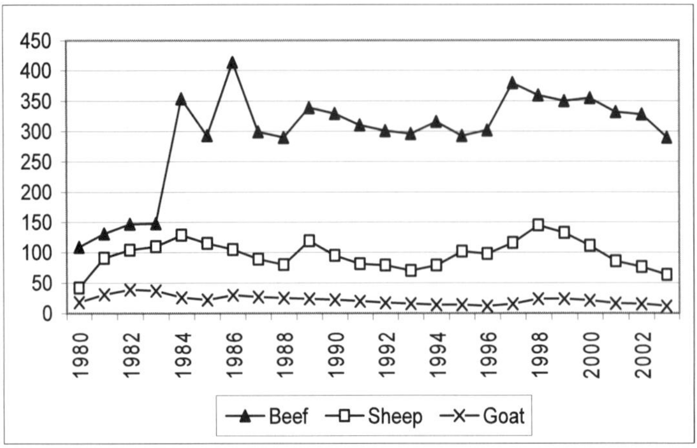

Figure 4.6. Meat production in Turkey, thousand tons, 1980-2003
Source SIS, 2004b and 2001

Agronomic potential

Agricultural production potential depends on both agro-ecological and socio-economic factors. This section examines evidence on Turkey's agro-ecological or physical agricultural production potential only, and covers the potential for increased agricultural area and increased efficiency of current use and the physical constraints that exist.

Almost all the land in Turkey with potential for arable cropping is cultivated (Keskin, 2001), so increases in production will need to come from more efficient use of current agricultural land. Fischer *et al.* (2000) estimated that around 250 thousand hectares of currently forested land is suitable or very suitable for crop production under rainfed conditions. However, conversion of forested land to agricultural land would have environmental consequences and may not be socially acceptable.

Various factors, such as the fertility and physical status of soils, climate and terrain (slope and elevation), may limit the production potential of Turkey. The fertility and physical status of Turkish soils indicate that the variability in soil properties is very high, which is not conducive to the uniform application of agricultural practices, such as fertiliser consumption. The majority of arable lands have problems with wind erosion. Only 14 per cent of total land area has a soil depth of 90 centimetres or more, and only 38 per cent of total land area has gradients that are highly suitable for farming (Keskin, 2001). A large proportion of soils have low organic content. Constraints vary considerably across the agricultural regions of Turkey and agricultural practices need to be adapted to individual situations in order to maximise production potential and minimise environmental damage.

Table 4.11. Share of total area (per cent) suitable for production of all crops under rainfed conditions

	High input[1]	Intermediate input[2]	Low input[3]
Very suitable, suitable and moderately suitable	32	23	10
Not suitable	57	62	87

[1] Farming system mainly market oriented, fully mechanised, high yield varieties, optimal fertiliser and pest control, large land holdings, capital intensive, low labour intensity.
[2] Production for subsistence plus commercial sale, intermediate capital intensity (credit accessible), medium labour intensity (incl. unpaid family labour), manual labour, some mechanisation, small sometimes fragmented landholdings, some fertiliser application/pest control. [3] Farming system largely subsistence based.
Source: Fischer et al., 2000

Fischer *et al.* (2000) developed a model for global agro-ecological zone analysis. Their results regarding crop production potential in Turkey are presented in Table 4.11. Currently, most of Turkey would be at the intermediate stage of input use, suggesting that 23 per cent of total area is suitable for production of crops. The model also estimates the climatic, soil and terrain (slope and elevation) constraints for crop production. Only 11 per cent of Turkey's land area has no climate, soil or terrain constraints. Severe terrain and soil constraints are present for 53 per cent of the land area. This study suggests a potential maximum wheat yield for Turkey of almost 5 tons per hectare, compared to the current 2 tons per hectare (FAO, 2003a). This yield, the maximum compatible with physical and agronomic constraints, may not be achievable in the given economic, administrative and environmental conditions.

Turkey's physically irrigable area is estimated at 26 million hectares, but only 8.5 hectares can be irrigated if economic considerations are also taken into account (DSI, 2004). Currently, 4.5 million hectares of agricultural land are irrigated and another 1.7 are expected to be irrigated under the Southeastern Anatolia Project (GAP) project. The average yield of irrigated land is 7.6 times that of dry-farming land, and the average value-added per hectare is 2.6 times that of one rain-fed hectare (Keskin, 2001). However, these figures might overstate the benefits of irrigation since often the more productive areas are irrigated.

Irrigation provides one avenue for increasing production. After completion of the GAP project, there appear to be a further 2 million hectares of potentially irrigable land. However, increasing the irrigated area may be constrained by water availability and potential environmental damage. Furthermore, given that the efficiency of current water use in irrigation has been estimated at 10-70 per cent (Keskin, 2001), it may be appropriate to focus more on increasing efficiency in current irrigation systems. To increase the efficiency of water use, various water pricing schemes have been proposed (Unver and Gupta, 2002).

The pasture resource has reduced from 56 per cent of total land area in 1940 to 16 per cent in 2000. Only 10-20 per cent of plants that are desirable in the botanical composition of pasture are present and plant cover ranges between 10-50 per cent (Keskin, 2001). Recovery of degraded pastures occurs slowly (Keskin, 2001) implying that the production potential of pastures will not improve dramatically in the short term. Research indicates a number of opportunities for improving pasture and forage productivity in Turkey (see Karagöz, 2003, for an overview) including fertiliser use, rotational grazing, artificial pasture establishment, forage production from fallow areas and fodder conservation. The establishment of the Pasture Law in 1998 has contributed to pasture improvement in registered areas, with some rehabilitation of degraded pastures.

It appears, from an agro-ecological perspective, that potential exits for increasing agricultural production despite a number of constraints relating to slope, soil types and structures and climate. First, agricultural practices need to become more adaptive to area-specific conditions, rather than a generic approach. This will allow producers to maximise production given area soil, slope and climate constraints. Second, a number of opportunities exist for improving the pasture resource, particularly forage production and fodder conservation. Third, current irrigation systems appear inefficient in terms of water use, which offers scope for increasing the productivity of existing systems. There is also further potential for increasing irrigation, though this is not without environmental consequences and may be constrained by water availability.

4.5 Agricultural prices

Producer and input price indices

Figure 4.7 illustrates the trends in real producer prices (1997 base year) for the period 1997 to 2003. Indices are the nominal farm price indices as published by SIS deflated using the GDP deflator. Agricultural prices in real terms declined after 1998

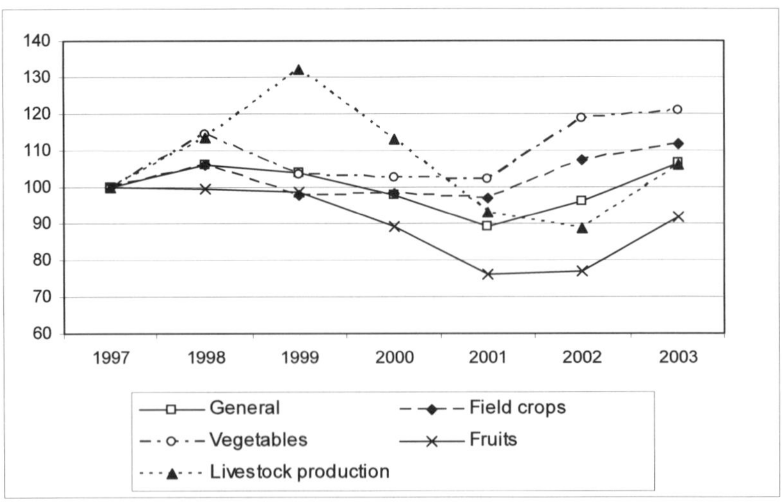

Figure 4.7. Trends in real price indices for crops, fruit, vegetables and livestock in Turkey, 1997-2003 (1997=100)
Source: SIS, 2004d; own calculations

(one year later for livestock products) but picked up again in 2002 and 2003. On average, real fruit prices were nearly 25 per cent below their 1997 level for two years (2001-2002), although field crop prices began to pick up in 2001. In 2002, livestock prices were at their lowest (one-third below their 1999 peak, and 10 per cent below their 1997 level). A large part of the decline in real prices in recent years has been the result of changes in the agricultural support programme (see Chapter 7).

Lundell *et al.* (2004) show the movement in the agricultural terms of trade (the ratio of nominal output prices to input prices) for the period 1997 to 2001. This ratio peaked in 1998 and fell steadily until 2001. Nominal input prices largely followed inflation over this period, so the fall in the terms of trade was due largely to output price changes.

Real price trends for selected products

Table 4.12 shows trends in real producer prices of selected agricultural products in Turkey over the period 1995-2003. Nominal producer prices (farm gate) were obtained from the OECD market support database (OECD, 2004b) and deflated using the implicit GDP deflator[7]. This provides only a rough idea of trends in real prices, since the GDP deflator captures price changes over the year, and does not therefore match the periods in which the products were marketed.

Table 4.12. Real prices for selected agricultural products 1995-2003 in Turkey, million TRL (at 1995 values) per ton

	1995	1996	1997	1998	1999	2000	2001	2002	2003
Wheat	7.4	10.4	10.9	9.4	8.6	7.4	7.7	7.8	9.1
Maize	6.8	9.5	8.8	8.3	7.4	6.8	6.8	7.2	8.4
Other grains	5.4	8.5	7.9	6.9	6.8	6.2	6.3	5.1	6.0
Oilseeds	18.0	19.7	20.1	19.4	18.1	15.9	19.5	16.5	16.0
Sugarbeet	3.0	2.8	3.8	3.1	3.2	2.7	2.4	2.6	2.7
Cotton	34.3	28.9	30.4	33.4	29.3	34.4	36.6	30.1	28.2
Tobacco	210.9	247.8	225.9	199.1	160.0	131.1	99.6	121.7	122.0
Potatoes	13.9	8.7	9.1	12.9	11.0	10.0	9.6	9.0	9.0
Tomatoes	10.2	15.1	13.5	12.5	10.1	12.5	10.9	9.7	13.2
Grapes	20.0	18.7	15.7	21.3	20.1	21.2	18.4	22.3	22.5
Apples	17.6	15.0	13.9	16.1	16.0	17.9	14.8	17.4	20.5
Milk	13.0	14.4	16.8	15.6	13.0	12.6	10.9	11.7	11.2
Beef	179.6	140.3	152.4	190.5	182.7	162.8	130.4	138.8	173.9
Poultry	79.7	68.0	65.1	78.2	66.4	56.6	61.1	56.1	49.6
Sheep meat	198.8	178.0	169.3	163.8	174.6	169.7	136.2	174.4	172.4
Eggs	72.4	70.8	66.2	56.5	50.3	64.2	61.0	57.7	59.7

Source: OECD, 2004b; own calculations

In general, it appears that the real prices of a number of products decreased prior to 2000/2001, consistent with changes in agricultural support under the ARIP

[7] Obtained from series for nominal and real (1987 prices) GDP published by SPO (2004).

programme. However, most prices have picked up again in the last two years. Fruit and vegetables are largely free from support and their prices (as illustrated by apples, grapes and tomatoes) vary significantly from year to year without showing any clear trends.

The prices of a number of crops (tobacco, wheat, and maize) decreased in the late 1990s but then increased in the last few years. This movement was particularly dramatic for the tobacco price, which by the end of the period was still about half its 1995 level. The price of sugarbeet reached a low in 2001 but has increased again in the last few years. An opposite trend is seen for cotton, whose price has been following world market trends. Prices for potatoes declined after 1998, remaining depressed in 2002 and 2003.

Recent real prices for milk appear low relative to prices received in the late 1990s. Prices for beef and sheep meat bottomed out in 2001 but have increased since then. Prices in 2003 were reasonably high when compared against prices since 1995. Prices for poultry reached their lowest level since 1995 in 2003.

Comparison of prices in the EU-15 and Turkey

Table 4.13 compares prices in Turkey and the EU-15 for certain agricultural products. Upon accession, Turkey will face prices in the EU-28. However due to unavailability of data a price comparison was only possible between Turkey and the EU-15. Producer prices at the farm gate are taken from the OECD market support database (OECD, 2004b) and converted into Euros using the reference exchange rates used throughout this book. Where price data was not available in the OECD database, other sources were used. In 1995, producer prices in Turkey were higher than in the EU-15, with the exception of wheat, maize, barley, cotton and milk. By 2000, only the producer price for milk was lower in Turkey than in the EU-15. Prices in Turkey 2000 ranged from 255 per cent of EU-15 prices (potatoes) to 101 per cent (cotton) and 95 per cent (milk).

Some prices moved closer to EU-15 price levels in 2003, but others diverged significantly. Prices for beef, potatoes, sugar, barley, maize and wheat showed some convergence over 2001 and 2002, but the trend was reversed in 2003. Prices for poultry, eggs, oilseeds, sheep meat and milk continued to converge towards EU-15 levels in 2003. In 2003 the prices of sheep meat and milk were substantially lower in Turkey than the EU-15, but all other prices were higher. Price differentials for wheat, maize and beef were exceptionally high at the end of the period shown, relative to their averages over the period as a whole. The price of potatoes remains much higher in Turkey (179 per cent of EU-15 level in 2003) as do prices for wheat (157 per cent) and beef (154 per cent).

The price convergence that has occurred partly reflects changes in agricultural support policies (see Chapter 7), although differences in agricultural support policies in the EU and Turkey are probably still important for explaining relative price levels. Persistent price differentials also suggest differences in quality and transport costs. The figures suggest that Turkey might be competitive in the production of milk, cotton, sugar and sheep meat. However, prices for processed dairy products are much higher in Turkey than in the EU, indicating high dairy processing margins in Turkey (Grethe, 2004: 50).

Table 4.13. Agricultural prices in Turkey and EU-15, EUR per ton, selected products

Producer Prices	1995	1997	1999	2000	2001	2002	2003
Wheat							
Turkey	124	204	170	169	143	159	193
EU-15	143	134	121	120	124	114	123
Turkish price as % of EU	*87*	*152*	*141*	*141*	*115*	*140*	*157*
Maize							
Turkey	113	166	146	157	127	148	178
EU-15	161	135	135	135	133	127	138
Turkish price as % of EU	*70*	*123*	*108*	*116*	*96*	*117*	*129*
Barley							
Turkey	90	148	134	142	117	105	128
EU-15	131	119	111	111	109	99	109
Turkish price as % of EU	*69*	*124*	*121*	*127*	*107*	*106*	*118*
Oilseeds							
Turkey	300	378	358	365	363	337	342
EU-15	220	251	198	215	230	255	257
Turkish price as % of EU	*136*	*151*	*181*	*170*	*158*	*132*	*133*
White sugar[1]							
Turkey	370	590	600	603	595	n.a.	n.a.
EU-15	n.a.	766	776	776	776	n.a.	n.a.
Turkish price as % of EU	*n.a.*	*77*	*77*	*78*	*77*	*n.a.*	*n.a.*
Potatoes							
Turkey	233	172	216	230	179	184	192
EU-15	216	90	167	90	121	113	107
Turkish price as % of EU	*108*	*190*	*130*	*255*	*148*	*163*	*179*
Milk							
Turkey	217	315	257	291	202	240	239
EU-15	306	304	300	307	326	316	309
Turkish price as % of EU	*71*	*104*	*86*	*95*	*62*	*76*	*77*
Beef							
Turkey	2997	2863	3605	3746	2422	2833	3710
EU-15	2723	2515	2533	2602	2229	2395	2413
Turkish price as % of EU	*110*	*114*	*142*	*144*	*109*	*118*	*154*
Poultry							
Turkey	1330	1223	1309	1302	1135	1145	1059
EU-15	996	1072	914	993	1082	987	1014
Turkish price as % of EU	*134*	*114*	*143*	*131*	*105*	*116*	*104*

[1] Turkish sugar prices are wholesale prices from Grethe (2004). EU price is an ex-factory price taken from Blume *et al.* (2002) and van der Linde, *et al.* (2000).

Source: OECD, 2004b; Grethe, 2004; Blume et al., 2002; van der Linde et al., 2000; own calculations

Table 4.13 contd. Agricultural prices in Turkey and EU-15, EUR/ton, selected products

Producer Prices	1995	1997	1999	2000	2001	2002	2003
Sheep meat							
Turkey	3318	3180	3444	3905	2530	3559	3678
EU-15	3138	3757	3299	3573	4112	4144	4187
Turkish price as % of EU	*106*	*85*	*104*	*109*	*62*	*86*	*88*
Eggs							
Turkey	1208	1244	993	1476	1133	1178	1274
EU-15	903	982	819	1033	1013	987	1141
Turkish price as % of EU	*134*	*127*	*121*	*143*	*112*	*119*	*112*
Cotton[2]							
Turkey	572	572	579	792	680	615	602
EU-15	834	836	631	784	637	765	n.a.
Turkish price as % of EU	*69*	*68*	*92*	*101*	*107*	*80*	*n.a.*
Tobacco[3]							
Turkey	3521	4243	3156	3017	1849	2483	2602
EU-15	2862	2904	2454	2354	2486	2936	n.a.
Turkish price as % of EU	*123*	*146*	*129*	*128*	*74*	*85*	*n.a.*
Olive oil[4]							
Turkey	2936	3103	2854	3949	2798	3705	3330
EU-15	2463	2382	2054	1849	1756	2012	1898
Turkish price as % of EU	*119*	*130*	*139*	*214*	*159*	*184*	*175*

[2] Cotton price for EU-15 is the average weighted minimum price in Greece and Spain taken from European Commission (undated).
[3] Tobacco price for EU-15 is the average weighted price of Basmas and Katerini varieties in Greece, taken from European Commission (2003a) and Eurostat (2004).
[4] Prices for EU-15 are the average of prices in Greece and Spain for extra virgin olive oil and taken from Eurostat (2004). The price for Turkey is from SIS wholesale statistics (SIS 2004e).
Source: European Commission, undated; European Commission,2003b; SIS, 2004e; OECD, 2004b; Eurostat, 2004; own calculations

The comparisons of prices for cotton, tobacco and olive oil should be interpreted cautiously. It appears that Turkey is competitive in cotton, and maybe also in tobacco in recent years. However, the price data we have been able to obtain may relate to different qualities and specifications of the products. Prices for olive oil are significantly higher in Turkey, however some of this difference could be explained by the stage of price collection: EU prices are selling prices of olive oil producers and Turkish prices are wholesale prices (Grethe, 2004).

Prices for fruit and vegetables are particularly difficult to compare, due to large variation in product specification and a lack of price data. Grethe (2004: 43) compared prices for some fruit and vegetable products in Turkey with the average of prices in Spain and Greece. In 2000, prices in Turkey were above the Spain/Greece average for apples, oranges, and lemons (112, 125 and 141 per cent of the Spain/Greece average price respectively), but well below for table tomatoes, cucumbers and grapes (34, 48 and 72 per cent respectively). It is highly probable that prices for fruit and vegetables differ significantly because of differences in quality and transport costs. Even within the EU-15, prices vary significantly across countries for most vegetables and fruits (Grethe, 2004: 43). As with milk, farm-gate prices for fruit and vegetables are significantly lower than fob prices in Turkey (40 to 85 per cent of fob price) indicating high marketing costs in Turkey. Furthermore, quality issues with Turkish produce occur, and only the high-priced range is currently eligible to enter the EU (Grethe, 2004: 44).

4.6 Agricultural incomes

Information and data on agricultural incomes are relatively scarce in Turkey. Cakmak (2004) reports some figures on disposable income (excluding transfer payments) based on the Household Budget Surveys conducted by SIS. These figures (see Table 4.14) provide evidence of the differences in incomes between the rural and urban population, and the agricultural and non-agricultural sectors.

Table 4.14. Rural and agricultural disposable income, 1994 and 2002 (in 1994 prices)

	Turkey	Rural[1]	Urban[2]	Agriculture[3]	Non-agriculture
Share in total					
1994 (per cent)	100	35.5	64.5	23.3	76.7
2002 (per cent)	100	34.6	65.4	19.3	80.7
Average income per employed household member (at 1994 prices)					
1994 (Turkey average=100)	100	60.4	156.4	46.4	154.2
2002 (Turkey average=100)	100	67.2	134.8	48.1	134.9
Percentage change 1994-2002	-3.1	7.8	-16.5	0.5	-15.3

[1] In settlements with population greater than 20 thousand. [2] In settlements with less than 20 thousand. [3] The main source of income is agriculture.
Source: Cakmak, 2004: 9

The rural population received one-third of total disposable income in 2002, whereas households dependent on agriculture received 23 per cent. In 2002, the average income per employed household member in rural areas was 67 per cent of the national average, while average income in urban areas was 135 per cent. The average income of agricultural households is only about half of the national level. However, relative average income for rural and agricultural households was higher in 2002 than in 1994. Cakmak (2004: 9) suggests that the impact of the 2001 crisis was more severe for urban areas.

Using gross agricultural value added as a proxy for agricultural income, Table 4.15 compares gross value added for Turkey, the EU-15, the NMS, Romania and Bulgaria. As well as agriculture, GVA includes the hunting, fishing and forestry sectors for all countries.

Agricultural income in Turkey is much lower than in the EU-15. GVA per person employed (including self-employed and unpaid family labour) was over eight times higher in EU-15 than in Turkey. GVA per person employed was also higher in the NMS and Bulgaria, whereas Romania had only half of the level in

Turkey. The very low GVA per employed also reflects significant hidden unemployment in agriculture in Turkey and probably also Romania. Turkey's relative position improves when the indicator GVA per hectare of utilised agricultural is used.

Table 4.15. Agricultural GVA in Turkey, EU-15, NMS and AC-2 (2002)

	Turkey	EU-15	NMS	Romania	Bulgaria
GVA (EUR million)	22,307	171,327	13,896	5,473	1,775
UAA (1000 hectares) data for 2000	38,883	130,809	36,167	14,819	5,325
Employment in agriculture (1000 persons)	7,458	6,537	3,880	3,683	299
GVA/employed (EUR/employed person)	2,991	26,209	3,581	1,486	5,936
GVA/UAA (EUR/ha)	574	1,310	384	369	333

Source: Eurostat, 2004; European Commission, 2004a

Agricultural incomes in Turkey have varied considerably over the period 1993-2002. Figure 4.8 shows indices of agricultural income in Turkey and the EU-15. Due to differences in data[8], these indices are indicative of trend only. There appears little similarity between agricultural income trends in Turkey and the EU-15.

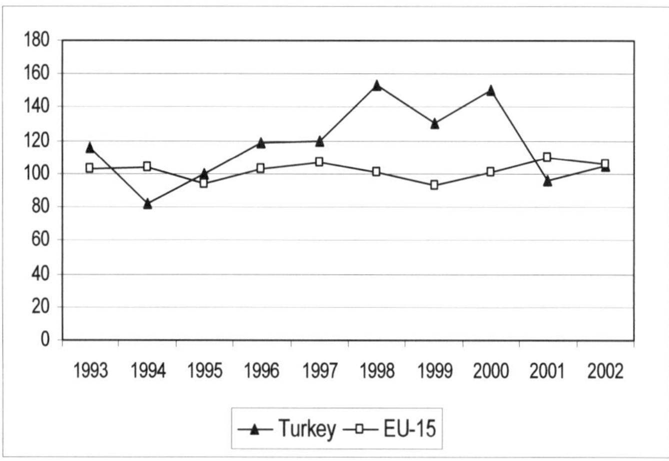

Figure 4.8. Real agricultural income (GVA) index (average of 1994-1996 = 100) for Turkey and EU-15, 1993-2002
Source: Eurostat, 2004; SIS, 2004a; European Commission, 2004a

[8] The EU-15 index is taken from European Commission (2004a). The index for Turkey was developed from GVA data and sector percentages provided in Eurostat (2004) and SIS (2004a). The base year for the EU-15 index was the average of 1994-1996, and the same base was used to construct the index for Turkey. The EU-15 index uses factor income (net value added at factor prices) and therefore provides a better income indicator than the index for Turkey.

4.7 Trade

In 2003, Turkey exported approximately EUR 4.3 billion of agricultural and food products and imported EUR 3.7 billion (Figure 4.9). Turkey typically has a trade surplus in agricultural products.

Raw products accounted for 78 per cent of Turkey's total agricultural exports and 92 per cent of imports in 2003. The share of raw products in imports has remained fairly steady since 1999, while raw products have decreased very slightly in importance for exports. Important export categories are fruit and vegetables, tobacco and cotton, and food preparations. In the category fruit and vegetables, 'edible fruit and nuts' and 'edible vegetables, roots, and tubers' accounted for 28 and 9.5 per cent respectively of total agricultural exports and for 9.5 per cent in 2003, and in the category food preparations, 'preparations of vegetables, fruits, nuts and plants' accounted for 13.7 per cent of total agricultural exports.

Agricultural imports are more diverse, with cereals, oilseeds, tobacco and cotton, and animal and vegetable oils and fats having relatively high import shares in 2003.

The EU-15 member states account for about 45 per cent of Turkey's raw agricultural exports. In particular, the EU-15 accounts for a significant share of Turkey's fruit exports. The EU-15 is less important as an export destination for the categories cereals, oilseeds and sugar. Relative to its total level, the NMS are an important destination for Turkish exports of oilseeds and cigarettes and tobacco, while the AC-2 countries are more important destinations for Turkish exports of fruits and nuts and preparations of vegetables and fruits.

Table 4.16. Self-sufficiency ratios[1], 2001

Commodity	Self sufficiency, %	Commodity	Self sufficiency, %
Wheat	97	Beef and veal	100
Rice	50	Mutton and goat meat	100
Barley	102	Poultry	101
Maize	76	Offals	123
Oil crops	84	Animal fats	54
Starchy roots	100	Milk, excluding butter	100
Sweeteners	114	Eggs	103
Pulses	117	Fish, seafood	80
Vegetable oils	53	Stimulants	80
Vegetables	105		
Fruit	129		

[1] Domestic production as percentage of domestic utilisation.
Source: FAO, 2003b

Table 4.16 summarises Turkey's self-sufficiency status for the main agricultural commodities. The situation for field crops is mixed: a clear deficit for rice, maize and oilcrops, and a clear surplus for sweeteners and pulses. Wheat, barley and starchy roots (including potatoes) are close to self-sufficiency level. Vegetables and fruit are surplus products. However, given the very large export flows from Turkey in these categories, it is surprising that these self-sufficiency ratios are not higher. As for livestock, most products are around self-sufficiency levels, except for animal fats.

4.8 Summary and conclusions

This chapter has provided an overview of Turkish agricultural production. Using the most recently available statistics, the agricultural sector has been summarised in relation to production and value, prices, agricultural incomes and trade. This final section provides a summary of the main findings and draws conclusions where appropriate.

Agriculture has a much larger role in the Turkish economy than in either the EU-15 or the NMS, with high shares in GDP and employment. Trends in agricultural land area are difficult to capture since estimates vary significantly according to data source. The following tentative trends can be identified in agricultural land use: fallow land has decreased, cultivated land has also declined slightly. Native rangeland pastures have decreased and cultivated pasture has probably increased. At this stage, it is difficult to reach a conclusion regarding trends in total agricultural area, since sources indicate conflicting trends. A best estimate of agricultural land in 2001 (the latest census year) in Turkey would lie between 22 and 27 million hectares for cultivated area and between 35 and 41 million hectares for total agricultural area.

Turkey is a major agricultural producer by world standards; it is one of the largest producers of many pulses, nuts, fruits and vegetables and the largest producer of apricots, hazelnuts and figs. Agricultural production varies across regions, according to climatic and topographic conditions. The highest production value occurs in the Mediterranean and Aegean regions, which are ideally suited to fruit and vegetable production. More extensive agriculture (crops and livestock) occurs in the more mountainous areas (particularly east and south-east).

The value of Turkey's agricultural production is similar to that of the NMS and around one-tenth of that in the EU-15. The crop sector is most important in Turkey and has a much higher share of total production than in the EU-15. However, the unreliability of livestock statistics suggests that official statistics may underestimate the true size of that sector. Even then, it is still likely that the livestock sector is less important in Turkey relative to the EU-15.

In terms of total agricultural production, field crops (cereals and industrial crops), fruit and vegetables and some livestock products (cattle, milk) have high shares. Some individual products such as wheat, tomatoes, olives, grapes and cows' milk have high output shares. The contribution of fruit and vegetables to total value of production is much higher in Turkey than in either the EU-15 or the NMS.

Crop production has grown slowly over the last twenty years (cereals, sugarbeet, cotton) or has shown very little trend at all (pulses, tobacco). In recent years, crop production areas and volumes show some significant changes, in response to the changing agricultural support environment and the Agriculture Reform Implementation Project. Grazing livestock numbers have fallen markedly. Over the same period, poultry production has increased threefold. In the last few years, a reversal in this trend is seen for sheep and goats but due to the short time frame, it is not yet clear whether this is truly a stabilisation of livestock numbers. Changes in livestock numbers are only indicative of the general trend, since livestock statistics are controversial.

Regarding future expansion possibilities for the sector, an analysis of the agro-economic conditions of Turkey indicates a number of climate, soil and slope constraints for crop production. Furthermore, the pasture resource appears to be fairly degraded. However, some potential exists for increasing production. Three key opportunities are identified: adapting agricultural practices to suit area-specific conditions, improving the pasture resource, and more efficient and possibly increased use of irrigation.

Prices in Turkey are declining in real terms for most agricultural products, mainly reflecting the reduction in agricultural support. Prices of outputs have fallen relative to inputs and also relative to prices in other sectors, indicating a reduction in profitability in the agricultural sector. Prices for agricultural products are generally much higher in Turkey than the EU-15; prices for some products indicate convergence to EU-15 prices while others have diverged in the last few years. Only the prices of sheep meat, milk, sugar, cotton and tobacco are significantly below those of the EU-15.

It is difficult to obtain a clear picture of agricultural income in Turkey, due to a lack of data. Average disposable income per person employed in rural households is around two-thirds of the Turkish average, but households whose main source of income is agriculture have an average disposable income of only half the Turkish average. Using gross value added as a proxy for income, it appears that agricultural income varies significantly from year to year. Gross value added per person employed in agriculture in Turkey is one-eighth that of the EU-15. In relation to the NMS and the AC-2 countries Turkish GVA/worker is about half the level in Bulgaria, only slightly lower than in the NMS and double the level in Romania.

Turkey maintains a trade surplus in agricultural and food products. The EU-15 is the main destination for Turkish agricultural exports and alongside the US, also a major source of Turkish imports. Key agricultural exports are fruit and vegetables and food preparations.

References

Blume, C., Strand, N. and Färnstrand, E. (2002) Sweet fifteen: The competition on the EU sugar markets. Swedish Competition Authority Report 2002:2, Stockholm.

Cakmak (2004) Structural change and market opening in agriculture. *CEPS EU-Turkey Working Papers,* No. 10, September 2004. Centre for European Policy Studies, Brussels.

DSI (2004) (General Directorate of State Hydraulic Works). Land use in Turkey. http://www.dsi.gov.tr/landuse.htm (accessed 16 August 2004).

European Central Bank (2004) Euro foreign exchange reference rates. http://www.ecb.int/stats/exchange/eurofxref/html/index.en.html (accessed 20 August, 2004).

European Commission (2002) *Agriculture in the European Union: Statistical and Economic information 2001*. Office for Official Publications of the European Communities, Luxembourg.

European Commission (2003a) Agricultural situation in the candidate countries. Country Report: Turkey. DG-AGRI. November 2003, Brussels.

European Commission (2003b) Raw tobacco: annex. DG-AGRI, 22 July 2003, Brussels.

European Commision (2004a) *Agriculture in the European Union: Statistical and Economic Information 2003*. Office for Official Publications of the European Communities, Luxembourg.

European Commission (2004b) 2004 Regular report on Turkey's progress towards accession. SEC (2004) 1201, Brussels.

European Commission (undated) The cotton sector. Working paper of DG-AGRI, Brussels.

Eurostat (2004) *Eurostat Database*. Eurostat, Luxembourg.

FAO (2003a) *World agriculture: towards 2015/2030. An FAO perspective*. Earthscan Publications Ltd, London.

FAO (2003b) Regional programme for food security in the Near East: Towards sustainable food security and poverty alleviation. Report from a High-level technical workshop, FAO-IDB, Jeddah 8-9/10.2003.

FAO (2004a) Major food and agricultural commodities and producers. http://www.fao.org/es/ess/top/country.jsp (accessed 25 August, 2004).

FAO (2004b) *FAOSTAT Database*. FAO, Rome

Fischer, G., van Velthuizen, H., Nachtergaele, F. and Medow, S. (2000) *Global Agro-ecological Zones*. CD-ROM. FAO and IIASA.

Grethe, H. (2004) *Effects of Including Agricultural Products in the Customs Union between Turkey and the EU: A Partial Equilibrium Analysis for Turkey*. CeGE-Schriften, Vol 9, Center for Globalization and Europeanization of the Economy, Georg-August-Universität Göttingen. Peter Lang Verlag, Frankfurt am Main.

Karagöz, A. (2003) Turkey. *Country Pasture/Forage Resource Profiles Database*. FAO Crop and Grassland Service, FAO, Rome.

Keskin, S. (2001) Gateway to land and water information: country report Turkey. FAO, Rome.

Lundell, M., Lampietti, J., Pertev, R., Pohlmeier, L., Akder, H., Ocek, E. and Jha, S. (2004) Turkey: A review of the impact of the reform of agricultural sector subsidisation. World Bank Working Paper, March 9, 2004.

OECD (1994) *OECD Economic Surveys: Turkey 1993-1994*. OECD, Paris.

OECD (2004a) *OECD Economic Outlook*, No 75, statistical annex tables. OECD, Paris.

OECD (2004b) *Producer and Consumer Support Estimates, OECD Database 1986-2003*. OECD, Paris.

Sarigedik, Ü. (2003a) Turkey: Livestock and products annual. *GAIN Report*, Number TU3019. USDA Foreign Agricultural Service.

Sarigedik, Ü. (2003b) Turkey: Tree nuts annual. *GAIN Report*, Number TU3020. USDA Foreign Agricultural Service, Ankara.

Sarigedik, Ü. (2003c) Turkey: Fresh deciduous fruit annual. *GAIN Report*, Number TU3023. USDA Foreign Agricultural Service, Ankara.

Sarigedik, Ü. (2003d) Turkey: Citrus annual. *GAIN Report*, Number TU3030. USDA Foreign Agricultural Service, Ankara.

Sarigedik, Ü. (2004) Turkey: Livestock and products annual report. *GAIN Report*, Number TU4022. USDA Foreign Agricultural Service, Ankara.

Sirtioglu, I. (2004) Turkey: Tomatoes and products annual report. *GAIN Report*, Number TU4019. USDA Foreign Agricultural Service, Ankara.

SIS (2001) *The summary of agricultural statistics 1980-1999*. State Institute of Statistics, Ankara.

SIS (2004a) *Agricultural structure 2002 (production, price, value)*. State Institute of Statistics, Ankara.

SIS (2004b) *The summary of agricultural statistics 1984-2003*. State Institute of Statistics, Ankara.
SIS (2004c) *General agricultural census: village information 2001*. State Institute of Statistics, Ankara.
SIS (2004d) Bulletin: Index of prices received by farmers, December 2003. SIS, Ankara.
SIS (2004e) *Wholesale Prices Index Database*. SIS, Ankara.
SIS (2004f) Household Labour Force Survey Bulletins, 6 April, 2004. SIS, Ankara.
SIS (various issues) Foreign trade statistics bulletins. SIS, Ankara.
SPO (2004) *Economic and social indicators 1950-2003*. State Planning Office: Ankara.
Tunali, I. (2003) Background study on labour market and employment in Turkey. Report prepared for the European Training Foundation, Italy.
Undersecretariat of Foreign Trade (2004) Turkey's foreign trade. http://www.byegm.gov.tr/references/turkeys-foreign-trade.htm (accessed 22 November 2004).
Unver, O. and Gupta, R.K. (2002) Water pricing: issues and options in Turkey. In Hamdy, A., Lacirignola, C. and Lamaddalena, N. (eds.) *Water Valuation and Cost Recovery Mechanisms in the Developing Countries of the Mediterranean Region*. CIHEAM-IAMB, Bari, pp. 123-148.
van der Linde, M., Minne, V., Wooning, A., Zee, F. van der (2000) Evaluation of the common organisation of the markets in the sugar sector. Report to the EU Commission. Netherlands Economic Institute (NEI), Rotterdam.

5

Turkey's Rural Population and Agricultural Workforce

Alison Burrell

This chapter concerns the rural population (those living in rural areas) and the agricultural workforce (people working in agriculture) in Turkey. These groups have considerable overlap, but are of course far from identical, since the rural population includes young and non-working elderly people, and rural dwellers who work outside agriculture. Moreover, some agricultural workers live in urban areas. The chapter examines urban-rural differences in labour force participation and unemployment. Education levels are reported, as well as the health and nutritional status, and the incidence of poverty, among the rural population.

5.1 Who is working in Turkish agriculture?

Some basic definitions and data

Data on the size of the rural and agricultural populations, employment in agriculture and the number of agricultural households in Turkey are given in Table 5.1. These data are not all mutually consistent, due to different definitions and measurement approaches used (Box 5.1).

Table 5.1. Rural population, agricultural employment and households (millions, percentage share of corresponding total in parentheses)

	Rural population[1]	Rural population[2]	Employment in agriculture[3]	Agricultural households[4]
1990	23.5 (41)	n.a.	8.7 (47)	4.1
2000	24.1 (35)	25.1 (39)	7.8 (36)	3.0-3.7

Source: [1] *UN Population Division (2003).* [2] *SIS (2001).* [3] *Percentage relates to total employment, SPO (2004).* [4] *From the Agricultural Census (1991 and 2001), census of village heads and agricultural household survey. As reported in Cakmak (2004).*

Despite differences in definition, the data in Table 5.1 all suggest that at least one in three of the Turkish population lives outside cities and larger towns, and that agriculture still provides about one-third of employment in Turkey. Moreover, the figures reveal an underlying dynamic process: as the urban population grew much faster than the rural population in the 1990s, the share of the rural population in total population fell although absolute numbers remained almost constant[1]. This movement was accompanied by a shift of economic activity out of the agricultural sector to non-agricultural sectors. According to the ARIP Quantitative Household Survey (Lundell *et al.*, 2004:30), farm households had on average 5.85 members in 2002.

Box 5.1 Measurement of rural population and agricultural labour

Turkey's population census covers the entire population, whereas the Household Labour Force Survey (HLFS) covers only the non-institutional civilian population of Turkey. It defines the urban population as persons living in a settlement of more than 20 thousand inhabitants; others (inhabitants of smaller towns, villages or the countryside) are classified as *rural population*. By contrast, Lundell *et al.* (2004) defined the *rural population* by classifying districts as urban if they are, first, city centres and, second, if their population density is more than 100 inhabitants per km^2. On this definition, the rural population was estimated to be 33.5 per cent of the total population in 2000.

The *agricultural workforce* is defined as people working in agriculture, hunting, forestry or fishing. They may belong to either the urban or rural populations. Thus, in the HLFS, the urban-rural distinction depends on the size of the settlement in which an individual lives, whereas the agricultural-non-agricultural distinction depends on the economic sector in which the individual is currently employed.

Most of the labour statistics in this chapter come from the Household Labour Force Survey (HLFS) of 2003, which collected information from over 78 thousand households (SIS, 2004). The HLFS uses the ILO definition of employment, which classes as employed anyone who worked at least 1 hour in the reference week. This definition has been criticised as over(under)-estimating (un)employment in Turkey (Tunali, 2003). Tunali reports that changing the threshold to 10 hours per week does not make much difference to unemployment rates. The effect on measured (un)employment, such as it is, is larger in rural areas.

Agricultural censuses and surveys usually cover the villages and district centres with less than 5,000 inhabitants. However, the 2001 agricultural census included the villages and districts with less than 25,000 inhabitants.

Labour force participation rates

Table 5.2 shows that Turkey's population is about 60 per cent urban and 40 per cent rural (according to the definition based on settlement size). However, of those working in 2003 only 53 per cent were from the urban population, whilst 47 per cent belonged to the rural population. This imbalance results mainly from differences in labour force participation rates. An adult male of working age (15 years or older) in a rural area is more likely to be in the labour force (that is, employed, or seeking work) than his counterpart in an urban area. For females, the difference is particularly marked: a woman of working age in the rural population is twice as likely to be in the labour force as one from an urban environment. It is striking, however, that, during the 1990s the urban female work-

[1] Between 1990 and 2000, the urban population had an annual growth rate of 2.68 per cent, compared with only 0.42 per cent for the rural population (Tunali, 2003). This difference is due to rural-to-urban migration.

Table 5.2. Labour force participation and employment by urban/rural status, 2003

	Turkey	Urban[1]		Rural[1]	
		Male	Female	Male	Female
Total non-institutional civilian population (thousands)	69479	21144	20991	13548	13796
Percentage of total population, %	100.0	30.4	30.2	19.5	19.9
Labour participation rate[2], %	48.3	68.9	18.5	72.9	39.0
Total numbers in work	21147	9025	2262	6231	3630
Percentage of total working, %	100.0	42.7	10.7	29.5	17.2
Agriculture[3] as % of total workers	33.9	2.9	9.5	55.4	89.0

[1] Urban = settlements with a population > 20 thousand. Rural = non-urban.
[2] Labour force (employed + unemployed seeking work) / population aged ≥ 15 years.
[3] Main economic activity in agriculture, hunting, forestry or fishing.
Source: SIS, 2004

force grew significantly faster than the urban male workforce.

Table 5.2 shows the importance of agriculture as a source of employment in rural areas, particularly for women: 89 per cent of employed women in the rural population are working in agriculture. Of these women, over 80 per cent work as unpaid family members (Table 5.3). As was suggested in Chapter 4, the proximity of house and work allows rural women to assist in agricultural work while maintaining child raising and household activities.

Table 5.3. Employment status by urban/rural status and gender, 2003

	Turkey	Urban[1]		Rural[1]	
		Male	Female	Male	Female
Agricultural workers					
All workers in agriculture[2] ('000s)	7,165	263	215	3,455	3,232
Share of agricultural workers by employment status, per cent					
regular & casual employees	5.4	21.3	15.3	6.1	2.7
self-employed & employers	43.7	66.9	30.7	68.8	15.9
unpaid family workers	50.9	11.8	53.5	25.1	81.4
Non-agricultural workers					
All workers in non-agriculture ('000s)	13,982	8,762	2,047	2,776	398
Share of non-agricultural workers by employment status, per cent					
regular & casual employees	73.8	72.9	88.5	65.1	78.6
self-employed & employers	22.7	24.4	6.8	30.9	10.1
unpaid family workers	3.5	2.7	4.7	4.0	10.8

[1] Urban = settlements with a population > 20 thousand. Rural = non-urban.
[2] Main economic activity in agriculture, hunting, forestry or fishing.
Source: SIS, 2004

By contrast, similar opportunities are not readily available for urban women. Their lower labour participation rate is due to both cultural and economic factors (World Bank, 2000). A major reason, however, is the lack of wage-earning opportunities for women in urban areas, particularly for those with low education (World Bank, 2000: 14). The World Bank study suggests that Turkey's falling total labour participation rate (from 64.5 per cent in 1975 to 47.9 per cent in 1997, based on Census data, and 48.3 per cent in 2003, based on Household Survey data) is due in considerable part to the withdrawal of women from the work force when the household moves from a rural to an urban area.

Table 5.3 shows that, among agricultural workers, self-employment or employer status prevail for men, whereas for women unpaid work predominates. Employment opportunities for rural women in the industrial and service sectors are particularly limited (Table 5.4). For non-agricultural workers, the breakdown by employment status is quite similar in both urban and rural areas, although in urban areas there is a higher probability of being a wage-earning employee rather than an employer or self-employed than in rural areas, and for women there is a greater probability of being an unpaid family worker in rural areas.

Table 5.4. Employment sector by urban/rural status and gender, per cent, 2003

	Turkey	Urban[1]		Rural[1]	
		Male	Female	Male	Female
All workers	100.0	100.0	100.0	100.0	100.0
Agriculture[2]	33.9	2.9	9.5	55.5	89.1
Industry	18.2	27.7	27.2	9.3	4.0
Construction	4.6	7.3	1.1	4.5	0.1
Services	43.4	62.1	62.2	30.7	6.8

[1] Urban = settlements with a population > 20 thousand. Rural = non-urban.
[2] Agriculture includes forestry, hunting and fishing.
Source: SIS, 2004

There is considerable regional diversity in agriculture's share of employment. Taking urban and rural areas together, Marmara stands out with only 14 per cent of workers employed in agriculture in 2003. Regions with the greatest reliance on agriculture for jobs were the Black Sea (61 per cent), East Anatolia (53 per cent) and Southeast Anatolia (44 per cent). Looking only at the rural areas within these regions, the share of jobs in agriculture is 78 per cent in the Black Sea region, 79 per cent in East Anatolia and 85 per cent in Southeast Anatolia, which shows the lack of economic diversification in the rural economy of these regions[2].

Unemployment

Not only are labour participation rates higher in rural areas, but also unemployment rates are lower. In 2003, urban unemployment was 13.8 per cent, as opposed to only 6.5 per cent in rural areas. The differential is especially great for

[2] East and Southeast Anatolia have the highest concentration of Kurdish inhabitants. Kurdish people comprise Turkey's largest minority ethnic group and are estimated to be 10-20 per cent of the population.

women (18 per cent unemployment for urban women, 4 per cent for rural women).

When interpreting these statistics, the question arises as to whether they include individuals working in the informal ('unregistered') economy. Tunali (2003) considers that the HLFS probably captures most employment in the informal economy (estimated by SIS in 2003 at 37.7 and 47.6 per cent of the male and female urban workforce, and at over 50 per cent of total employment by OECD, 2004), although some types of work that are typical of the informal economy (such as domestic service) are excluded from the HLFS coverage. OECD (2004: 146), based on SIS data, gives an estimate of over 91 per cent of agricultural workers as unregistered.

The phenomenon of hidden unemployment, or under-employment, is often attributed to traditional or semi-subsistence agriculture. In countries where labour markets are not functioning well, it is possible that employment statistics mask considerable under-employment. The HLFS seeks to shed light on this, by investigating whether those working less than 40 hours per week are doing so by choice, and also by identifying individuals who, although working full time, are seeking a job change because of insufficient income[3].

According to HLFS definitions and results, only 4.8 per cent of the Turkish labour force was under-employed in 2003, with under-employment lower in the

Box 5.2 How young is Turkey's working population?

In 1991, Turkey joined the ILO-initiated International Program on the Elimination of Child Labour (IPEC). In 1998 it signed the ILO Convention 138 that establishes the minimum age of employment at 15 years. Prior to 2000, the relevant population for calculating labour force participation in the HLFS included those aged 12 years or more. Since 2000, the HLFS has used 15 years of age as the age threshold. The labour force participation rate measures the proportion of those working, or available for work, from the population aged 15 and older. Therefore, economic activity at ages below 15 is not recorded in the regular employment statistics.

However, HLFS supplementary data indicate that in 1999, 511 thousand children (4.2 per cent) in the 6-14 age group were working (down from 8.5 per cent in 1994). Tunali (2003, Table 3.6.1) reported that in 1999, in the age group 6-17, there were 1.64 million economically active children. Tunali's figures show that child employment rates were much higher in rural areas, and 58 per cent of all working children worked in agriculture. However, child employment in agriculture tends to be part-time; a far greater proportion of working children worked over 40 hours per week in urban areas than in rural areas. Participation in work within the 6-17 age group increases with age in both urban and rural areas, and is higher for males than females at each age in urban areas. However, the gender differences are small in rural areas. Overall, more than half the children working in 1999 worked as unpaid family labour, and in rural areas, five out of six working children had this status. The latest HLFS results (fourth quarter, 2003) estimate 654 thousand children working in the 12-17 age group, which represents a fall of 22 per cent relative to the same quarter in 2002.

---

[3] Implicitly, this defines under-employment from the respondent's subjective viewpoint, rather than in terms of economic output relative to the amount of time spent (the economist's viewpoint).

rural population than in the urban population. This average result masks a large gender difference: under-employment rates for men were slightly higher for rural men than for urban men, whereas in all rural areas – including the low-income south-east and east regions – under-employment rates were particularly low for rural women[4].

The urban-rural differentials in unemployment rates measured over all workers also hold, at national level, for the rate of unemployment amongst the 'educated youth' (persons aged 15-24, educated to high school level or above): in urban areas in 2003, these rates were 27 and 33 per cent for men and women respectively, and 22 and 25 per cent respectively in rural areas. At the regional level, the urban-rural differential is reversed for some regions (see Table 5.5).[5]

Table 5.5. Educated youth unemployment, Turkey and selected regions, 2003

	Total unemployment %	Educated youth unemployment, %				Agriculture's share in employment %	Under 15 share of population %
		Urban[1]		Rural[1]			
		Male	Female	Male	Female		
Turkey	10.5	27	33	22	25	34	30
Marmara	11.8	24	30	17	32	14	25
Black Sea	4.8	33	37	16	21	61	26
Southeast Anatolia	21.6	36	51	69	33	44	43

[1] Urban = settlements with a population > 20 thousand. Rural = non-urban.
Source: SIS, 2004

Table 5.5 compares the educated youth unemployment rate for Turkey as a whole with three selected regions: Marmara (high-income region, industrial), Black Sea (medium-income region, but predominantly agricultural) and Southeast Anatolia (low-income region, with above average reliance on agriculture). The high rates of educated youth unemployment in Southeast Anatolia, together with the high proportion of under-15-year-olds in the population of this region, are striking.

The World Bank Report (2000) estimated that 22 per cent of income inequality between Turkish households could be explained by differences in the educational status of the household head, and a further 25 per cent by his employment/unemployment status. However, the figures in table 5.5 indicate that, for young people, better education does not guarantee employment. Moreover, except in the case of Southeast Anatolia, these figures suggest that a young, more educated person, who wants to remain in his region, has a better

[4] A striking exception is East Anatolia, where in 2002 and 2003 the under-employment rate for rural males was nearly 20% of the labour force.

[5] Males who do not enrol for higher education are enlisted for compulsory military service at age 20 (Tunali, 2003). Therefore, the 15-24-year age group may be slightly under-represented in the non-institutional civilian labour force.

chance of finding employment if he remains in a rural environment rather than moving to a larger conurbation.[6]

The relatively low unemployment rate for the Black Sea region shown in Table 5.5 indicates that reliance on agriculture *per se* does not condemn a region to lag behind the others. Agriculture in this region has good productivity, and specialises in the high-value crops tobacco, hazelnuts and tea. Moreover, each year the region attracts several hundred thousand seasonal workers from other parts of the country. World Bank (2000) argues that it is *low-productivity* agriculture (low mechanisation, low capital/land ratios), with its weak multiplier effects in the local economy, that is the decisive factor for unemployment in rural regions. Arguably, low human capital in the agricultural work force also contributes to this low productivity.

The current situation presents a challenge for labour market policy. As Tunali (2003: 80) concludes: "Turkey will have to train its labour force so that it can improve productivity, but currently it is unable to provide jobs for a large segment of those who are trained". This is part of an even larger challenge: due to the age structure of the Turkish population, where nearly 20 per cent of the population is in the age group 15-24 (see Figure 5.2), many new jobs will have to be created in the coming years for young people at all levels of educational attainment (Taymaz and Özler, 2003).

5.2 Human capital in agriculture

The human capital invested in the agricultural workforce is of interest for several reasons. It determines the efficiency with which given agricultural resources and technologies can be exploited, as well as the speed with which new techniques, regulations and opportunities can be communicated and are likely to be adopted. It is likely to be a factor in the rate of 'spontaneous' (endogenous) innovation engendered by farmers themselves. Furthermore, in periods of rapid structural change, when agriculture is shedding workers in considerable numbers, their level of human capital is likely to influence the rate at which they can be absorbed in other sectors.

The measurement of human capital is fraught with difficulty. For agriculture, various indicators are used, such as years of formal schooling, years of work experience or age distribution. The following section focuses on literacy, years of education and specialised training in agriculture, while acknowledging that they cannot tell the whole story.

Literacy and schooling

In 1990, overall literacy rates for 15 years and older were 89.8 per cent for males and 67.4 per cent for females; these rates had risen to 94.4 and 78.1 per

---

[6] Tunali (2003) reports that the south and south-east were among the regions badly affected by the 2000-2001 crisis, which may help to explain these very high figures in 2003.

Box 5.3 Disincentives to acquire education

In a very large study of Turkish children using data for 1994 and 1997, Tansel (1998) found that the number of years of school completed at each of three levels (primary, secondary and high school) was positively related to the educational attainment of both parents, to household income, and to an urban location. In addition, from middle school onwards, self-employment status of the father had a negative effect on the number of years' schooling completed by both male and female children. This indicates that the cycle of low educational attainment amongst agricultural families, where self-employment is the norm and adult levels of education are very low, tends to be self-perpetuating.

Economic incentives also play a role. Tunali (2003) provides econometric evidence that, although the economic return (in terms of lifetime earnings) to an extra year of schooling is positive all the way through to university, it is lower for extra years beyond primary school and yet lower again for extra years after high school. Moreover, the incentive (in terms of *incremental* return) to continue past primary school had fallen in 1994 compared to 1988 (unless the child also went on to high school). Finally, the additional economic incentive to proceed to university after high school was lower than the incentive to complete high school for both men and women, regardless of their future status (wage earner, civil servant, etc). The evidence on rates of educated youth unemployment in Table 5.5 confirms the lack of labour market incentives for individuals to invest in additional years of education, in both urban and rural areas.

Finally, transaction costs and expectations influence the choice for more education. According to World Bank (2000), most of Turkey's 30 thousand villages have a 5-year primary school only, so that to continue beyond primary level, village children would have to commute daily or attend a boarding school. This report also cites evidence (from East and Southeast Anatolia) that parents may decide not to keep their children in school because they (rationally) consider their children's chance of success in secondary school to be small, given the poor quality of village primary schools (World Bank, 2000: 73).

cent by 2000. The literacy rate in the 15-24 year age-group was 97.8 per cent for males and 93.2 per cent for females (UN, 2004).

Figures for literacy rates in Turkish agriculture relative to other economic sectors are given in Table 5.6. It is striking that nearly one in five of the agricultural workforce is illiterate. Among male agricultural workers, one in twelve is illiterate, and 85 per cent have received no education above primary school

Table 5.6. Literacy levels in the Turkish workforce by sector, 2003 (percentages)

	Illiterate	Literate, no school	Primary	Lower secondary	Higher secondary	Higher education	Total
Agriculture	18.1	6.1	65.0	6.0	4.4	0.4	100
Male	8.5	6.5	69.7	8.0	6.7	0.6	100
Female	28.5	5.8	59.9	3.8	1.9	0.1	100
Manufacturing	1.2	1.1	51.9	15.1	23.5	7.2	100
Construction	2.6	2.6	58.2	13.8	15.8	7.2	100
Services	1.4	1.1	34.2	13.9	28.2	21.3	100
All sectors	7.1	2.9	48.8	11.4	18.8	11.0	100

Source: Derviş et al., 2004: 87

level. Among rural women, more than one in four is illiterate and over one-third have not completed primary school. Less than 6 per cent of rural women have been educated to lower secondary school standard or higher.

Low rates of educational attainment in the adult population may prove to be quite persistent. Evidence for 1999 (Tunali, 2003: 36) shows that over 8 per cent of children aged 6-11 were not enrolled at school, despite the extension of compulsory schooling in 1997 from 5 to 8 years (from age 6-14). The net primary education enrolment ratio was still only 92 per cent in 2003.[7] The net secondary enrolment ratio in 2003 was 46.5 per cent (50.2 per cent for males and 42.4 for females). The primary school non-enrolment rate was the same in both rural and urban areas. At levels beyond primary school, enrolment rates were lower in rural areas compared with urban areas, particularly for girls. The reasons for low levels of schooling, and particularly for the high drop-out rate after primary school, are complex (Box 5.3).

In the early 2000s, Turkey was spending only 2.1 per cent of GDP on education, despite having over 20 per cent of the population in the 5-14 years age group. For comparison, average expenditure on education for OECD countries is around 3.5 per cent of GDP. Turkey's expenditure per pupil was 20 and 12 per cent of the OECD average for primary and secondary school pupils respectively (European Commission, 2001). Law 4702 of 2001 provides a framework for extending compulsory basic education from 8 to 12 years (i.e. beyond the current ceiling at 14 years of age), but the deadline has been moved for lack of resources. Currently, high school (starting at age 14) consists of 3 years of general or vocational high school, or 4 years of technical high school. The participation rate in vocational programmes in 2002/3 for the 15-19 age group was 19.4 per cent of the age group (Corradini and Fragoulis, 2004).

In December 2003, the State Planning Organisation prepared the first Preliminary National Development Plan as a basis for the financial assistance towards economic and social cohesion to be provided by the EU during 2004-2006. The plan identifies four development targets, the second of which is 'developing human resources and increasing employment'. Within this second target area, labour market policies and reinforcement of the education system to strengthen links between the labour market and education are specifically mentioned. Twelve NUTS II regions have been identified as having priority under this project. The total budget allocation to this second target area for 2004-2006 for these 12 regions is EUR 70 million (EUR 48.5 million from the EU, EUR 21.5 million from the national budget) (Corradini and Fragoulis, 2004).

These initiatives are positive. However, the challenge is enormous. Although the increase in numbers of urban school children peaked with the age cohort that was aged 15-19 in 2000, and 5 years later in the rural population (Figure 5.1), there are millions of children in the education system and this will continue. In order to extend the years of schooling and improve the quality of education, very large increases in expenditure are required.

---

[7] See Population and Development Indicators: http://nkg.die.gov.tr/.

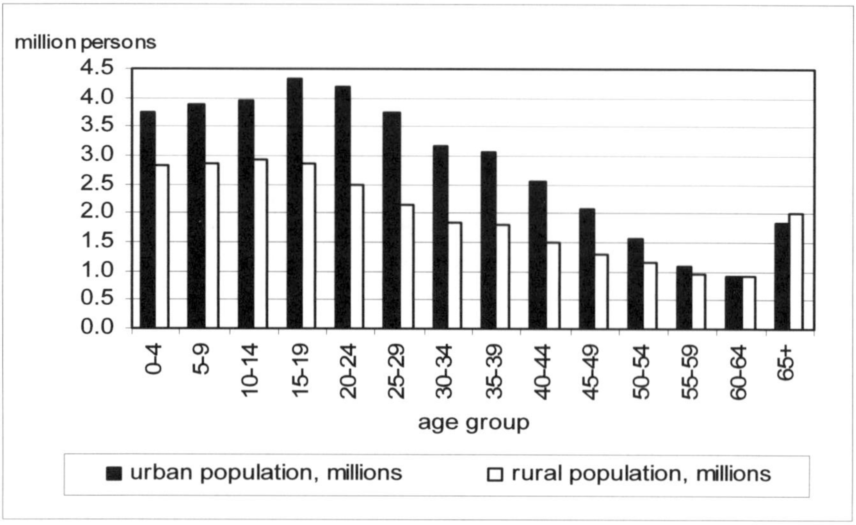

Figure 5.1. Rural and urban populations by age group, 2000
Source: SIS population census data 2000, in Tunali, 2003

Agricultural education and training

After finishing secondary school at age 14, students may attend a vocational high school. This involves a 3-year diploma programme aimed at training qualified people for various professions (for example, workers in health, agricultural extension, land surveying) and also at preparing students for higher education. In 1998-99, 536,317 students were enrolled in 1,129 Vocational High Schools[8]. There are 18 Agricultural Vocational High Schools, which were operated under the responsibility of MARA until their transfer to the Ministry of National Education in September 2004. Table 5.7 gives a breakdown of the total educational enrolment in 1999-2000.

Table 5.7. Overview of education, 1999-2000 school year

Levels of education	Number of schools, '000s	Number of students, '000s	Number of teachers, '000s
Pre-school	9.88	251.60	15.70
Primary schools	43.32	10,053.13	324.92
Secondary schools	6.17	2,444.41	143.47
General high schools	2.66	1,506.38	70.25
Vocational and tech. high schools	3.51	938.03	73.22
Total formal education	59.37	12,749.13	484.09
Non-formal education	6.53	2,978.80	48.52
TOTAL	65.91	15,727.93	532.60

Source: Ministry of Foreign Affairs, 2004

[8] Turkish Education System 2004, see http://karatekin.cmyo.ankara.edu.tr/minneso/s_2.htm.

Non-formal education, the other main component of the national education system, includes training, education, guidance and practical activities outside formal channels. It is designed for those who have never entered the formal education system, or who have dropped out early. Its objectives are adult literacy, development of knowledge and skills, and vocational training. Apprenticeship schemes and distance learning are among the formats used for non-formal education. In this context, MARA provides training on agriculture and home economics to villagers through its local organisations (Tunali, 2003).

At the level of higher education, there were in 1999 19 Faculties of Agriculture (25 thousand undergraduate students), 14 Faculties of Veterinary Medicine (5.8 thousand undergraduates), and 9 Faculties of Forestry. Since then, 4 new agricultural faculties and 5 new veterinary faculties have been established. All these faculties offer a 4- or 5- year bachelor's programme, and most offer higher degrees. The research effort of these faculties is constrained by high teaching loads, lack of coherent research strategies, weak links with other research institutions and inadequate research resources (Uzunlu *et al.*, 1999). The first role of postgraduate programmes in agricultural and veterinary faculties is seen as producing researchers for Turkey's various research institutes.

5.3 Quality of life in rural areas

Health

In Turkey, life expectancy at birth for someone born in 2000-2005 is 68 years for a male and 73.2 years for a female. This represents an increase of about 10 years since the late 1980s, but still lags about 10 years behind western Europe.

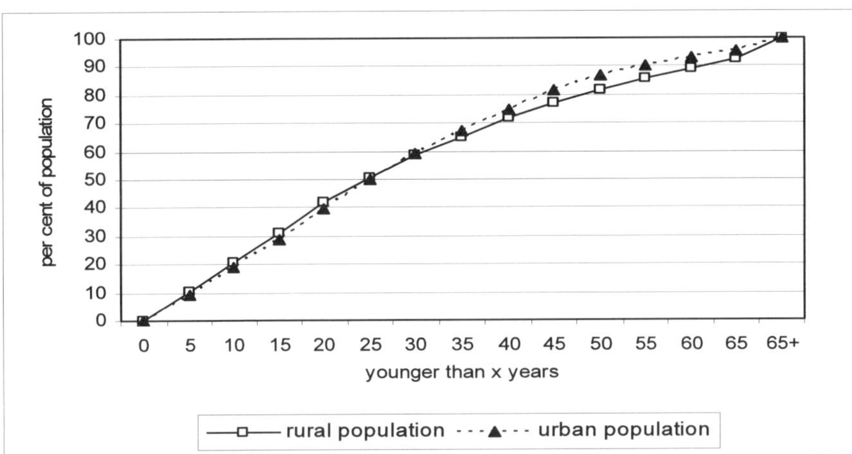

Figure 5.2. Age distributions of rural and urban populations, 2000
Source: SIS population census data 2000, in Tunali, 2003

Moreover, World Bank (2000: 38) cites evidence from the late 1990s of a 10-year gap in average life expectancy between Turkey's poorest and most developed areas.

Figure 5.2 shows the cumulative age distributions in the rural and urban populations for 2000. In both populations, about 60 per cent were younger than 30 years of age. However, compared with the urban population, the average age of those above 30 was higher in the rural population, whereas the average age of those below 30 in the rural population was younger. This suggests a movement of individuals in the middle age groups to an urban environment[9].

Turkey's infant mortality rate fell from 53 per thousand in 1993 to 36 per thousand in 2002[10]. However, large differences in public health levels and access to health services still exist between western and eastern Turkey, and between urban and rural areas (Ergör and Öztek, 2000). Behind the average infant mortality rate of 42.7 per thousand in 2000 were rates of 35.2 and 55 per thousand in urban and rural areas respectively (Savas *et al.*, 2002). The difference in infant mortality between 'west' and 'east' was even more marked: 32.8 and 61.5 per thousand respectively. Similar regional differences are observed for child immunisation rates (Savas *et al.*, 2002; Ministry of Health, 2004). The incidence of people with disabilities registered in the 2000 census was 1.69 per hundred in the urban population and 2.0 per hundred in the rural population as a whole, with a rate of 3.19 per hundred for men aged over 20 in rural areas. Figures for the number of inhabitants per health care professional (doctors, dentists, pharmacists, midwives, etc) show large differences between metropolitan, 'developed' and 'underdeveloped' provinces. For example, the average number of inhabitants per doctor in each of these types of region in 2000 was 879, 1294 and 2299 respectively (Savas *et al.*, 2002: 81). Expenditure on health care increased as a proportion of GDP from about 3.7 per cent in 1992 to about 4.8 per cent in 1998 (of which 3.4 per cent was publicly funded). However, it is clear that a further increase is needed just to bring lagging areas up to the standards prevailing in the more favoured parts of the country.

Other public health indicators show a more mixed picture of the urban-rural 'divide': for example, in 2000 86 per cent of rural inhabitants had access to an improved water source, compared with 81 per cent in urban areas. By contrast, only 70 per cent of the rural population had access to improved sanitation as against 97 per cent of the urban population (UNICEF, 2004)[11].

Malaria was a major disease in Turkey in the first part of the twentieth century. Massive malaria control activities began in 1926, and the disease virtually disappeared from the late 1940s onwards. The early 1970s, however, saw a resurgence in Southeast Anatolia. In 1976 and 1977, nearly 31 and 116

---

[9] The ratio of males to females in the 15-30 age group was 1.06 in urban areas compared to 1.01 in rural areas, whereas the ratios over all age groups were rather similar. This suggests an out-migration of young men from rural to urban areas in search of work.

[10] Rates of well under 10 deaths per thousand births prevail in western Europe, compared with 12 per thousand in Hungary and Poland, 16 per thousand in Bulgaria and 23 per thousand in Romania.

[11] For definitions, see http://earthtrends.wri.org/pdf_library/country_profiles/Pop_cou_792.pdf

thousand cases, respectively, were reported. Large–scale control programmes were re-introduced, but despite this, cases have been reported each year throughout the 1990s, with a peak in 1998 (over 35 thousand cases). The disease has remained concentrated in Southeast Anatolia, and the small number of cases that are observed in other parts of the country are attributed largely to migrant workers from that area. Just over 10 thousand malaria cases were reported in the year 2002 (Ministry of Health of the Republic of Turkey, 2004).

Nutrition

According to FAO (2003), per capita calorie intake in Turkey in 2001 was 3343 k-calories per day, the highest among the countries of the Economic Cooperation Organisation (ECO). Yet in Turkey only 10 per cent of daily energy intake was coming from animal products, the third lowest rate for the region. Although Turkey's per capita *total* protein consumption was one of the highest in the region, both the share and the absolute level of *animal* protein in the total was below the ECO regional average[12].

FAO (2001) reported that per capita total calorie intake remained constant during the 1990s, although the proportions of carbohydrate and protein both fell slightly (to 64.3 per cent and 11.5 per cent of calories, respectively, in 1998) whereas the share of fat increased (to 24.2 per cent). Per capita intake of meat, dairy products and eggs actually fell slightly over this period (FAO, 2001: 10). This study cites a number of surveys, spanning the period 1974 to 1999, reporting micronutrient deficiencies in the Turkish population, including significant levels of anaemia due to iron deficiency in adolescent girls, pregnant and lactating women. Results of several surveys undertaken in the mid- and late 1990s indicate significant proportions of underweight children, especially in rural areas and in particular in the eastern region, as well as stunting[13].

As is suggested in Section 8.4, agricultural and trade policies in recent years have restricted supply for animal products, especially meat, and kept prices high. This is likely to be an important factor behind the poor nutritional status of a significant proportion of the population. However, increasing meat imports and allowing domestic meat prices to fall is unlikely to improve the situation for the poorest households.

Poverty and security

The results of the 2002 Poverty study (SIS, 2002) reveal that nearly 20 per cent of the rural population had consumption levels 50 per cent or more below the national

---

[12] In 1999/2000 Turkish consumption of meat, eggs and milk equivalents stood at 22, 75 and 51 per cent, respectively, of EU-15 levels (Grethe, 2004: 49).

[13] In a sample of nearly 1 thousand rural children, 22 per cent had height/age ratios that were at least 2 standard deviations below the median of the international reference population, and over 8 per cent were at least 3 standard deviations below. Comparable figures for urban children (in a sample of 1700 children) were 12.6 per cent and 4.7 per cent.

average. In the rural and urban populations respectively, 4.1 and 2.4 per cent of individuals were living below USD 2.15 per day (converted from TRL 1329 thousand using the PPS exchange rate). When this poverty threshold is doubled, the proportions become 39 and 25 per cent respectively. The urban-rural difference is partly due to the greater share of large households (7 persons or more) in the rural population. The inverse correlation between poverty and education is particularly striking (Table 5.8).

Table 5.8. Percentage of urban and rural individuals in poverty by education level, 2002

Education level	Percentage below the poverty line[1]		
	Turkey	Urban	Rural
Total	30.0	22.0	34.5
Illiterate	41.1	35.9	46.4
Literate without a diploma	34.6	30.0	41.1
Primary school	26.1	21.8	31.1
Elementary school	26.5	21.2	34.3
Secondary school	18.8	13.8	30.1
High school	9.8	7.1	17.7
University	1.6	1.1	4.4

[1] Defined according to a 'basic needs' criterion.
Source: SIS, 2002

Rural seasonal workers and the rural unemployed had the highest rates of poverty (45 per cent and 63 per cent respectively). In rural areas, the poverty rate among the economically inactive was 30 per cent, but this was lower than the rate for both the rural self-employed and rural unpaid family workers. The poverty rate among agricultural workers was 36 per cent (33 per cent if they lived in an urban area, 37 per cent in a rural area). Lundell *et al.* (2004: 41) report that for the rural poor, wages were a smaller share (19 per cent) of household income than the non-poor in 2002 (24 per cent).

Unlike developed economies, income distribution in Turkey is determined largely by market outcomes. State transfers typically play a negligible role in income redistribution (World Bank, 2000: vii). Social insurance and pensions are primarily linked to employment in the formal sector and mainly benefit the middle class (World Bank, 2000: 57). In 2003, only 9 per cent of agricultural workers had social security coverage (compared with 36 per cent of construction workers and 58 per cent of workers in the service sector) (SIS, quoted from Cakmak, 2004). Prior to the recent agricultural reforms, poor agricultural workers did not benefit from redistribution to agriculture. The World Bank has estimated that 37 per cent of total agricultural subsidies went to the 5 per cent of farmers with the largest farms, whereas just 22 per cent went to the two-thirds of farmers with less than 5 ha.[14] Even then, these figures over-estimate the support

[14] Lundell *et al.* (2004: 41) report that for both the rural poor and non-poor in 2002, DIS payments represented on average 7-8 per cent of household income. However, as these payments are linked to land, their distribution will follow the inequality of land distribution over households up to the 50 hectare ceiling for payment.

received by the poorest, because small farmers tend to use less of inputs that were subsidised (chemical fertiliser, machinery).

Regarding the elderly, although old-age income assistance accrues to the poor and non-poor equally, it too is mainly linked to previous employment in the formal sector. A study of over 1 thousand people over 65 years of age in 1998 (representative of both the urban and rural populations) found that 93 per cent had no old age pension, just 56 per cent had water piped into their dwelling and "the majority were living in undesirable conditions" (Celik and Celik, 2001).

5.4 Discussion, implications and conclusions

The agricultural work force represents about one third of total employment in Turkey, has low levels of educational attainment, and in some regions very low productivity. Almost half of the agricultural workforce are women, who work mainly as unpaid family labour. Agricultural workers experience a high incidence of poverty, as well as low provision of public services and social transfer payments. Although the evidence shows that employment prospects are positively related to educational attainment, unemployment among the 'educated youth' is in fact higher than for the population as a whole.

Two conclusions emerge from this overview. First, improving education provision, access and attainment levels must become a top priority for Turkey in the coming years. As the World Bank concluded (World Bank, 2000: 45): "No other policy measure can have an equally deep and lasting effect as eradicating illiteracy and increasing the level of education for all." This same report goes on to say that policies must begin by redressing the current inequalities in educational provision, quality and attainment, and this means directly targeting the agricultural – and rural – groups in Turkey.

This situation represents a particular challenge for Turkey in a pre-accession period. Given that, up to now, education levels within candidate countries have been much more comparable with levels among existing EU members[15], relatively less attention has been given in accession discussions to this area. Although the EU is working towards greater harmonisation and comparability between member states in terms of education, education policy is still largely a matter for each individual member state. By contrast, the *acquis* focuses particularly on regulations aiming to support the single market and to impose common (i.e. EU-level) policies. This means that in a pre-accession phase, the attention of Turkey's policy makers will be drawn to these specific areas, at the risk of diverting attention away from national education policy as a top priority area. And yet the longer-term performance of Turkey as an EU member, and the success of its economy within a competitive single market, depend crucially on the human capital of young Turkish people.

Second, the duality of Turkey's economy is epitomised by the third of its population that remains in agricultural employment. Although some parts of Turkish agriculture are successful and competitive, there are large pockets, in

[15] See http://www.nationmaster.com/cat/ Education.

particular in livestock production and in the south and east of the country, where performance and living standards are very low. A pre-condition for regenerating these parts of agriculture is a restructuring of the small-scale and fragmented farm-size structure. This will put huge pressure on socio-economic groups for whom there is currently no safety net.

It follows that preparing for greater exposure to competitive agricultural markets, and promoting job creation in non-agricultural sectors, should be central to Turkey's pre-accession strategy. This should be accompanied by liberalisation of the labour market so as to improve access, incentives and rewards for better educated and more mobile young people. It appears that, on becoming an EU member, the biggest adjustment would be asked of the most vulnerable in the population. Current low levels of literacy in the most vulnerable areas, together with the particular age structure of the population, mean that new strategies are needed to manage structural adjustments that go beyond the experience gained in recent enlargements.

References

Cakmak, E. (2004) Structural change and market opening in agriculture. *CEPS EU-Turkey Working Papers,* No. 10, September 2004. Centre for European Policy Studies, Brussels.

Celik, S.S. and Celik, Y. (2001). Ageing in Turkey. Hacettepe University. http://employees.csbsju.edu/jmakepeace/perspectives2001/Celik2001.jmm.html (accessed October 2004).

Corradini, M. and Fragoulis, H. (2004) Achieving the Lisbon goal: The contribution of vocational education and training systems. Country report: Turkey. European Training Foundation, Italy.

Derviş, K., Emerson, M., Gros, D. and Ülgen, S. (2004) *The European Transformation of Modern Turkey.* Centre for European Policy Studies, Brussels.

Ergör, G. and Öztek, Z. (2000) The development of national health policy in Turkey. In: *Exploring Health Policy Developments in Europe.* WHO Regional Publications, European Series No. 86. WHO, Copenhagen, pp. 194-208.

European Commission (2001) Regional features in Turkey. In: *Unity, solidarity, diversity for Europe, its people and its territory.* Second report on economic and social cohesion. European Commission, Brussels, pp. 157-158.

FAO (2001) Turkey: Nutrition Country Profile. FAO, Rome.

FAO (2003) Regional programme for food security in the Near East: Towards sustainable food security and poverty alleviation. FAO-IDB Report, Jeddah 8-9/10/2003.

Grethe, H. (2004) Effects of Including Agricultural Products in the Customs Union between Turkey and the EU: A Partial Equilibrium Analysis for Turkey. CeGE-Schriften, Vol 9, Center for Globalization and Europeanization of the Economy, Georg-August-Universität Göttingen. Peter Lang Verlag, Frankfurt am Main.

Lundell, M., Lampietti, J., Pertev, R., Pohlmeier, L., Akder, H., Ocek, E. and Jha, S. (2004) Turkey: A review of the impact of the reform of agricultural sector subsidisation. World Bank Working Paper, March 9, 2004.

Ministry of Foreign Affairs (2004) The Turkish National Education System. http://www.mfa.gov.tr/grupc/cg/01.htm (accessed 26 October 2004).

Ministry of Health of the Republic of Turkey (2004) Malaria in Turkey. www.saglik.gov.tr/eng/malaria/malaria.htm (accessed 23 October 2004).

OECD (2004) OECD Economic surveys: Turkey 2004. OECD, Paris.

Savas, B.S., Karahan, Ö. and Saka, R.Ö. (2002) Health care systems in transition: Turkey. European Observatory on Health Care Systems, Copenhagen, 4(4) 2002.
SIS (2001) Household Labour Force Survey Bulletins, 13 April 2001. State Institute of Statistics, Ankara.
SIS (2002) Results of 2002 Poverty Study. Press release, 13 April 2004. State Institute of Statistics, Ankara.
SIS (2004) Household Labour Force Survey Bulletins, 6 April, 2004. SIS, Ankara.
SPO (2004) Economic and social indicators 1950-2003. State Planning Organisation, Ankara.
Tansel, A. (1998) Determinants of school attainment of boys and girls in Turkey. Economic Growth Center Discussion Paper No 789, Economic Growth Center, Yale University, Connecticut.
Taymaz, E. and Özler, S. (2003) Labor market policies and EU accession: Problems and prospects for Turkey. Paper presented at *Turkey: Towards EU accession*, Bilkent University, 10-11 May 2003, Ankara.
Tunali, I. (2003) Background study on labour market and employment in Turkey. Report prepared for the European Training Foundation, Italy.
UN (United Nations) (2004) Demographic and Social Statistics. http://unstats.un.org/unsd/demographic/products/socind/illiteracy.htm (accessed October 2004).
UN Population Division (2003) World Population Prospects: The 2002 Revision Population Database. UN Population Division, New York.
UNICEF (2004) At a Glance: Turkey. Statistics. http://www.unicef.org/infobycountry/turkey_statistics.html (accessed 19 October 2004).
Uzunlu, V., Düsünceli, F., Talug, C., Uyanik, M., Yalvac, K. and Istanbulluoglu, E. (1999) NARS of Turkey. In Casas, J., Sohl, M. and Hafaz, H. (eds.) *The National Agricultural Research Systems in the West Asia and North Africa Region*. Icarda, Aleppo, Syria.
World Bank (2000) Turkey: Economic reforms, living standards and social welfare study. Report no. 20029-TU. World Bank, Washington DC.

6

Agricultural and Food Industry Structure

Siemen van Berkum

6.1 Introduction

During the 1960s and 1970s the state had a major influence on the economic process through its central planning philosophy and state-owned enterprises. In the early 1980s Turkey made major economic changes and adopted a more market-oriented economic policy. Economic liberalisation paved the way for the private sector and both domestic and foreign investments took over existing state enterprises and established new companies. State-ownership, however, still dominates some sectors of the agrifood chain. This chapter describes the structure of all components of the agribusiness cluster, which includes the upstream and downstream sectors as well as the primary agricultural sector. In addition, the structures of the wholesale and retail sector are presented. Structural features of the Turkish agrifood chain have implications for institutional arrangements in the fields of agricultural production and trade. This chapter concludes with a section on how these linkages between farmers, processors, traders and retail can be characterised in Turkey.

6.2 Structural features of the upstream industries

This section presents the structural features of industries delivering to the agricultural sector: fertilisers, plant protection products, feed concentrates and seeds. In addition, financial services delivered to the agriculture and food sector are discussed.

Agro-chemicals, animal feed and the food processing machinery industries

The Turkish fertiliser industry is dominated by a small number of private companies and one state-owned enterprise, TÜGSAŞ (Table 6.1). Total fertiliser production capacity is 5.3 million tons/year, but total production is 3-3.5 million tons/year, implying a two-thirds utilisation of the national production capacity. Domestic demand has been around 5 million tons/year in recent years, implying that imports amount to 1.5-2 million tons/year. The performance of the Turkish fertiliser industry went down after 2001 due to increasing production costs and a drop in demand. The industry depends heavily on imported materials (natural gas, phosphate rock) and on intermediates (ammonia, phosphoric acid). Prices of these imported raw materials and intermediates rose sharply, due to the strong decline of the currency in 2001 and 2002. Furthermore, the industry fertiliser subsidy was phased out in 2001. As a large part of the subsidy went to the fertiliser processors, this meant a loss for the industry. Furthermore, domestic demand decreased by 25 per cent as prices went up. As a result of these developments, Lundell *et al.* (2004:18-19) report a significant decline of the capacity utilisation and increasing concentration in the Turkish fertiliser sector.

Two fertiliser companies have 50 per cent of their capital in foreign ownership (Turkish Treasury, 2004). The state-owned companies TÜGSAŞ and IGSAŞ, of which TÜGSAŞ is the larger, together account for 40 per cent of the total national fertiliser production capacity and 20 per cent of total sales (SPO, 2004). According to the government's privatisation programme, TÜGSAŞ will be privatised before the end of 2005. TÜGSAŞ has four subsidiary companies. By mid-2004, two asset sales have been agreed already, one is in contract stage (block sale) and the other (asset sale) is in approval stage.

Table 6.1. Concentration of Turkish agrifood upstream industries

Activity code (ISIC Rev 3)	Name of manufacturing activity	No. of companies	CR4[1]	CR8[1]
2412	Fertilisers and nitrogen compounds	14	61.5	82.1
2421	Pesticides and other agro-chemical products	15	83.0	95.8
1533	Prepared animal feeds	130	33.0	44.6
2921	Agricultural and forestry machinery	83	79.2	84.7
2925	Machinery for food, beverage and tobacco processing	82	31.1	45.6

[1] The concentration rates CR4 (CR8) are measured as the ratio of the sales of the 4 (8) largest companies to the total domestic sales of the branch of industry.
Source: SIS, 2004

The domestic production of pesticides and other agro-chemicals (excluding fertilisers) is in the hands of 15 companies. Despite this substantial number, the four biggest companies dominate the industry, with a market share of 85 per cent (see Table 6.1). The agro-chemical sector suffers from significant over-capacity, with production capacity 50 per cent in recent years (SPO, 2004).

In other agro-chemical markets also, a small number of firms dominate domestic market sales. Whether these large firms have market power is difficult to say without having company- and sector-specific information on, for instance, price margins. Such information is not available. Yet it seems that the import regimes for these inputs allow foreign suppliers to come in on competitive terms. This would imply a competitive market without price-setting dominance by any individual company with activities in these input markets.

Since the government sold the state-owned feed mills in 1994, the feed industry has been privately owned. The number of feed mills increased to over 350 in the mid-1990s (Kindap, 1998). Apparently, there has been a rapid process of concentration in recent years as, according to SIS data, the animal feed industry had 130 companies in 2001 (Table 6.1). Still, there is much scope for further concentration as most of the feed mills have little production capacity: the 8 largest feed mills account for 45 per cent of total compound feed production in Turkey.

The food processing machinery and equipment manufacturers in Turkey range from small to medium-sized companies mainly located in the bigger cities. The line of products manufactured by these companies varies substantially from highly automated equipment to manual and basic models. The Turkish food processing machinery industry is able to produce every kind of machinery and equipment necessary for the local food processing industry. Manufacturing firms in this field are also serving export markets, mainly in the Middle East, Balkans and Central Asia (Royal Netherlands Embassy, 2004a).

Seed industry

Presently, 93 private companies and 31 public sector entities produce, import, mediate or distribute seeds for agriculture and horticulture (Royal Netherlands Embassy, 2004b). The private companies established after the economic reforms of the 1980s formed the Turkish Seed Industry Association (TÜRK-TED). TÜRK-TED is a lobby group that also provides the national seed sector services aimed at upgrading knowledge and skills of its members and improving market transparancy by providing statistical data on variety improvement, seed production, quality, distribution, trade, etc.

Table 6.2. Some features of the seed industry

Products	Share of private sector in seed industry (%)		Share of certified seeds (%)
	1986	2000	2000
Wheat	0.2	13.5	25
Barley	0.0	13.3	10
Soybean	41.2	94.4	25
Maize (hybrid)	91.2	99.8	100
Sunflower	100.0	100.0	100
Potatoes	48.9	99.9	20
Vegetables	94.4	99.8	25
Fodder crops	10.9	52.7	50
Cotton	0.0	23.4	n.a.

Source: TÜRK-TED, Royal Netherlands Embassy, 2004b

The share of the private sector in seed production is increasing, although publicly-owned enterprises still dominate the production of seeds for wheat, barley, cotton and fodder crops (see Table 6.2). For these products, the General Directorate of Agricultural Enterprises (TIGEM – State Farms) has been the major public organisation dealing with seed propagation and distribution to farmers. Turkish farmers are increasingly using certified seeds, although in certain subsectors (e.g. barley, potatoes) the proportion of certified seeds used is still particularly low. Approximately 40 per cent of seed used each year is produced by the formal seed industry. The high percentage of uncertified seed purchased and/or farm saved seed generally affects the quality of the produce negatively.

Seed imports are allowed only for companies that produce, procure and distribute seeds in Turkey. The main goal behind this policy is to encourage both local and foreign investors to invest in Turkey's seed industry. Presently, 13 foreign seed companies are active in Turkey (Royal Netherlands Embassy, 2004b:40). However, the import procedures are complicated and time-consuming (Royal Netherlands Embassy, 2004b:15). All seed companies have to get their varieties included in the Annual Seed Programme (ASP) of MARA (DG TÜGEM). For that application, they prepare their import, production and distribution programmes at least 1 or 2 years in advance. TÜGEM publicises its list at the beginning of the year, creating a lot of stress for the international seed houses, as they supply considerable amounts of imported lots of seed exactly during that period. These imported lots are paid for in December and run a risk of rejection if not included in the next ASP. Furthermore, imported varieties have to be labelled in Turkish, based on specific requests of the Ministry. For that reason, rejected lots can not be re-exported to other countries. Import rejection causes tremendous financial losses to all international companies involved.

Financing the agriculture and food sector

The Agricultural Bank of Turkey (Ziraat Bank) was established in 1888 and is the oldest bank in Turkey. Moreover, it is the largest bank in the country, and extends more than 90 per cent of agricultural credit. Since 2000, the Agricultural Bank has operated as a joint stock company with all shares belonging to the Turkish Treasury. The Bank's main sources of finance consist of (interest-bearing) deposits and borrowing from the central bank. Before the agricultural policy reforms started in 2001, the Agricultural Bank was also used as financial intermediary in government support policies, extending short-term loans to agricultural sales cooperatives for commodity price support (see also Section 6.5). The agricultural policy reforms had important implications for the sector's access to credit.

The Agricultural Bank provides loans to farmers through several channels. One channel is the Agricultural Credit Cooperatives (ACCs). Almost every village in Turkey is served by an ACC. Until 2002, ACC farmer members had the right to use credit up to a pre-defined ceiling (TRL 2.25 billion in 2002, being roughly USD 1,500) uniform for all members. Agricultural credit was given 80 per cent in kind, and mostly as fertiliser.[1] The Bank also provides loans

---

[1] The extensive network of ACCs represents 38 per cent of registered fertiliser distributors reaching a market share of 30 per cent in 2000. However, due to agricultural policy reforms (see chapter 7) and the loss of preferential treatment (access to concessionary funds for credits) by the government, ACCs' market share dropped to 12 per cent in 2002 (Lundell *et al.*, 2004: 19).

to individual farmers directly. These loans are mainly of a medium- to long-term nature. Although its lending is in principle (by law) targeted at small farms, the Bank's loan requirements restrict these credits to farmers who own agricultural land or have other properties needed as collateral. Land titles are not always clearly defined (see Lundell *et al.*, 2004:57), which potentially hinders farmers' access to credit. In practice, the Bank's direct lending activities have focused mainly on larger farms and state-owned enterprises, while ACCs serve smaller-scale farmers[2] (see also Table 6.3). The Agricultural Bank also provides loans to upstream and downstream enterprises in the agrifood sector.

Since 2001, credit provision to the agricultural sector has declined significantly. As part of the 2001 agricultural reform programme, credit subsidies were phased out in 2002[3]. Furthermore, in 2002, the Treasury ceased supplying funds to the Agricultural Bank and the ACC system. In addition, the new banking law prevents the Bank from providing funds to those ACCs with outstanding debts. Lundell *et al.* (2004) estimate that this will cut off 40 per cent of all ACCs from Agricultural Bank financing. Since flows of credit resources from the Treasury have ceased, the two main agricultural sector lenders, the Bank and the ACCs, have reduced their loan portfolios by about three-fold from the peak level of USD 7.3 billion in 1997 to reach USD 2-3 billion in recent years (of which half is renewed every year). This amount is estimated at one-third of all credit used by farmers (Lundell *et al.*, 2004:21). Farmers also obtain credit from merchants, wealthy farmers and moneylenders. However, compared to formal loans provided by the Bank and ACCs, the conditions of these informal loans are much worse, as these sources offer credits often at exorbitant interest rates (interest rates on the Bank's agricultural loans were 39 per cent in 2003 (TCZB, 2004). These changes resulted in reduced access to credits by farmers and this will have a serious negative impact on the possibilities for the sector restructuring and modernisation.

6.3 The structure of primary agriculture

Farms in Turkey are generally family-owned, small and fragmented. The 2001 agricultural census recorded 3 million farms, against 4 million in 1991 (Table 6.3). The average cultivated area per holding increased during the 1990s to reach about 6 ha in 2001, which is about one-third the average size (19 ha) in the EU in 1999/2000. Irrigated land is distributed slightly more evenly than cultivated land (Cakmak, 2004).

About 65 per cent of agricultural holdings are smaller than 5 ha. The majority of these holdings are vegetable producers, which typically cultivate an area of 0.2-1 ha (Royal Netherlands Embassy, 2004b:20). Eighty-three per cent of holdings (41 per

---

[2] In 2001, the Agricultural Bank made loans worth USD 539 million to 374 thousand farmers (average size of loan: USD 1441), while the ACCs lent a total of USD 502 million to 800 thousand farmers (average size of loan: USD 627). Thirty per cent of the ACCs' funds came from the Agricultural Bank (Lundell *et al.*, 2004:22).

[3] These credit subsidies amounted to around USD 1.1 million to farmers (Lundell *et al.*, 2004:4).

Table 6.3. Size distribution of land, 1991 and 2001 (per cent)

	1991		2001	
Size of holdings (ha)	Farm HHs[2]	Cultivated area	Farm HHs[2]	Cultivated area
No Land	2.50		1.77	
< 0.5	6.19	0.29	5.78	0.26
0.5 - 0.9	9.37	1.08	9.44	1.02
1 - 1.9	18.49	4.28	17.54	3.82
2 - 4.9	31.33	16.28	30.91	15.48
5 - 9.9	17.53	19.80	18.21	20.41
10 - 19.9	9.42	21.21	10.64	24.05
20 - 49.9	4.27	20.23	5.00	23.69
50 - 99.9	0.59	6.49	0.57	6.32
100 - 249.9	0.25	5.63	0.14	3.07
250 - 499.9	0.05	2.88	0.01	0.40
500 +	0.01	1.83	0.00	1.50
Total	100.00	100.00	100.00	100.00
Gini Coefficient[1]		0.60		0.59
	(1000 HHs)	(1000 ha)	(1000 HHs)	(1000 ha)
Village Head Census	4,092	21,103	3,698	22,156
HH Survey	4,068	21,449	3,076	17,164

[1] Calculated by the author of the table from grouped data.
[2] Farm households (HH)
Source: Cakmak, 2004:3

cent of total agricultural land) were smaller than 10 ha. Fifteen per cent of holdings were from 10 to 50 ha (nearly half the cultivated land). A relatively high number of more specialised farms are located in the Aegean and Mediterranean regions. These two coastal regions focus largely on fruit and vegetable production while the predominantly rural and mountainous areas in the central and eastern parts of the country specialise in livestock and animal products. The share of irrigated land increased from 14 per cent in 1991 to 20 per cent in 2001 and is much higher in the west than elsewhere in Turkey. A third of the holdings smaller than 1 ha are irrigated (Cakmak, 2004). These holdings produce fruit and vegetables.

Subsistence and semi-subsistence farming is an important characteristic of Turkish agriculture, which is similar to the situation in some regions in the new member states of the EU-25 (e.g. Poland), as well as in Bulgaria and Romania.[4] This type of farm is characterised by very low productivity, high hidden unemployment and low competitiveness. These farms, however, are crucial for providing income security and livelihood for millions of the rural population in Turkey. Subsistence farming in Turkey, though, should not be confused with small-scale farming: the

[4] Compared to Bulgaria and Romania, there seem to be more medium-sized farms in Turkey, which generally have better development perspectives than the very small-scale farms.

agricultural sector is characterised by a relatively large horticultural sub-sector, where production for the market on relatively small plots can be profitable.

6.4 Trends in the food industry structure and performance

Importance of the food industry

According to 2002 data, the Turkish food industry contributes around 5 per cent of GNP and accounts for 20 per cent of total production of the manufacturing sector. The Turkish food industry has kept a stable share in total manufacturing production over the last few years, from 20.1 per cent in 2000 to 20.9 per cent in 2002. During the 1990s this share was increasing, with the production of processed foods growing by about five per cent per year. From 1990 to 2000, the share of the food industry in total manufacturing industry value added increased from 13 to 16 per cent (Rehber, 2004:87 and TÜSIAD, 2003). However the share in manufacturing industry export has declined from 6 per cent in 2000 to 4.9 per cent in 2002 (SPO, 2004).

The food sector employs more than 100,000 registered workers and technical staff in more than 28,000 enterprises (SPO, 2004). Most of them are small to medium-sized enterprises. The State Planning Organisation estimates that around 10 per cent of these enterprises are relatively modern and large. Hanson (2004) reports that only one out of six firms uses modern technology for production and quality control. As a result, one may expect only a small share of firms to meet the EU quality norms. Azabagaoglu *et al.* (2003), for instance, estimate that only 6 to 7 per cent of Turkish total milk supply is processed by dairies meeting EU norms and having ISO 9000 quality assurance certificates.

Branch composition

Table 6.4. Key figures on the food industry structure

	Share in total number of enterprises, 2000 %	Share in total food output value, 2002 %
Meat and meat products	n.a.	13.8
Milk and milk products	11.0	14.4
Processed fishery products	n.a.	1.5
Cereal and starch products	65.0	40.7
Processed fruits and vegetables	11.5	7.2
Vegetable oils and oil products	3.5	6.1
Sugar, confectionary, etc	3.0	12.4
Others	n.a.	4.1
Total	100.0	100.0

Source: SPO, 2004

The distribution of the number of enterprises among sub-sectors of the food industry has not changed much since 1990. In 2000, the cereal and cereal-based sub-sector accounted for 65 per cent of the total number of food enterprises. Processing enterprises in the fruit and vegetable sector were in second place with 11.5 per cent, followed by the dairy enterprises (see Table 6.4).[5] The other categories identified in the table account for a lower percentage of the total number of enterprises. Sub-sectors with the highest production values are cereals, meat, dairy and sugar processing.

Ownership in the food industry

After the economic policy changes of the early 1980s, economic liberalisation stimulated the private sector, and both domestic and foreign investments took over existing state enterprises and established new companies. Today, Turkey's food sector is dominated by the private sector, but in a few branches of the food industry, such as the sugar, meat and tea industries, there are still state-owned enterprises. These state enterprises very much dominate the branch, although they do not have monopoly power as private firms coexist at the production and marketing stages.

In the mid-1990s, state-owned factories in the milk and feed industries, and a number of meat combines, were privatised. As part of the general economic reform programme, the Government intends to privatise the surviving state enterprises in due time. For instance, for 2004 the remaining meat processing units from EBK Meat and Fish Production Inc. are scheduled to be privatised. EBK was a major player in the market in the past, but now handles less than 3 per cent of production (Sarigedik, 2004). Also in the sugar sector, privatisation of state processing units is foreseen. However, the stage this process has reached is not so clear: while the State Planning Organisation reports that the privatisation of 26 state-owned sugar factories is already under way (SPO, 2004), the responsible Privatisation Administration reports that there is not yet a schedule for privatising TÜRKŞEKER (Privatisation Administration, 2004). TÜRKŞEKER (Turkish Sugar Company) dominates the sugar industry: the company had a market share of approximately 70 per cent in 2001.

Government monopolies existed for decades in the beverage and tobacco industry, but this situation has changed now. The government monopoly of the production of alcoholic beverages (wine, beer distilled beverages) ended in 2003. Beer and wine are mainly manufactured by the private sector (SPO, 2004). The beer industry consists of two private firms and one (relatively small) state company (TEKEL). As of mid-2003, there were 10 foreign-owned companies in the beverage (alcohol and non-alcohol) industry, with average 50 per cent foreign capital.

In the tobacco industry, the state monopoly was abolished already in 1991. Soon afterwards, some foreign tobacco manufacturers entered the Turkish market. For example, Philsa, the Corporation of Phillip Morris and Sabancõ Holding

---

[5] In contrast to the SPO source used for table 6.5, the Royal Netherlands Embassy states that 18 per cent of all food processing industries are active in the milk and dairy industry. The source of this information is, however, not mentioned (Royal Netherlands Embassy Ankara, 2004c).

Company, and RJ Reynolds started to manufacture cigarettes in 1993. By 2003, there were 12 foreign companies in the tobacco industry, with 93 per cent foreign capital (Turkish Treasury, 2004). TEKEL, the state company, is the only manufacturer of oriental cigarettes that are made from 100 per cent oriental tobacco produced domestically. However, TEKEL competes with private firms in the manufacturing of blended cigarettes. Domestic production of non-filtered cigarettes, which are produced only by TEKEL, is small. Cigarette production is carried out in 9 factories, 6 of which belong to TEKEL. The privatisation of TEKEL's activities in the tobacco industry is expected to be completed in 2004.

Although the privatisation process (combined with market size and growth prospects) offers many opportunities to foreign investors, Turkey has not attracted high inflows of foreign investment. Reasons for this low performance include structural barriers, heavy bureaucratic requirements, macroeconomic instability, corruption and political instability (TÜSIAD and YASED, 2004). In the 1990s Turkey was not able to attract more than USD 1 billion on average annually, of which only a small percentage has been invested in the food industry (Loewendahl and Ertugal-Loewendahl, 2001:5).[6] In 2003, there were 155 foreign-owned companies in the food manufacturing sector, with an average of 64 per cent foreign capital. This accounted for almost 5 per cent of total foreign direct investment (Turkish Treasury, 2004). Foreign investment has relatively large shares in vegetable oils and fats, candy and chewing gum, dairy products, confectionary and artificial sweetener industries (Sirtioglu, 2002). As reported above, foreign investors have also found the beverage and tobacco industry attractive.

Concentration in the food industry

The concentration of companies in the food industry is highest in the branches of starch production and beverages (beer, wine, spirits) (see Table 6.5). In these branches, the four biggest companies have a market share of 70 per cent and more. However, it should be noted that also in other branches of the food industry a small number of companies have a significant market share. In the fish, tobacco, cereal processing and sugar confectionery branches, the four largest companies have more than 60 per cent market share. Also in branches with relatively high numbers of enterprises, some companies dominate the branch. For example, the dairy industry consists of over 100 enterprises, yet the four largest dairies have a 50 per cent market share. SIS counts almost 400 manufacturers of bakery products and the largest four have a market share of one-third.

These numbers indicate that the food industry (including beverages and tobacco) is fragmented in many industry branches, yet that in some branches a few companies dominate the market while in other branches the largest four companies have a significant market share. Whether these tendencies towards increasing

[6] According to generally accepted international standards in FDI inflow ranking of 2000, the minimal annual FDI attraction potential of Turkey would have been USD 35 billion (UNCTAD, 2002). This potential is almost equal to FDI inflows in Brazil in 2000.

concentration in certain branches have had implications for efficiency and/or price formation is not known.[7]

Table 6.5. Concentration in the Turkish agrifood industry, 2001

Activity code[1]	Name of manufacturing activity	No. of companies	CR4 [2]	CR8[2]
1532	Starches and starch products	6	95.8	100.0
1553	Malt liquors and malt	8	77.2	100.0
1554	Soft drinks and mineral waters	54	75.0	84.7
1552	Wines	13	73.5	91.5
1551	Distilling, rectifying, blending of spirits, etc.	13	71.3	95.5
1512	Processing and preserving of fish and fish products	16	68.1	84.6
1600	Tobacco products	25	66.7	88.5
1544	Macaroni, noodles, couscous and similar farinaceous products	19	61.6	81.8
1543	Cocoa, chocolate and sugar confectionery	85	61.4	82.1
1520	Dairy products	114	51.8	66.1
1549	Other food products n.e.c.	113	38.3	51.8
1542	Sugar	39	35.9	53.4
1541	Bakery products	372	35.5	54.5
1514	Vegetable and animal oils and fats	95	35.1	48.9
1511	Production, processing and preserving of meat and meat products	99	34.7	50.3
1513	Processing and preserving of fruit and vegetables	234	20.0	29.4
1531	Grain mill products	264	18.1	27.5

[1] ISIC Rev 3.
[2] The concentration rates CR4 (CR8) are measured as the ratio of the sales of the 4 (8) largest companies to the total domestic sales of the branch of industry.
Source: SIS, 2004

Performance of the food industry

Data on turnover, value added and profits of the food sector are hard to find. SPO (2004) reports that while the food industry accounts for 20 per cent of total production in the manufacturing sector, its share in value added is 16 per cent. To the author's knowledge, details providing insights into profitability of specific sub-branches are not systematically published by accessible public sources, and one has to rely on indirect indicators.

The degree of capacity utilisation in food sub-sectors could be used as an indication of the rate of profitability at sub-sector level. According to SPO (2004), capacity utilisation in most of the sub-sectors of the food industry has been at

[7] Note that SPO reports 28,000 enterprises in the food industry, of which very many belong to the category of small-scale manufacturing (SPO, 2004). SIS, however, reports about only 1,569 companies in the food (related) industry. The reason might be that SIS reports only companies with a certain minimum number of employees. However, the criteria for selecting the companies are not explained in the statistics provided on the SIS website.

approximately 50 per cent in the years 2001-02, while the utilisation rate was 70 per cent on average between 1995 and 1999.[8] The 2001 economic crisis in Turkey had a significant downward impact on the utilisation rate in the food industry. Furthermore, the reduction of government support in the form of administered prices and subsidies (see Chapter 7) may also have contributed to the decline in the profitability and subsequently the capacity utilisation rate in certain sub-sectors. Anyway, structural weaknesses in the food sector have aggravated the impact of the cyclical problems on the capacity utilisation rate in the sector. SPO refers to the weak financial structure of SMEs in the sector, wrong investment decisions, instability in export markets, seasonality of agricultural production and insufficient integration or coordination between agriculture and industry, which are factors that contributed to a lack of flexibility of the food sector in responding to economic downturns and changing consumer preferences. The low level of capacity utilisation indicates that a significant number of companies in the food sector produce rather inefficiently and would have rather low levels of profitability. What is promising, however, is that when the Turkish economy showed its first signs of recovery at the end of 2002, SPO noticed that capacity utilisation in the entire food sector also showed signs of improvement (SPO, 2004). Data on capacity utilisation that would reveal whether these improvements have continued in 2003 are not yet available.

The food industry is characterised by duality, with many small and medium-sized companies but only a limited number of large-scale, modern companies quoted on the stock exchange. Food and beverage companies listed on the Istanbul Stock Exchange (ISE) provide some information on the performance and structure of large companies in this industry. The ISE lists 24 food and beverage manufacturing industry companies; for 22 of these companies more detailed data were available in the ISE 2003 yearbook of companies[9]. The majority of companies were specialised in a few products, but a few large companies had a very diversified product range. Most companies that produced tomato paste and tomato products did so in conjunction with other products, such as canned and frozen fruit and vegetables. Sixteen of the companies were established between 1960 and 1989 and five during the 1990s. Companies that were more orientated towards the domestic market (i.e. with exports less than 20 per cent of sales) produced mainly meat, milk products and beverages. Very export-oriented companies (i.e. with exports more than 60 per cent of sales) produced mainly frozen and canned fruit and vegetables, other fruit and vegetable products and vegetable oils. Five of the 22 companies recorded a loss in 2003; however, no pattern emerges amongst these companies.

[8] This might still be low compared to EU standards. For comparison: capacity utilisation in the major branches of the Dutch food industry has been between 80 and 90 per cent throughout the period 1990-2004 (CBS, Statline statistics). The generally low capacity utilisation in the Turkish food industry may be due to the fact that state ownership is still significant in a number of branches. State-owned companies do not have to bear the consequences of inefficient production when their losses are covered by the government budget.

[9] Available at: http://www.ise.org/company/companies_yb_2003.htm.

6.5 Wholesale and retailing

Wholesale structures

The wholesale structure was dominated by the state or parastatal enterprises and quasi-state organisations for many years until this started to change as part of the 2001 economic reform programme. In the grains sector, the Turkish Grain Board (TMO) happened to be the key player, acting as a buffer stock agency to stabilise producer and consumer prices in wheat production. TMO provided signals to merchants about the future directions of the market by announcing purchasing prices, which were later re-determined based on market conditions. As part of the reform programme, TMO prices will be increasingly linked to the world price. In doing so, state procurement functions only as a 'buyer of last resort', as is now the case in the EU. As a first result of the reforms, TMO reduced its volume of intervention purchases to about 800,000 tons in 2002, which is only a third of 1999-2001 purchases (Lundell *et al.*, 2004:16). In 2002, TMO also ceased announcing minimum purchases. The prices paid by TMO dropped by 13 per cent over 1999-2001, and by an additional 10 per cent in 2002.

In the marketing of agricultural commodities and inputs, agricultural sales cooperatives (clustered in 16 unions, called ASCUs) have been a major player in collecting and distributing a wide range of agricultural commodities for a long time. Established in the 1920s and 1930s to serve farmers in the purchase and processing of export crops such as cotton, hazelnuts, sunflower and olives, the ASCs/ASCUs were given a major role in the implementation of the government programme in the 1960s. Through the network of ASCUs the government was able to bolster producer prices through subsidies and market intervention. State control was further tightened in 1984 when the Ministry of Industry and Trade obtained extensive power to direct the operations of the ASCs/ASCUs. These cooperatives were used to purchase a maximum amount of agricultural produce with the aim of maintaining high prices for these products. In fact, the cooperatives acted as a state intervention agency. As part of the 2001 economic reform programme, ASCs/ ASCUs are in a process of being transformed from parastatal organisations into financially autonomous and sustainable cooperatives that can compete with private traders while operating for the benefit of the farmers who formally own them. Presently 330 ASCs have around 750,000 members (Lundell *et al.*, 2004:60). Whether the restructuring of the cooperatives and their reorientation will be successful remains to be seen. Inadequate revenue structures, overstaffing and little business orientation are just a few of the problems that need to be tackled before these organisations can play a role in improving the functioning of the market.

Whereas special laws govern ASCs and ACCs, the roughly 5,000 Agricultural Development Cooperatives (ADCs) with some 500,000 members operate under the general cooperative law. ADCs tend to focus on activities not covered by the ASCs and ACCs, such as dairy and livestock, handicrafts, consumer articles and the marketing of fruits and vegetables. ADCs claim that they market 50 per cent of all milk and rice produced for the market in Turkey (Lundell *et al.*, 2004: 63). The increasing numbers of newly established cooperatives and the expansion of the business volume of existing ADCs illustrate the increasing popularity of ADCs, which are currently mainly located in the western and central parts of the country.

As well as the cooperative structures, private wholesale traders act as important intermediaries between the producer (farmer), processing and/or retail stage. The wholesale of perishable products such as fresh fruits and vegetables is largely in the hands of so-called commissioners. By law, the wholesale marketing of fresh fruits and vegetables has to go through recognised wholesale markets, where the tax office charges 5 per cent VAT, before products can be sold to the retail. The commissioners, appointed by the government, are the key intermediary party between the producer and the buyer. Growers are obliged to sell via commissioners, but are free to choose and may change from year to year. Services provided by commissioners in terms of grading and sorting are generally low, and commissioners tend to mix supply from different small-scale growers to create enough volume (see, for instance Wijnands *et al.*, 2004). On the other hand, many commissioners finance growers and offer them credits. Yet, such relationships condition sales and therefore work against transparent price-making. As a consequence, Turkish wholesale markets are not playing an important role with regard to the development of quality standards and economic transparency. The wholesale markets are established and controlled by main cities or municipalities and/or regional municipalities. Both the commissioners and the municipalities receive a certain percentage of commission from the trade of the products.

Retail market structures

Modern retail in the form of super- and hypermarkets presently has over 40 per cent market share of Turkish consumer food expenditure (see Table 6.6).

Table 6.6. Retail food sector trends (market share in per cent according to outlet)

	1999	2000	2001	2002	2003[1]
Hypermarkets (over 2,500 m^2)	6.5	8.0	9.5	9.6	10.0
All supermarkets (100-2,500 m^2)	17.0	20.0	25.0	27.5	31.0
Markets (50-100 m^2)	12.5	10.0	9.5	9.4	9.0
Bakkals (< 50m^2)	49.5	48.0	42.0	40.5	36.0
Others (convenience stores, kiosks)	14.5	14.0	14.0	13.0	14.0

[1] Estimate.
Source: Sirtioglu, 2004:3

The structure of the retail sector is strongly influenced by the type of food consumed by the bulk of the population. For low-income groups, which comprise the majority of the Turkish population, it is estimated that 55 per cent of the diet is made up of bread with an additional 15 per cent consisting of rice, potatoes and pasta products (AAF Canada, 2004). People buy food products mainly on open markets (bazaars) and in local neighbourhood stores where the majority of products are made with local ingredients. For the entire population, processed products are only 15-20 per cent of consumption. The latter are mainly bought in the supermarkets, which are situated in urban areas and cater to those who have benefited most from Turkey's rising prosperity. Now that they have the income to afford it, this clientele has developed a penchant for Western, imported products. The first modern supermarkets in Turkey date back to the mid-1950s, but the true take-off of super-markets occurred in the country only in the 1980s (Codron *et al.*, 2004). Since then, the retail sector has developed rapidly with a sharp increase in the larger supermarkets and discount segment from the mid-1990s onwards. The share of these modern food stores in the overall food retail market is expected to grow from 42 per cent (2003) to more than 50 per cent of the retail market by the end of 2005 (Sirtioglu, 2004).

Modern supermarkets and discount stores are increasingly replacing traditional stores. The latter include the small grocery retailers, called *bakkals*, that up to 1999 had a 50 per cent market share in the food retail sector (see Table 6.6). This transformation has mainly occurred in the larger cities, but recent investments have targeted medium-sized cities where shopping habits are changing or cities where tourism is intensive. In the future, hypermarkets, supermarkets and hard discount chains are expected to dominate the sector in the wealthier and larger urban areas of the country.

So far, the involvement of foreign investment in the retail sector is rather limited. Retail chains are largely in the hands of Turkish investors (like the companies Migros SOK, Gima, BİM, Tansas Makro, Yimpaş), with only some investment from Germany (Metro, Real), UK (Booker) and France (Carrefour).

The economic recovery and increased food sales (in real terms) in 2003 also attracted investments from new international chains buying local chains. For example, Tesco (UK) bought majority shares in Kipa in 2003. Carrefour already entered Turkey in 1993 but was not very aggressive in gaining market share until 2001 when it rose to be a leader in the Turkish market.

The rather late entrance and cautious operations of foreign investors in the Turkish retail sector are due to the generally unattractive economic environment for private investors in the food chain, which persisted until recently. Although investment policy liberalisation started in the early 1980s, foreign investment was stimulated only by the customs union with the EU (1996) and the initiation of an EU membership procedure in 2001 (see also TÜSIAD and YASED, 2004). As a result of these decisions, trade barriers were reduced, inflation has been brought down, consumer price support has been almost totally removed, government control over strategic exports has been significantly reduced and FDI has been stimulated. The new law on FDI, ratified in 2003, may further encourage investments from abroad in the food sector.

6.6 Relationship of farms with markets

Linkages between farms, processors and traders

While in the preceding sections the structural features of the different components of the agrifood chain have been described, this section focuses on the relationships

Box 6.1 Food consumption patterns

Turkey may be considered a very promising market for food products. Turkey's population of almost 71 million is growing at over 1 per cent annually. Over 27 per cent of the population is under the age of 15, and this young population will continue the trend that has seen an increase in the demand for Western products and lifestyles. On the other hand, the extreme inequality in the distribution of family income throughout Turkey[1] limits Turkey's potential of being a major consumer market.

Turkish consumers spend around 30 (in urban areas) to 45 per cent (in rural areas) of their income on food, beverages and tobacco (AAF Canada, 2004). Food consumption patterns over the recent decades have been affected by rapid urbanisation, growing participation of women in the labour force and tourism. Consequently, demand for processed products and for livestock (milk, meat) products has grown over time. Dairy consumption estimates range between 15 (Sirtioglu, 2002) and 32 (OECD, 2002) kg of milk per capita. The highest figure mentioned is, however, still far below west European consumption levels. Meat consumption has shifted in recent decades from lamb, mutton and goat to beef, veal and poultry due to changing tastes and costs. Domestic beef consumption decreased in recent years due to reduced animal supply, high rates of inflation, increased beef prices, reduced incomes because of economic crisis, and trade policy (see Chapter 8). Annual beef consumption is approximately 9 kilograms per capita in Turkey (Sarigedik, 2004), while total meat consumption is estimated at 17 kg/capita. The principal share of the diet remains flour and flour-based products, estimated to be around 200 kg of cereals per capita. Furthermore, the diet is rich in fruit and vegetables. Per capita consumption amounts to 230 kg fruit and 100 kg vegetables, or 330 kg total as compared with 180 kg in France. The Turkish consumption figures, which are already high, are underestimates as they exclude consumption from own gardens. This, and the significant share of semi-subsistence farming in Turkey and its attendant unregistered production and consumption, generally obscure the available consumption data.

[1] The wealthiest 10 per cent of the country account for one-third of its consumption, while the poorest 10 per cent only account for 2-3 per cent of consumption (AAF Canada, 2004).

between the market actors. Linkages between farmers, traders and processors can be either through the market or through contracts. Contract farming is not yet widespread in most agricultural sub-sectors in Turkey. For instance, in the dairy, beef, sunflower, olive and vegetable oil industries, most processors rely on open market purchase to provide raw materials (Rehber, 2004). The market structure of these products is highly fragmented, with many small-scale suppliers and many middlemen. The marketing channels are diverse, from local markets and local collectors to regional and municipal wholesale markets, traders, and many small-scale processors. Consequently, costs of collection, storage, marketing and processing are relatively high, while the quality and prices of the agricultural commodities may differ widely in time and place. Some details of the milk marketing system illustrate this. The low supply and low quality of the milk are the major problems in the sector. Approximately 80 per cent of all dairy milk produced is not offered to processing units but is sold as milk and dairy products on local and regional markets, not cooled and under unsuitable hygienic conditions (Azabağaoğlu *et al.*, 2003). The processors are predominantly small-scale, have a low capacity usage (only one-third in 2000) and have hardly invested in cold chains. They compete for the raw milk on price; due to high inflation rates in recent years, farmers were reluctant to agree on a settled price and just waited for the best deal. Processors sell their products mainly through the mediation of wholesale markets. Processing firms sell to traditional and small groceries but very few firms work with modern retail chains as they are not attractive to that outlet, due to their small scale and insufficient quality performance.

On the other hand, vertical coordination through contract farming occurs in some sectors in Turkey, when agricultural products have to go through a processing stage. Examples are certain horticultural products like tomatoes and peas, where the importance of contract farming has increased over time, and hops where 60 per cent of production is under contract (Rehber, 2004). Sugarbeet growing (a minor crop in the agricultural sector) is grown only under contract, while fruit and vegetable processors rely partly on contract farming and partly on spot market purchases (as, for example, in the citrus sector).

Most traded volumes of fresh produce are sold via (a large network of many small-scale) traders through wholesale markets. As described in Section 6.5, a commissioner is the key intermediary between the producer and the buyer. The most common method for determining the price is by negotiation on the spot and/or at auction. Retail shops, *bakkals* and open bazaars all purchase most of their fresh products from a commissioner at the wholesale markets. A substantial share – approximately 35 per cent - of Turkey's citrus crop is processed, graded and packed for the high-quality domestic and export markets. About a dozen large packing companies dominate this part of the market. These companies purchase their raw material through contracting. Packers generally begin contracting in August and purchase the crop on the tree. Farmers are paid after the harvest, which is normally in September/October. This implies that contract farming does not include elements of pre-financing or access to inputs, yet the advantage for growers is certainty of payment. Packers estimate that about half the crop will be first or second grade, destined for the upscale local market and/or export market. The remainder receives minimal processing and is sold through a series of regional wholesalers and local retailers (Sarigedik, 2003).

Retail procurement systems and vertical integration in the agrifood supply chain

The development of the modern retail chain has a significant impact on the purchase systems in the food chain in Turkey. Codron *et al.* (2004) illustrate this with a case study of fresh fruits and vegetables (FFV). This group of perishables is largely bought by final consumers on open markets (bazaars) and *bakkals*. These traditional outlets provide the basic attributes of low price and freshness. Yet, modern Turkish retailers place high priority on the FFV section and want to gain market share, by providing high and consistent quality at prices equivalent to those at the bazaars. To achieve this aim, several supermarkets in Turkey have been shifting over the past years from the old system based on terminal wholesale markets towards the use of more integrated channels.[10] Codron *et al.* (2004) categorise and position the Turkish supermarkets according to the level of backward integration and the main retailer-supplier governance structure: market or contract. Some retailers still procure fresh fruit and vegetables in the local wholesale market. These are the smaller chains with only a few stores. Other retailers centralise the purchase of their FFV procurement with backward integration into the sorting function, yet they buy on the central wholesale market from numerous mono-product wholesalers. These retailers are located in the larger cities within a short distance of the central wholesale market. Both types of buyers rely on the market as the major governance structure. Some supermarkets, however, are turning to contractual agreements with shippers, i.e. those who buy from the field. For instance, Kipa-Tesco has established contractual relationships with many efficient local shippers who deliver products requested to the different retail stores, at short notice and with their own refrigerated trucks.

The wholesale market system in Turkey generally does not provide much service in terms of grading and sorting, while supply is often in small volumes that are difficult to combine. Retailers accept that quality standards are generally low, because they perceive consumer quality awareness as still too low to warrant charging a premium. Contracts allow for complying with the private standards or requirements imposed by the retailer. Kipa-Tesco has an explicit strategy to compete on quality, and demands guaranteed quality goods from their suppliers. These guarantees are enforced by contract. When the Turkish economy grows and a larger group of consumers become more discerning and quality aware, more supermarkets are expected to emphasise the quality attribute of their FFV products and may want to use contractual arrangements to realise that. A major challenge to the farming sector, then, is to meet the growing quality requirements and standards all along the chain. Failure to meet consumer requirements may result in further import penetration.

6.7 Conclusions

The structural features of the Turkish agrifood chain vary along and within the different levels of the supply chain. In the industries upstream of farming a few large enterprises dominate the fertilisers and pesticides industry, whereas the animal feed industry is characterised by many smaller firms. At primary level, farm structures are highly fragmented with a large part of the sector being (semi) subsistence. The structure of the food processing industry also varies between branches. Although generally fragmented, there is significant structural

[10] As described in the previous section, fresh fruit and vegetables must pass through wholesale markets by law. Currently, when a supermarket circumvents this through vertical coordination, its contracts are illegal. However, a law to address this is in process.

concentration in a number of branches of the food industry, such as in starch, beverages and tobacco. Market power does not seem to exist although hard evidence to verify this is not available.

The government used to intervene in the agricultural market through state-owned enterprises and government-controlled marketing organisations but it has reduced its role during the last two decades. For instance, the private sector is increasingly involved in the seed and meat sectors while the state has completely privatised state-owned factories in the dairy and feed industries in the 1990s. Further privatisation of state-owned enterprises is, however, still to come in the sugar, meat, fish, beverages and tobacco industries. Agricultural marketing cooperatives (in previous times controlled by the government) are in the process of being transformed into fully independent organisations. The success of this transformation is, however, questionable, as these organisations are characterised by inadequate revenue structures, overstaffing and little business orientation. The wholesale market system for fresh products is dominated by commissioners, appointed by the government. This system does not contribute to the development of quality standards, while low economic transparency limits opportunities for tracing products in the food chain.

The performance of the food industry is hard to assess, due to a lack of information. A bad sign is that in general, the food industry suffers from over-capacity. The low level of capacity utilisation indicates that a significant number of companies produce rather inefficiently and would have low levels of profitability. On the other hand, there is a small number of large, modern food companies, some of them quoted on the stock exchange. Information on the performance of the latter indicates that some of those companies recorded high profits in recent years.

Processors purchase most agricultural commodities on the market: contract farming is not widespread in Turkish agriculture. Supermarkets, on the other hand, are increasingly shifting from buying fresh products at wholesale markets towards the use of more integrated channels in order to purchase guaranteed quantities and quality against competitive prices. The dynamic in the retail sector is impressive: supermarkets are developing rapidly and are increasingly replacing traditional stores. Foreign investment in the retail sector as well as in the food processing industry is rather limited. The new law on foreign direct investment, ratified in 2003, may further encourage investments from abroad in the food sector.

References

AAF Canada (Agriculture and Agri-Food Canada) (2004) Agri-food country profile Turkey. AAF Canada, Ottawa, March 2004.

Azabağaoğlu, M.O., Gaytancioğlu, O., Kubaş A. and Erbay, R. (2003) Analysis of the marketing structure of the dairy industry in the Trakya region and the determination of emerging issues with multidimensional scaling. *Turkish Journal of Agricultural Forecasting* 27, 117-122.

Cakmak, E. (2004) Structural change and market opening in agriculture. *CEPS EU-Turkey Working Papers,* No. 10, September 2004. Centre for European Policy Studies, Brussels.

Codron, J.M., Bouhsina, Z., Fort, F., Coudel, E. and Puech, A. (2004) Supermarkets in low-income Mediterranean countries: Impacts on horticulture systems. *Development Policy Review* 22 (5), 587-602.

Hanson, R. (2004) Turkey exporter guide annual 2004. *GAIN report*, Number TU4008. USDA Foreign Agricultural Service, Ankara.

Kindap, A. (1998) Current state of the mixed-feed manufacturing in Turkey in relations with livestock and poultry sectors and raw material supply position, In: Brufau, J. and Tacon, A. (eds.) *Feed Manufacturing in the Mediterranean Region: Recent Advances in Research and Technology*. CIHEAM-IAMZ, Zaragoza, pp. 337-345.

Loewendahl, H. and Ertugal-Loewendahl, E. (2001) Turkey's performance in attracting foreign direct investment. ENEPRI Working Paper, No. 8, November 2001. ENEPRI, Belgium.

Lundell, M., Lampietti, J., Pertev, R., Pohlmeier, L., Akder, H., Ocek, E. and Jha, S. (2004) Turkey: A review of the impact of the reform of agricultural sector subsidisation. World Bank Working Paper, March 9, 2004.

OECD (2002) OECD Economic Surveys: Turkey 2001-2002. OECD, Paris.

Privatisation Administration (2004) Privatisation implementations between 1985 and 2004. http://www.treasury.gov.tr/english/ybs/ybsyeniing.htm (accessed 16 August 2004).

Rehber, E. (2004) Vertical integration in the food industry and contract farming: The case of Turkey. *Outlook on Agriculture* 33(2), 85-91.

Royal Netherlands Embassy (2004a) Food processing machinery in Turkey. Office of the Agricultural Counsellor, Ankara.

Royal Netherlands Embassy (2004b) Turkish horticultural sector. Office of the Agricultural Counsellor, Ankara.

Royal Netherlands Embassy (2004c) Market scan – Turkish cheese sector. Office of the Agricultural Counsellor, Ankara.

Sarigedik, Ü. (2003) Turkey: Citrus annual. GAIN Report, Number TU3030. USDA Foreign Agricultural Service, Ankara.

Sarigedik, Ü. (2004) Turkey: Livestock and products annual report. *GAIN Report*, Number TU4022. USDA Foreign Agricultural Service, Ankara.

Sirtioglu, I. (2002) Turkey food processing ingredients sector, Market update 2002. *GAIN report*, Number TU2047. USDA Foreign Agricultural Service, Ankara.

Sirtioglu, I. (2004) Turkey retail food sector report 2004. *GAIN report*, Number TU4005. USDA Foreign Agricultural Service, Ankara.

SIS (2004) Concentration in Turkish manufacturing industry, 2001. Concentration in Turkish Manufacturing Industry News Bulletin, 27-08-04. SIS, Ankara.

SPO (2004) Sector profiles of Turkish industry, a general outlook. Industry Department, SPO, Ankara.

TCZB (Agricultural Bank of Turkey) (2004) Annual Report 2003. TC Ziraat Bank, Ankara.

Turkish Treasury (2004) Foreign investment statistics. http://www.treasury.gov.tr/english/ybs/ybsyeniing.htm (accessed 16 August 2004).

TÜSIAD (2003) Output and industrial structure. In: *European Union and Turkey: Towards Economic Integration*. TY/222/2003, TÜSIAD.

TÜSIAD and YASED (2004) FDI attractiveness of Turkey: A comparative analysis. *Insight YASED* 6, 40-46, April 2004.

UNCTAD (2002) *World Investment Report 2002.* UNCTAD, Geneva.

Wijnands, J., Deneux, S.D.C., van Paassen, R.A.F., Poot, E.H. and van Woerden, S.C. (2004) Internationalisation and competitiveness of the Dutch greenhouse vegetable sector. LEI Report 5.04.06. LEI, The Hague, (*in Dutch*).

7

Overview of Agricultural, Food, Rural and Structural Policies

Arie Oskam

7.1 Introduction

Turkish agricultural policy dates from the 1930s, when a protectionist policy was introduced in line with similar developments in various industrialised countries (Hathaway, 1963). In this respect, advice from the Soviet Union played an important role (Zürcher, 2004: 197). The Kemalist principles of a strong central state coincided with the socialist principles of a centrally planned society. The five-year planning system, however, started only in 1963.

Not surprisingly for a country with a strong nationalist orientation, agricultural policy has often been rather protectionist. In certain periods (e.g. 1980-84 and again after 2001), however, there has been a strong reduction in support and protection (OECD, 1994: 63-92: OECD, 2004; Lundell *et al.*, 2004). In the economic reform of 1980-84, Turkey switched from a growth strategy based on import substitution to an export-orientated strategy. The reform at the turn of the century was more focused on reducing budget costs and starting institutional reform. Because reforms were introduced in an unstable macro-economic environment, it is difficult to distinguish intended policy changes and side effects of macro- economic changes: inflation, large changes in exchange rates, interest rates, etc. The combination of unstable governments, large fluctuations in economic conditions, and the so-called five-year planning system resulted in a situation where many groups and organisations were profiting from agricultural policies[1]. Agriculture was often used as a political football by politicians seeking votes. A clear and consistent line in policy was lacking. According to the World Bank (2001: 6), the package of government initiatives as a whole was even counter-productive. This suggests that the combination of high support prices and input subsidies, and their inconsistent use over time, slowed the agricultural sector down rather than stimulating it.

---

[1] An example is the fertiliser industry, which (except in 1995-97) received a subsidy that was not passed on to farmers (Lundell *et al.*, 2004, Fig. 3.2). Farm prices for fertiliser did not follow changes in the import price or the fertiliser subsidy.

Since 2001, and mainly under pressure from the IMF and World Bank, some important agricultural policy changes have been introduced. The so-called 'Agriculture Reform Implementation Project' (ARIP) is an experiment in major agricultural policy change, and can reveal how Turkey manages important policy adjustments that also require institutional changes. The World Bank appears to be very confident that the changes brought in by ARIP are working well (Lundell, *et al.*, 2004).

Policies dealing with food quality and food safety started to develop in the mid-1990s, due to the customs union with the EU. These developments were strengthened after 2000 when Turkey took a more outward-looking position with respect to agricultural production and food. Exporting to developed market economies requires greater concern for food safety and food quality. The penetration of supermarkets into domestic retail markets is another driving force behind food quality and safety. With respect to food policy, the government of Turkey has been following rather than leading.

There was virtually no rural development policy over the last part of the 20th century. As a country with a long history as a transport nexus between Europe and Asian countries, Turkey had a well developed infrastructure of roads and – later – electricity. Investments in rural development were and are mostly related to irrigation schemes. Irrigation water is provided to farmers using a strong top-down management approach, and often without pricing according to use, which makes good resource management at micro level difficult. The Eastern Anatolia Watershed Rehabilitation Project is an example of a more integrated rural development project for improving the resource base and changing the production structure of a backward region of Turkey (World Bank, 2001). Currently, attempts are being made to get farmers more involved in the micro-development of this project, although so far it is difficult to foresee whether they will be successful. What is required for success is institutional change, which - as argued in Chapter 3 – is a process that takes time.

This chapter considers three partly related policy areas: agriculture, food, and rural development/structural policy. Most attention will be given to agricultural policy (Section 7.2), because of its recent reform in Turkey and its relevance for the EU. The agricultural policy section starts with the policy-making process, because this is seen as one of the explanations for the inconsistent use of policy instruments over time. The Agriculture Reform and Implementation Project (ARIP) will be discussed in detail.

Food policy (Section 7.3) receives less attention in this chapter, because government involvement started later and is still limited. We deal with two aspects of food policy: food safety and food quality. Other aspects are discussed in Chapter 6 in the context of the food chain. Rural and structural policy (Section 7.4) has a different tradition in Turkey compared to the EU. The concept 'rural policy' is quite unknown, although 'rural affairs' and 'rural development' are more familiar concepts. Structural policy in relation to agriculture and the food industry has been rather limited and has taken a special form, namely as part of the policies relating to State Economic Enterprises (SEEs) and Agricultural Sales Cooperative Unions (ASCUs). Rural development policy takes the form of rural development plans, which have sometimes covered infrastructural elements such as electrification, drinking water provision, roads and communication. Discussion and conclusions are in Section 7.5.

7.2 Agricultural policy

Rather than describing the chronological evolution of agricultural policy in Turkey, or analysing the succession of legal decisions marking that evolution, here we focus on the policy process and ask: Who are the actors involved? What positions do they typically take? And what have been the outcomes of the process?

The long history of agricultural policy, the central position of the government, the heavy involvement of many interest groups and farmers' perception that 'the government is responsible' are good reasons for starting with the policy process.

Agricultural policies in Turkey have so far been shaped by expected political benefits of governments. Various studies (e.g. Cakmak, 2003) show the association of support price changes with elections. Budgets were allocated in favour of short-term price protection, input subsidies and the operational costs of maintaining employment in organisations subsidised by the state. Turkey belonged to the group of developing countries that protected its agricultural sector and related activities rather than taxing them, but high protection levels have not stimulated an efficient agricultural sector. Beef and dairy, cereals, oilseeds and some industrial crops have been protected at the expense of the more efficient fruit and vegetable sectors. This diagnosis of Turkish agricultural policy is widely shared (World Bank, 2001; Togan *et al.*, 2003; Lundell *et al.*, 2004).

Policy process

Here, we identify and discuss the interest groups and state organisations that have been driving Turkey's agricultural policies.

The first and most powerful player has been the government and the leading political party.[2] They granted market power to the state-owned economic enterprises (SEEs) and the cooperative unions, whilst also underwriting their budgets. As a result, these organisations and to an extent farmers also benefited at the expense of consumers and taxpayers. Agricultural policy was implemented mainly by the Ministry of Agriculture and Rural Affairs (MARA), the Ministry of Industry and Trade, the Agricultural Bank of Turkey and the Treasury. The Agricultural Bank was not only the principal supplier of credit to crop and livestock producers. It also functioned as intermediary in channelling money to the agricultural sales cooperatives (ASCs) and to their own retail network of Agricultural Credit Cooperatives (ACCs). Given the close relationship between

[2] There is little evidence that the Parliament has acted as an independent player in the policy-making process.

the government and the Agricultural Bank, there was no pressure for the Bank to operate in a cost-effective way (Togan *et al.*, 2003: 28-29).

Second, a number of SEEs played a prominent role in the process. SEEs are government-installed monopoly enterprises that organise and regulate particular markets. In this category are TÜRKŞEKER (Turkish Sugar Company), TEKEL (Turkish Alcohol and Tobacco Company), and TMO (the Turkish Grain Board), which had nearly complete market power. SEEs were powerful organisations, as long as the government protected their power in the market and supplemented their operations from the budget (see also Chapter 3).

Third, one can identify the large group of Agricultural Sales Cooperatives (ASCs), each of which specialised in a particular product. The ASC sector included 16 unions (ASCUs), circa 330 ASCs and around 750 thousand farmer members. ASCUs' directors were appointed by the government and in that respect they were not an independent player in the policy process. Initially, cooperatives were established in selected regions for strategic export crops such as cotton, sunflower, raisins and olives. In the 1960s, the cooperatives became channels for increasing producer prices, organising market intervention or providing and subsidising inputs (fertiliser, credit, and irrigation water). Some cooperatives had a supply-limiting role (tobacco, hazelnuts and tea). They received government financing not only for crop purchases and input subsidies, but also for investments in industrial installations, storage facilities and administrative buildings. Farmers had very limited control over the activities of their cooperatives. Moreover, maintaining employment in these overstaffed organisations became an important objective for the government (Lundell *et al.*, 2004: 60).

Incidentally, the international funding organisations (IMF, World Bank) and recently also the European Union have had considerable influence over the direction of policy developments through the leverage that they exert as donors on the Turkish government.

Farmers in Turkey have been mainly represented by the 'Chambers of Agriculture'. These Chambers are semi-public institutions, which makes it difficult for them to take a position independent from the government. They have been rather weak players in the process. The same holds for the few 'real' farmer-controlled organisations and other non-governmental organisations (NGOs).

Given the main players in the process, it is clear that the countervailing power of consumers and taxpayers in the policy process has been very limited.

The agricultural sector and the functioning of agricultural policies

Turkey's agricultural sector has been a rather closed and domestically oriented sector (see Chapter 4). Until about 2000, policy was strongly supportive to agriculture, via price support and input subsidies. There was a gradual increase in real budget outlays for output price support and input subsidies over the period 1985-2000 (OECD, 1994; World Bank, 2001) (see Table 7.3). These support policies gave Turkey a special position among developing countries, many of which maintain a low level of agricultural prices by taxing the agricultural sector (World Bank, 1986). Turkey's agricultural protection did not, however, result in strong growth of agricultural output and Gross Value Added (GVA) in agriculture (see Figure 7.1). A declining trend in agriculture's GVA can be observed, although instability is great which makes it difficult to reach clear conclusions.

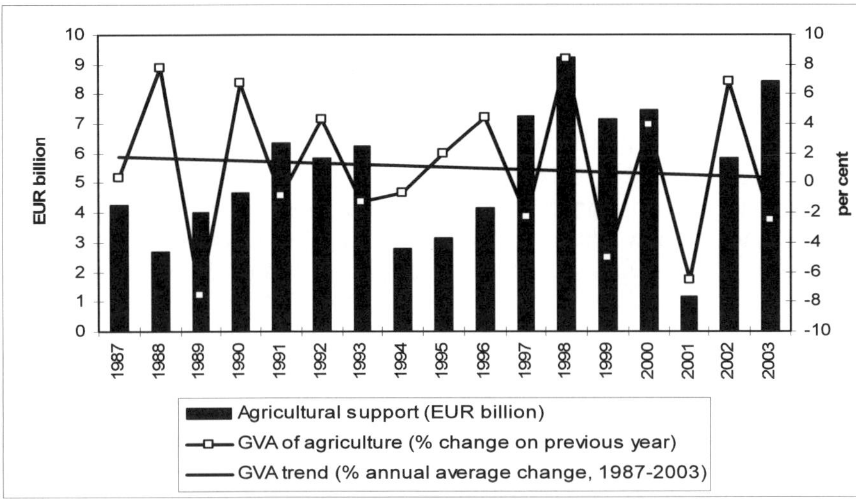

Figure 7.1. Agricultural production development in Turkey 1987-2003
Source: SPO (2004a and b) and OECD (2004)

Price support was the most important part of Turkish agricultural policy (Flam, 2003: 19-20). SEEs and ASCs were commissioned to buy commodities such as cereals, tobacco, tea and sugarbeet from farmers at prices determined by the government. Deficiency payments were first used in 1993 for cotton. After that, no deficiency payments were made until 1998. In 1998, olive oil was introduced into the deficiency payment programme alongside cotton, and in the following years sunflower, soybeans and maize were added. Markets were also protected by import tariffs. Moreover, the output of tobacco, hazelnuts, tea and sugarbeet has been controlled in different ways. This illustrates a policy regime that was strongly driven by the groups who profited from the system.

Input subsidies were the second most important component of agricultural policy. Various subsidies, grants and exemptions were provided with the purpose of reducing the cost of inputs, including, credit, fertiliser, seed, pesticides and water (Flam, 2003: 20).

The development of rural areas and agriculture in particular has been impeded by heavy government intervention in the sector, which was often counter-productive. Trade controls, government procurement, strong government involvement in marketing, input subsidies (especially for credit and fertiliser), and heavy investment in irrigation infrastructure on a fully subsidised basis have created a net inflow of resources from the government to agriculture. All these elements either required payments from the state budget or implicit transfers from consumers, which had negative effects for the whole economy. Within the agricultural sector, policies discouraged production of products in which Turkey has a comparative advantage, squeezed out private sector marketers and subsidised inefficient production technologies (World Bank, 2001: 6).

Until the new Agricultural Sales Cooperative and Agricultural Sales Cooperative Union (ASC/ASCU) Law came into force in June 2000, cooperatives were mainly channels for implementing government programmes rather than member-owned cooperatives. Funded by government, the cooperatives were put under the supervision and direct control of the Ministry of Industry and Trade via the ASCUs. The Ministry controlled the ASCUs through the appointment of management staff and operational directives. The ASCUs controlled the primary societies (ASCs) which operated as *de facto* branches of the union rather than independent cooperatives responsible for their own finances, management and operations (World Bank, 2001: 9).

The dominant role played by the government prevented the cooperatives from developing as organisations providing services for their members. Controlled from the top and financed by funds coming from government, the cooperatives had no incentive to develop their operations, nor were they given the opportunity to do so. Having to work as agencies for delivering price subsidies to farmers, the cooperatives could not become profitable. Their losses were aggravated by the fact that they were compelled to employ staff in excess of their requirements, at wages far above comparable private sector levels (World Bank, 2001: 9).

Another negative consequence of government intervention in the operations of the cooperatives was that members had no real feeling of ownership of their cooperatives. This is illustrated by the very low rate of participation in general assembly meetings in nearly all societies. There was a widespread perception among members that, although they were officially the owners, in reality the cooperatives were government entities (World Bank, 2001: 9).

The above criticisms all centred on the inefficiency and unsustainable budget cost of the support to agriculture. Most of the support was not reaching farmers, and they did not benefit. In the late 1990s, these criticisms were used by the World Bank when designing and justifying a reform proposal for Turkey's agricultural policy.

The Agriculture Reform and Implementation Project (ARIP)

Following several severe financial crises starting in 1997, the IMF and World Bank promoted major change in macroeconomic and agricultural policies as part of a recovery package. The crises not only resulted in a large devaluation of the Turkish lira, but also in a severe drop of real GDP in 1999 and 2001 (see Chapter 2). The devaluation helped Turkey to adjust agricultural prices in the direction of world market prices, but made imported inputs (pesticides, fertiliser and fuel) even more expensive.

The agricultural reforms (ARIP) are described in detail in several accessible documents (World Bank, 2001; Togan, *et al.*, 2003; Cakmak and Kasnakoglu, 2002; Togan, 2004). It is useful to go over them here, however, given their importance for Turkish agriculture. The main philosophy of ARIP is to 'liberalise' Turkish agricultural markets and market organisations, to remove input subsidies and to compensate farmers by means of non-distorting direct income support.

Reduction of price support

The ARIP focuses on smoothing the transition from a heavily government-influenced agricultural sector to a partly market-driven one. This shift, however, is taking place in the context of substantial import tariff protection. Although it is difficult to identify real price movements under conditions of very high inflation, it appears that agricultural prices fell about 15 per cent in real terms from 1999 to 2001, but by 2003 had returned to their previous levels (see Section 4.5). Market forces (higher input prices, increased demand after the crises and higher export prices in Turkish lira) gained importance, but still in a situation with substantial import protection.

Restructuring the SEEs and ASCUs

Other key elements of ARIP are the privatisation of SEEs and the reorganisation of the agricultural cooperatives. Because these are institutional reforms, it may take some time before these organisations perform in line with their newly formulated objectives. A number of formal steps have been taken with respect to the SEEs (see Chapter 3). For sugar, tobacco and tea, privatisation is taking place. For cereals the situation is less clear. There is a risk that, for these organisations, market power, inefficient operation and lack of competition will continue despite formal privatisation. The SEE for agricultural inputs has been discontinued.

It is a priority of the present government to strengthen farmers' organisations. The 'Action Plan' of MARA recognises the need for urgent reforms, including: (i) the drafting of a single cooperative law replacing the three different existing laws and aiming for the independence of cooperatives by firmly integrating them into the private sector, (ii) the development of a programme that would facilitate the consolidation of rural cooperatives by merging many small and narrowly-focused units in order to develop larger units capable of providing comprehensive services to farmer members, and (iii) a new law on farmers' unions as well as amendments to the law on chambers of agriculture, both of which would strengthen the representation of farmers and improve non-commercial farmer services. This reorganisation is also based on the FAO-IFAP Technical Co-operation Project, which analysed farmers' organisations in Turkey and compared them with existing farmer institutions and structures in Europe. A three-legged system was recommended for Turkey, based on cooperatives, producers' unions and chambers of agriculture. Comprehensive restructuring and development of each leg was deemed necessary. These recommendations are now implemented in the current MARA Action Plan (Lundell *et al.*, 2004: 59).

The intentions of the reform are quite ambitious and their implementation will take time. According to Lundell *et al.* (2004: 61), many important restructuring steps still need to be taken, such as turning the unions into efficient service organisations for the primary cooperatives. An interim situation appears to have arisen where old mechanisms do not work any more but new approaches (and their embedding) are not yet operating either.

Farmers are not well organised and a new law on farmers' unions is not helping to remedy this. An environment that is conducive to the development of farmer-led organisations has been lacking for many years. One simple result of this is that the participatory approaches now being promoted cannot be effective. Whether effective producers' organisations will be established in the future is not clear.

Decreasing and abolishing input subsidies

The reduction and, for some items, complete removal of input subsidies is one of the clear achievements of ARIP. Reduction in the fertiliser subsidy started in 1997 and was completed in 2001. Although the subsidy was nearly 50 per cent of the fertiliser price in 1997, the (real) price to farmers certainly did not double over the period of abolition. Part of the subsidy used to go to the fertiliser industry and the organisations distributing the fertiliser, and presumably its abolition was partly absorbed by them. Moreover, competition, including from foreign suppliers, increased. Following the subsidy removal, fertiliser use has fallen by 25 to 30 per cent (Lundell *et al.*, 2004).

Removing interest subsidies, which used to be paid via the Agricultural Bank and the ACCs, was another significant change in the environment of agricultural producers. There was a dramatic change in real interest rates from about *minus* 20 per cent in the 1990s to 30 per cent in 2001-2002 (Lundell *et al.*, 2004: 22, 23). The previous credit policy of the Bank and the ACCs amounted to a soft budget constraint due to high inflation. The present situation, however, goes fully in the opposite direction and makes it nearly impossible to use bank credit. Credit use must have fallen considerably, although good figures are not available. Informal credit has possibly taken over part of the market.

The reduction in subsidies for pesticides, seeds and water should also be mentioned. Here it is less clear whether all subsidies for pesticides and seeds have already been removed. The OECD database on Agricultural Support shows increasing input subsidies between 2002 and 2003, particularly for feed and diesel. Quantity reductions for seeds are small and for pesticides more long-term statistics are required to provide a good indication. Water prices are much more determined institutionally and the price subsidy for water still continues.

Direct income support

As partial compensation for the removal of output support and input subsidies, the Turkish government introduced direct income support (DIS), starting in 2001.

This involves a flat-rate payment per hectare, amounting in 2003 to about TRL 160 million (about EUR 92), paid up to a ceiling of 50 hectares per farm. The programme started in 2001 with a cap at 20 hectares, but this created an incentive to split up larger farms in order to receive more direct income payments. Land registration is not yet complete (see Chapter 3). However, all land users (owners, tenants and share-croppers) are eligible to apply for DIS, as long as they can demonstrate that their land is legally cultivated and is registered in the land registry, or that they can show a document from the village head that they are legal users of the land (Lundell, *et al.*, 2004; Cakmak, 2004).

Table 7.1 shows the stepwise increase in DIS payments. Although the payment of DIS is slightly behind schedule, it is expected that this part of the reform will be easily accepted, as it provides substantial amounts of cash to farmers. According to Cakmak (2004: 16) a per hectare payment was the only feasible tool to compensate for price decreases and reduced input subsidies.

Table 7.1. Direct Income Support payments, 2001-03

For the year[1]	Farmers registered for DIS (1000)	Area registered for DIS (1000 ha)	DIS Payments (TRL mn)	
			(TRL mn)	(EUR 1,000)[2]
2001	2,193	11,821	1,182,095	946,685
2002	2,593	16,080	2,170,831	1,279,994
2003	2,765	16,650	2,664,023	1,535,911

[1]The payments for the intended years were delayed and made in two instalments.
[2] The conversion to euros is made according to the periods of actual payment at market exchange rates.
Source: Cakmak (2004: 16)

The 16.65 million hectares registered for DIS in 2003 falls short of the expected maximum. Factors explaining this discrepancy include the incompleteness of the land register (although farmers legitimately occupying unregistered land may obtain a document form the village head to certify that they farm a given area), and the transaction cost disincentive to register a very small holding.

Restructuring agricultural production

The World Bank also looked at the opportunities for Turkey to reduce production and increase prices for certain products. An obvious product is hazelnuts for which Turkey supplies 70-75 per cent of the world market. The hazelnut area of Turkey had grown substantially in the past, due to high subsidies (Lundell *et al.*, 2004:51-58). A dominant world market position gives scope for increasing prices by restricting supply, and this led to the idea of offering a subsidy for grubbing-up part of the hazelnut area. Such measures, however, are not easy to enforce. Individual farmers are not interested in giving up or reducing their production. This element of ARIP is not working at the moment. Although there is less information available for tobacco, where supply control has also been envisaged, one can expect similar problems for this product.

Observing the effects of policy change in household surveys

It is quite clear that the ARIP reforms brought major changes to Turkish agriculture, but also more generally to households. This was the reason for carrying out an extended household survey where farm households are included. First indications are available with respect to the years 2001 and 2002. Evaluations by the World Bank have been made on the basis of a quantitative household survey in 2002.[3] Policy-induced price reductions can be observed at farm level. Due to those price reductions and reduced input subsidies, agricultural production fell by about 4 per cent (2 per cent for crop products and 10 per cent for livestock products). The direct income support compensates for about 40-50 per cent of the revenue loss at farm level (Lundell *et al.*, 2004). Compensation goes mostly to crop farmers, but this cannot be derived from the data overview.

Given the basic principle that the effect of price changes is smaller in the short run than in the long run, a larger production effect will result in the future. Over the longer term, however, weather effects, technology change and continuing price adjustments due to exchange rate changes and price adjustments arising from changing import regimes, might make it difficult to isolate the long-term effects of ARIP.

Evaluation

Table 7.2 provides an overview of the main items of ARIP. The World Bank's evaluation of ARIP is quite optimistic: "By international standards, the magnitude of this fiscal adjustment from agriculture (agricultural transfers were cut by over two-thirds, or USD 4.3 billion) and its quality (since the adjustment squarely focused on subsidies rather than investments) are impressive" (Lundell *et al.*, 2004: vii). The evaluation concludes that the programme is largely on track with respect to lowering output and input subsidies (see Table 7.3).

The implementation of direct income support has encountered problems regarding land registration and the enrolment of farmers, but it is expected that this part of the programme will be fully implemented by the end of 2004 (Lundell *et al.*, 2004). If so, this would illustrate the ability of the Turkish administration and farm sector to change policies under high internal and outside pressure.

The continuation of the project in 2003 and 2004 should be assessed carefully before a realistic judgement of the project can be made. OECD data for 2003 signal increased domestic support and remaining input subsidies. Also the effect of policy changes on consumer prices is still unclear. Already for decades, the Turkish government has tried to reform agricultural policies, but time and again the country has failed to sustain initial policy changes.

---

[3] See Lundell *et al.* (2004: 40; in particular footnote 22) for a description of the methodology used to gather the data. The data set contained 5508 village households surveyed in 500 rural villages in November-December 2002.

Table 7.2. Implementation of ARIP, starting 2001

Item	Targets	Achieved	Comments
Reduction in price support	Phasing out of price support and linking up to world market prices and reducing import tariffs.	Procurement prices have decreased (ca. 20% in real terms). This was mostly done by keeping price increases below inflation rates. Some reductions in import tariffs have been introduced. Not all export subsidies were abolished.	Price reductions, although they might raise difficulties from a political perspective, are easier than institutional changes. But even the price reductions have not been sustained.
Restructuring - State Economic Enterprises (SEEs); - Agricultural Sales Cooperatives (ASCs) /Unions (ASCUs)	Complete restructuring of SEEs by either privatisation or discontinuation. Transformation of the ASCs into real cooperatives.	State-owned organisations still exist and many restructuring steps still remain to be taken, e.g. turning the unions into efficient service providers for the village-level cooperatives (Lundell *et al.*, 2004: 67).	It is difficult to say whether internal organisational changes will be sufficient to achieve a market-oriented system.
Decrease input subsidies	Abolition of fertiliser subsidy by 2002. Reduced seed, pesticide and water subsidies. Reduced credit subsidies.	Fertiliser subsidies have been reduced and according to Lundell *et al.* 2004: 18) they have been discontinued. Credit subsidies have been abolished.	Water, seed and pesticide subsidies (although reduced) still exist.
Introduction of direct income support	Introduction of direct income support starting in 2001.	The introduction of direct income support had some delay (due to land registration problems), but seems to run quite well.	Land registration was quite successful. Direct income support was attractive for farmers, given price decreases.
Restructuring agricultural production	Relocating hazelnut and tobacco production by 2002.	The grubbing-up of hazelnut trees has not been successful.	Destroying 'capital goods' for financial compensation is difficult with high inflation rates and free-rider behaviour.
Recording policy impacts in a household survey	Annual surveys of households over the period 2001-2005.	Lundell *et al.* (2004) use the results of the household survey of 2002.	Survey for 2003 has not been carried out.

Table 7.3. Trends in fiscal transfers by type, 1999-2004, EUR million

Type of Transfer	1999	2000	2001	2002	2003[1]	2004[2]
Transfers to agricultural SOEs	2994.0	1800.9	1187.8	135.4	92.6	-6.4
Transfers to ASCUs	908.9	786.6	134.4	163.4	166.1	165.3
Direct output subsidies	222.4	356.9	371.3	275.6	242.6	302.7
Input subsidies	272.8	208.2	91.1	0.7	0.0	0.0
Credit subsidies	1301.9	241.2	318.9	0.0	0.0	50.8
Direct income support	0.0	2.9	76.3	1318.4	1376.1	1526.2
Total	5700.1	3396.7	2179.7	1893.4	1877.4	2038.8

[1] Revised – final figures can be slightly different. [2] Provisional estimate.

Source: Lundell, et al., 2004: 4; M. Lundell, personal communication, 2004

Is ARIP a preparation for the EU?

Clearly, the basic principles of the ARIP are broadly consistent with the long-term policy direction of the Common Agricultural Policy. A striking difference is that the European Union has taken about two decades to change policy (starting in 1992), while ARIP is a five-year programme. Strong external pressure has induced a faster, albeit less fundamental, policy change in Turkey. ARIP started, however, in a situation where many input subsidies were used as well as output price support, and where state organisations and obligatory co-operatives played a powerful role.

The World Bank (2001: 10) argues: "This overall program of agricultural policy reforms will assist the Government in its aspirations for accession to the EU by increasing the efficiency of the sector and the economy at large". This statement was, however, later clarified as follows. "Originally, the Bank discussed with the Government a multi-component single loan, which would have the components currently envisioned in the ARIP, plus several others as well. These other components included commodity market development, land management, agricultural services, and preparation for EU accession (harmonisation of legislation and institutional improvements). However, it is clear that some of these tasks are more urgent than others to ensure the success of the reform program. Thus, the Government requested that we prepare a project to support these more urgent tasks on a priority basis, leaving some of the others for a later operation. The ARIP focuses only on the tasks that must be started in 2000-2001 to support the transition from a heavily government-influenced to a market-driven agricultural sector" (World Bank, 2001: 20).

It is quite probable that the World Bank was not interested in the discussions on the shape of future EU policy and preferred a clear policy change that would fit into a rather wide spectrum of future EU policies. This, of course, raises the issue of whether changes that are difficult for the Turkish agricultural sector to absorb and which are quite outside the future EU policy range will be questioned as soon as negotiations with the EU start.

Agricultural policies for specific products

As in the EU, there are differences in policies between products. It does not make sense to discuss individual products in detail here. The interesting question, however, is whether in the future large differences between Turkish and EU agricultural policies are expected for particularly important or sensitive products.

Olives and olive oil

For olives, a typical Mediterranean product, EU member countries Spain, Italy and Greece have dominant positions in the world market. Compared to the EU, Turkey is a small producer with about 11 per cent of the combined production of Spain, Italy and Greece.[4] In Turkey, market regulations and processing for olives and olive oil are in the domain of the Agricultural Sales Cooperatives (ASCs). Before the recent policy reforms, olives and olive oil were under a deficiency payment system (Togan *et al.*, 2003: 4). As in Italy, a large share of the olive trees are old and not very productive. Given the high real interest rates, the removal of interest subsidies and the reduction in prices, it is not attractive for producers to invest in new trees.

In the EU, olive oil and olives have been protected by high tariffs, direct premiums and export subsidies. Moreover, Turkey also applies import tariffs to olives and olive oil, which are maintained for olive oil under the Customs Union between Turkey and the EU (Grethe, 2004: 60), and provides export subsidies to olive oil. Tariff bindings at the WTO for these products are lower for Turkey than the EU. Grethe (2004: 44) compares the price-supporting policies of the EU (Greece and Spain) and Turkey over the period 1990-2001. The lower price of Turkish olives (compared to Greece and Spain) and the higher price of olive oil, illustrate an inefficient oil processing sector for Turkey or quite different product qualities.[5]

Even though there are some differences between Turkish and EU policies for olives and olive oil, the basic instruments are not too different, except for the EU's area premium. According to Grethe (2004: 324), full integration of Turkey into the EU in 2006 would lead to only small changes in supply and demand of table olives and olive oil, although Turkey would become a slightly larger net exporter of olive oil: circa 9 thousand tons.[6]

Sugar

The sugar policies of Turkey and the EU show various similarities. Turkey is a substantial producer (12 per cent of EU-25 output) and for several years has been a next exporter of sugar, with high import tariffs and an implicit export subsidy (Grethe, 2004:37, 38; OECD, 2004). Sugar quotas have been set for a

---

[4] According to Grethe (2004: 140) the Statistical Institute of Turkey might underestimate Turkish production substantially. Turkish olive production fluctuates between high and low production years (see Figure 4.4, Chapter 4). This fluctuation is much larger than for the EU producing countries, and is sometimes out of phase with them.

[5] The very large price difference between Greece and Spain signals important quality differences.

[6] This is about 7 per cent of total supply in Turkey and 1 per cent of the supply of EU-25.

five-year period by the Sugar Agency, which was established in 2001. During the period 1999-2001, sugarbeet factories cut purchases by 26 per cent. During this period the strong involvement of the government via TurkŞeker was also reduced (Lundell *et al.*, 2004; Grethe, 2004). By 2002 and 2003, however, the implicit or applied level of protection of the Turkish sugarbeet sector increased again (OECD, 2004). These implicit export subsidies or special arrangements with countries are in operation to promote the export of sugar to third countries.

The Sugar Law imposes strict quotas at processing plant level. The quota classification follows the current EU structure (Cakmak, 2004: 14). According to Lundell *et al.* (2004: 61) there are some 1.5 million sugarbeet farmers registered as members with the Sugarbeet Cooperatives. This very high number would imply that nearly every second farmer is allowed to grow sugarbeet, with on average 0.2 hectare. This indicates a highly fragmented primary sector.

Cotton

Turkey is a substantial producer and also an importer of cotton. It has been classified as a largely deregulated crop, although with some deficiency payments (Cakmak, 2004:14) or production premiums (Grethe, 2004: 39). This is related to both its production conditions and to its textile and garment industry. The strong price decreases for cotton on the world market, however, made importing (of cotton lint) more attractive and led to a lower domestic price with a slightly expanding cotton area and production, because other products were even less attractive (Lundell *et al.*, 2004: 25-28). Moreover, cotton production in Turkey profits from large-scale irrigation areas that are coming on stream.

Bringing Turkey under the EU cotton regime would mean putting a rather efficient production sector under a direct income support system (Council Regulation 864/2004). Here, a specific amount per hectare might be chosen: for Greece it is EUR 550 per hectare, of which 65 per cent is incorporated into the single farm payment, the rest being coupled to area.

Dairy

In Turkey, dairy production is protected by high tariffs (between 67 and 150 per cent according to the product, which isolate the industry from international competition (Grethe, 2004: 50). Although milk prices are below EU prices, prices of butter and skimmed milk powder are much higher and this reflects a relatively inefficient dairy industry compared to that of the EU. Farms with dairy cows are much smaller in Turkey and not specialised, which leads to high milk collection costs.

Real prices for dairy farmers have fallen over recent years. Both in cow numbers and production, dairy is a declining sector of Turkish agriculture. It is difficult to say how dairy farming will benefit from the DIS scheme within ARIP.

7.3 Food policy

Food policies began in Turkey in the 1980s with a more open trade policy, increasing urbanisation and growing attention from the general public for food quality and food safety. This increased attention is partly due to the higher income levels during the last two decades. As is shown in Chapter 6, the same factors driving the development of food policies have also been influencing the Turkish food sector. Because of the low level of Foreign Direct Investment (FDI), the introduction of new technologies and methods has not contributed very much to the process of increased food quality and food safety.

Food safety

Food safety concerns depend on a society's level of development. In subsistence agriculture, food safety is mostly related to habit formation and survival. Even in such societies, however, certain habits had or have a negative influence on food safety, mostly arising from lack of knowledge or maintaining old practices. In poor urban environments, food safety is even more at risk because primary products and food are not grown and processed by consumers themselves. This can permit extremely low standards of food safety. In highly developed rich societies, science and technology largely drive food safety: consumers require high standards and pay for them. In subgroups of highly developed society, food safety *and* food quality are part of the culture (for example in organisations like the 'Slow Food Movement'). Given the existence of all four consumer groups in the highly diversified Turkish society, it is not very easy to describe the food safety situation in Turkey from the demand side of the market.

The food safety situation in Turkey

According to Togan *et al.* (2003: 34), food legislation in Turkey has been updated continuously since 1985. The harmonisation of 'Good Agricultural Practices' has been completed, and the regulation on Agricultural Quarantine has been in force, and regularly strengthened, since 1991. A Food Act was passed in 1995, according to which all stages of food production are targeted for inspection. Turkey has formally adopted a number of typical elements of food safety regulations and control systems. Here, the EU has served as the main model for rules and legislation with respect to food. One example concerns upgrading or establishing monitoring laboratories. Accreditation by the Accreditation Council, set up in 2001, has been initiated for some of the laboratories, which are involved in tests organised by the Food Analyses Performance Assessment Scheme (FAPAS) and by the Turkish Scientific and Technical Council (TÜBITAK). Another important set of standards relating to food safety is the CODEX 1997 adopted by Turkey in 1997. Since 1994, the Turkish Standards Organisation (TSE) has been using internationally recognised guides in certification. This signals some of the more formal approaches to dealing with food safety.

One of the few available empirical analyses of food safety applications in Turkey is that of Alpay *et al.* (2002). This study is based on face-to-face interviews with 100 firms in 5 different food sub-sectors in Turkey in the period 1997-99. The study is not focused on safety and quality standards *per se*, but on their effects on the export performance of firms. The analysis addressed the following three questions: (i) whether compliance with safety and quality standards is rewarded in the EU market, (ii) whether the impact of strong vertical integration and care for environmental quality is positively related to export performance, and (iii) whether or not there are differences across sub-sectors. The authors also sought firms' views on Hazard Analysis at Critical Control Points (HACCP), a concept that was at that time still rather unfamiliar in the Turkish food industry. The study collected data and constructed indices measuring the compliance with quality and safety standards, vertical integration and environmental performance. As might be expected, it confirmed that higher quality and safety standards have a positive effect on firms' export performance.[7] The paper does not provide any information on food safety levels achieved.

Although Decree 560/1995 made both the Ministry of Agriculture and Rural Affairs (MARA) and the Ministry of Health responsible for implementing food safety legislation, both ministries agreed in 2003 that MARA would take over this responsibility. Law 5179, giving the responsibility of food safety to MARA, has been in force since 5 June 2004, creating a delicate situation where a producer-related organisation is responsible for the food safety of consumers. The General Directorate of Protection and Control within MARA administers 81 Provincial Directorates, 39 Provincial Control Laboratories and one Food Control and Research Institute. Food control inspection services were being carried out by about 1400 food inspectors as of the end of 2003. Food analysis services were carried out by about 1000 food analysts. This does not say very much about the level of food safety in practice. One of the main difficulties seems to be the shift away from an inspection approach to a system audit approach for food safety.[8]

Certification of fruits and vegetable growers is important for Turkey because they are the important agricultural export products. By June 2004 there were 102 farmers certified by EUREPGAP[9] with a total area of 2,905 ha and certification was growing rapidly (EUREPGAP, 2004).

[7] The same holds for vertical integration and environmental performance. The use of HACCP seems related to both food quality and food safety indicators. As always, it is difficult to distinguish cause and effect.

[8] Information derived from the Twinning Project Proposal 'Restructuring and Strengthening of the Food Safety and Control System in Turkey' within TR 0403.03.

[9] EUREPGAP started in 1997 as an initiative of retailers belonging to the Euro-Retailer Produce Working Group (EUREP), and has developed into an equal partnership of agricultural producers and their retail customers aiming to establish widely accepted standards and procedures for the global certification of Good Agricultural Practices (GAP).

Food quality

This area shows some parallels with food safety, but there are also differences. Food quality is more difficult to define, because as well as safety characteristics, it also refers to whether the particular food item fits into a healthy diet for 'average' consumers, and to the perceived taste and attractiveness of the food. So far, we have not found any accessible study dealing with food quality in Turkey.

Conclusions on food safety and food quality

Only very limited information is available on the actual functioning of legislation and standards with respect to food safety and food quality in Turkey. Legislation is increasingly oriented towards EU standards. The ongoing legislative and other efforts have focused on establishing an effective food safety system that responds to the 'from-farm-to-table safe food' objective of the White Paper prepared by the European Commission (Republic of Turkey, 2003). The effectiveness of governance and control, however, cannot be assessed, and would also be difficult to categorise in a society with weak institutions and considerable corruption.

Two remarks are relevant. Due to the growth of exports, the increase in supermarkets and expanding tourism, food quality and food safety standards should increase. This linkage between food safety, quality and export development is confirmed by Alpay *et al.* (2002). Moreover, the preparation for EU accession may have a positive influence on the further adoption of legislation. The consequences of big food safety scandals in Turkey become more visible and more serious as a society and its trading position become more open. This is illustrated by the increased attention on food safety in Turkey.

Lower economic growth over the last few years and a relatively low level of foreign direct investment in the food industry are factors that have inhibited rapid development of food safety and food quality standards.

7.4 Rural development policy and structural policy

Structural policy and rural development policy aim at increasing the productivity of rural areas, regions or industries. This can be done by policy instruments directly targeted to increasing productivity (e.g. education, research, extension) or by policy instruments directed at the allocation of production factors (e.g. reducing labour in agriculture, investment in infrastructure, etc.). Often policies contain both elements: increasing productivity and the reallocation or increase of input use.

Rural development policy is quite different in Turkey and the European Union in several aspects, beginning with the philosophies of rural development policy (Turkey) and rural policy (EU). Turkey is strongly oriented towards developing the basic conditions for agricultural production (e.g. irrigation) and improving basic infrastructure. It is sometimes highly influenced by the typical World Bank projects centred on irrigation schemes. EU rural policy is intended to improve: (i) the competitive position of agriculture, (ii) the environment in relation to agriculture, and (iii) the living conditions in local communities. Rural policy in the EU has a strong bias towards agriculture, because the funding of rural policy comes from the agriculture budget.

Turkey's rural development policy in the past has shown more similarity with EU structural policy, where large infrastructural projects are used to stimulate the development of a region. Structural policy in relation to agriculture and the food industry is presently not used in Turkey. Before ARIP, the SEEs and the ASCUs had a highly protected market position and could use part of their funds for investment in their production facilities. Whether this will continue in the partly privatised industries and cooperatives under transformation is not yet clear.

Within Turkey's National Development Plan 2004-06 (Republic of Turkey, 2003:vii) the approach to rural development is more in line with that of the EU. Clearly, some EU terminology has been copied. This could be due to the fact that 75 per cent of the budget (a total of EUR 200 million over the period 2004-06) will be provided by the EU pre-accession funds.

Rural development projects

Rural development in Turkey concentrates on long-term investment projects that are highly subsidised by the government. Various rural development projects have been completed, others are currently under way or under consideration. The Master plan for the Eastern Anatolia Project (DAP) was completed in 2000. The Eastern Black Sea Regional Development Plan, the Southeastern Anatolia Project (GAP), the Aegean Regional Development Plan and the Western Mediterranean Regional Development Plan are all under way. Studies have begun for the Central Black Sea Regional Development Plan and the Yesilirmak Basin Development Plan. Preparatory studies are under way for the Central Anatolia Regional Development Plan. The Eastern Mediterranean Regional Development Plan was launched at the end of 2002. Several of these projects are scheduled to be completed by the end of 2005 (CIHEAM, 2004). Whether this can be achieved is not possible to say.

The Eastern Black Sea Regional Development Plan (DOKAP) is in the implementation phase with technical assistance from the Japan International Cooperation Agency (JICA) under the coordination of the State Planning Office (SPO). This plan for the Eastern Black Sea Region covers eight provinces, and its target year is 2020. Its major objective is to prepare a medium- and long-term integrated regional development plan, and to define priority sectors and possible investment projects towards this plan – a typical situation of 'planning to plan' (Republic of Turkey, 2003: 165).

The World Bank (2001: 21) mentions the Eastern Anatolia Watershed Rehabilitation Project, which is classified as a Natural Resource Management Project. Although the project is classified as 'developing satisfactorily', it is difficult to say what this means. For all these plans, information is lacking about both the type of expenditure and the sources of the funding. Evaluations of previous and present plans are not available. This is not unusual for Turkey, where most effort has been put into developing five-year plans without making evaluations afterwards. This is also true of the National Development Plan 2004-06 (Republic of Turkey, 2003), which contains many attractive aims for rural development[10]. However, our institutional analysis (Chapter 3) suggests that many of them will be difficult to realise in such a short period.

One of the advantages of land registration (see Chapter 3) is that the ownership of land becomes clear. Undefined property rights, which make it difficult to involve participants in the ownership and the costs of projects, often create problems in rural development.

The Southeastern Anatolia Project (GAP) Institution is a national agency under the Prime Ministry. It is responsible for managing socio-economic development in the less developed south-eastern region of Turkey. Lundell *et al.* (2004) describe this project as an example of an integrated rural approach. The main infrastructure is already in place (electricity, communication facilities, roads and feeder roads), but they need upgrading. About 90 per cent of the rural zones have drinking water, but only 52 per cent of this water is supplied through the distribution network. Rural housing is considered to be one of the bottlenecks. Here, loans for house building are used.

There is no experience in Turkey with projects similar to LEADER[11], where a bottom-up process generates initiatives and stimulates the search for co-financing.

Structural policy

Until recently, structural policy as such was an unknown concept in Turkey. Now, however, new documents are starting to copy the EU terminology. The National Development Plan 2004-06 is very focused on the development of infrastructure: communication, roads, railways, electricity, and also on social infrastructure (Republic of Turkey, 2003). Although Turkey's physical infrastructure provision tends to be lower than in the New Member States of the EU, this does not hold for all types. Infrastructure is often developed in relation to urbanisation.

Although structural policy as a concept is quite new in Turkey, several elements of Turkey's rural development projects could be likened to typical structural policy in the European Union. There is a strong tendency to expand resources, while targeting the increase of productivity receives less priority.

7.5 Discussion and conclusions

The structural features of Turkish agricultural policy can be characterised as

[10] These aims include "to increase employment and support income generating activities in rural areas", "to strengthen capacity towards efficient utilisation of the agricultural lands", "to increase the living standards of rural population by means of modern agricultural techniques", and "to contribute to the reduction of interregional and intra-regional disparities by stabilising the migration from rural areas to urban areas". (Republic of Turkey, 2003: 129).

[11] Liaisons Entre Actions de Développement de l'Économie Rurale, EU initiative aiming to foster integrated development strategies within very small districts.

strong government involvement either directly or indirectly via state enterprises and cooperatives. It has been argued on various grounds that this type of policy was not fostering an efficient agricultural sector. Given the small and fragmented farm structure, a more rational approach would have been to encourage education and extension to transfer new and suitable technologies. However, as has been shown in Chapter 3, the institutions for such an approach were and still are seriously lacking. Moreover, the chain approach, in which industries or retailers take the lead in transforming agriculture, is underdeveloped (see Chapter 6). Although the shift to private industry is encouraged through ARIP, it is not yet clear whether these policies really encourage the development of competitive private industries that will function well in the Turkish environment. A good example is credit policy, where subsidies have been removed. According to the new rules, collateral is required, which indirectly limits small farmers' access to the credit. Agriculture in Turkey has kept its role as a major employer. However, most of the agricultural sector is quite inefficient compared to the EU, although there are exceptions (e.g. cotton).

For many years, the Turkish agricultural sector has been a political football in the competition to gain votes, rather than a sector receiving serious attention from the government with a view to improving its efficiency and adjusting to the present and future needs of society. But that situation changed with the ARIP (2001-05). From an analytical perspective, ARIP is an interesting policy experiment: price support has been reduced, subsidies have been removed and direct income support has been introduced. Clearly, ARIP should make a contribution to preparing Turkey for the EU, although markets for some products still enjoy higher levels of tariff protection than in the EU (see Chapter 8). It is too early to observe the full impact of ARIP. However, a short-term production fall of 4 per cent has already been observed.

At the product level, many conclusions could be drawn regarding Turkish policies and their compatibility with EU commodity regimes. Here we mention only a few of them. For example, Turkish and EU policies for table olives and olive oil are not very different, except for the EU's area premium. Sugar policies in Turkey and in the EU are both highly protective and include a quota system. Turkey operates a relatively efficient cotton sector, but an inefficient dairy chain.

Food policy in Turkey mainly consists of various measures to impose international food safety standards. Domestic income growth has so far not been sufficient to encourage demand-led increases in food safety and food quality standards.

Rural development policy in Turkey has been dictated by international donors, and by the extent to which clusters of rural population around big cities have created demand for public infrastructure. To slow down the trend of migration to urban centres, investments have periodically been made in rural areas, which are usually water resource development projects for irrigated farming. Therefore, rural policy in Turkey is partly an endogenous outcome of the flow of population from villages, towns and cities in the east and south-east of Turkey to the western part of Turkey. Rural development policy in Turkey is more focused on large-scale investments in, for example, irrigation. Structural policy would be a new concept for Turkey. The National Development Plan 2004-06 already introduces the terminology, which has yet to be made operational.

In summary, both food policy and structural policy in Turkey are quite remote from the present policies of the European Union.

References

Alpay, S., Yalcin, I. and Dolekoglu, T. (2002) Export performance of firms in developing countries and food quality and safety standards in developed countries. Paper presented at *ERF's 8th Annual Conference*, Cairo, January 2002.

Cakmak, E.H. (2003) Evaluation of the past and future agricultural policies in Turkey: are they capable of achieving sustainability ? In: Jacquet, F. and Lerin, F. (eds.) *Libre-échange, agriculture et environnement: L'Euro-Méditerranée et le développement rural durable : état des lieux et perspectives*. CIHEAM-IAMM, Montpellier, pp. 155-165.

Cakmak, E. (2004) Structural change and market opening in agriculture. *CEPS EU-Turkey Working Papers,* No. 10, September 2004. Centre for European Policy Studies, Brussels.

Cakmak, E. and Kasnakoglu, H. (2002) Assessment of agricultural policies and the impact of EU membership on agriculture in Turkey. Unpublished manuscript.

CIHEAM (2004) Agri.Med: Agriculture, fishery, food and sustainable rural development in the Mediterranean region. CIHEAM Annual Report 2004.

EUREPGAP (2004) Statistics on EUREPGAP certified growers, 18 August 2004. http://www.eurep.org/documents/infoletter/EUREPGAP_certification_status_half_2004.pdf (accessed 2 November 2004).

Flam, H. (2003) Turkey and the EU: Politics and economics of accession. Seminar Paper, No. 718, Institute for International Economic Studies, Stockholm University.

Grethe, H. (2004) *Effects of Including Agricultural Products in the Customs Union between Turkey and the EU: A Partial Equilibrium Analysis for Turkey*. CeGE-Schriften, Vol 9, Center for Globalization and Europeanization of the Economy, Georg-August-Universität Göttingen. Peter Lang Verlag, Frankfurt am Main.

Hathaway, D.E. (1963) *Government and Agriculture*. Macmillan, New York.

Lundell, M., Lampietti, J., Pertev, R., Pohlmeier, L., Akder, H., Ocek, E. and Jha, S. (2004) Turkey: A review of the impact of the reform of agricultural sector subsidisation. World Bank Working Paper, March 9, 2004.

OECD (1994) *OECD Economic Surveys: Turkey 1993-1994*. OECD, Paris.

OECD (2004) *Producer and Consumer Support Estimates, OECD Database 1986-2003*. OECD, Paris.

Republic of Turkey (2003) Preliminary national development plan (2004 - 2006). Ankara, December 2003.

SPO (2004a) Economic and social indicators 1950-2003. State Planning Organisation: Ankara.

SPO (2004b) Main economic indicators. State Planning Organisation, Ankara.

Togan, S. (2004) Turkey: Toward EU accession. *World Economy* 27 (7), 1013-1045.

Togan, S., Bayaner, A. and Nash, J. (2003) Analysis of the impact on agricultural markets and incomes of EU enlargement to Turkey. Paper presented at *Turkey: Towards EU accession*, Bilkent University, May 10-11 2003, Ankara.

World Bank (1986) World Development Report. World Bank, Washington DC.
World Bank (2001) Project appraisal document on a proposed loan in the amount of US$ 600 million to the Republic of Turkey for an Agricultural Reform Implementation Project/Loan. Report No: 21177-TU. June 6, 2001.
Zürcher, E.J. (2004) *Turkey: A Modern History,* 3rd edn. Tauris, London.

8

Turkey's Foreign Trade Position

Alison Burrell

8.1 Overview

Total merchandise trade and balance of payments

Foreign trade has been growing in importance for the Turkish economy for several decades. In 1990, Turkey's total imports and exports of goods were 14.6 per cent and 8.5 per cent of GNP, respectively. These shares grew steadily during the 1990s, levelling off in the early 2000s (Table 8.1). In 2003, Turkey's deficit on merchandise trade amounted to over 9 per cent of GNP.

Table 8.1. Turkey's merchandise trade flows, EUR bn, selected years 1990-2003

	1990	1995	2000	2001	2002	2003[P]
Exports[1]	10.18	16.54	30.07	34.99	38.13	41.61
as % of GNP	*8.50*	*12.60*	*13.80*	*17.80*	*19.70*	*19.70*
Imports[2]	17.51	27.30	59.01	46.22	54.52	60.83
as % of GNP	*14.60*	*20.80*	*27.00*	*23.50*	*28.20*	*28.80*

[1] Excluding entrepot and shuttle trade. [2] Excluding gold imports. [P] Provisional.
Source: Undersecretariat of Foreign Trade, 2004

Table 8.2 shows the main sources and destinations of Turkey's merchandise trade flows. The EU is by far both the largest exporter to, and importer from, Turkey. For comparison, about half the foreign trade of North African countries is also with the EU (CIHEAM, 2000), which emphasises the dominant role of the EU as trading partner for the whole Mediterranean region.

On 1 January 1996, a customs union between the European Union and Turkey came into force. Agricultural products remained outside the customs union, although both the EU and Turkey awarded significant trade preferences to each other's agricultural products. There is surprisingly little evidence of this policy change in the figures: Turkey's exports as a share of GNP are roughly the same in 2000 as in 1995, and the importance of the EU as a trade partner – for both imports and exports – has remained unchanged. On the export side, this is partly explained by the fact that already for many years the EU had granted considerable concessions to imports from Turkey.

Table 8.2. Direction of trade in goods, selected years, 1990-2003

Trading bloc	1990	1995	2000	2001	2002	2003[P]
	Imports by source, % share					
EU	44.4	47.2	48.8	44.2	45.2	45.8
EFTA	2.7	2.5	2.1	3.6	4.9	4.9
USA	10.2	10.4	7.2	7.9	6.0	5.0
Middle East	12.2	7.6	5.7	6.8	5.8	5.9
CIS	5.6	9.3	10.4	11.2	10.8	11.2
of which Russian Federation	*0.0*	*5.8*	*7.1*	*8.3*	*7.5*	*7.9*
Others	24.9	23.0	25.7	26.4	27.3	27.2
Total	100.0	100.0	100.0	100.0	100.0	100.0
	Exports by destination, % share					
EU	55.4	51.2	52.2	51.4	51.2	51.8
EFTA	2.6	1.4	1.2	1.0	1.1	1.1
USA	7.5	7.0	11.3	10.0	9.3	7.9
Middle East	13.8	10.8	8.0	9.2	8.6	10.5
CIS	4.1	9.5	5.9	6.3	6.3	6.3
of which Russian Federation	*0.0*	*5.7*	*2.3*	*2.9*	*3.3*	*2.9*
Others	16.7	20.2	21.4	22.1	23.4	22.3
Total	100.0	100.0	100.0	100.0	100.0	100.0

[P] Provisional.

Source: Undersecretariat of Foreign Trade, 2004

Typically, Turkey has a deficit on merchandise trade with the EU (EUR 2.2 billion in 2002). Within the EU, Germany is Turkey's main export destination, taking about one-sixth of its exports in 2002. The UK, Italy and France took 8.5, 6.4 and 6 per cent respectively. Germany, Italy, Russia and the USA are the most important import sources. As a trade partner of the EU, Turkey's ranking in 2002 was eleventh as an import source and tenth as an export destination, with shares in EU trade of 2.13 per cent for imports and 2.47 per cent for exports (AAF Canada, 2004).

The Commonwealth of Independent States (including Russia) has gained importance as a trading partner since 1990, whereas the Middle East has declined slightly. The Middle East is a traditionally important source for Turkey's crude oil imports, whereas Russia is increasingly important for both oil and natural gas imports. The USA has recently lost import share, but its importance as an export destination is about the same as it was in 1990.

The composition of Turkey's imports in the early 2000s is rather similar to that of 1990: about one-sixth of imports by value consists of investment goods, nearly three-quarters of intermediate goods and the remainder of consumption goods. The contraction in imports due to the 2001 economic crisis (see Table 8.1) restored the import share of intermediate goods at the expense of consumption goods, which in 2000 accounted for over 13 per cent of imports.

As well as Turkey's persistent merchandise trade deficit, a second negative item in Turkey's foreign current account is the interest paid on foreign debt. These items are partly offset by a positive services balance (where tourism makes a large contribution) and transfers (workers' remittances from abroad). Nevertheless, the current account was in surplus in only four years over the period 1990-2003. After recording current account deficits of 0.7 and 4.9 per cent of GNP in 1999 and 2000, respectively, Turkey adopted a floating exchange rate at the start of 2001. The ensuing 21 per cent depreciation of the Turkish Lira and the contraction in imports produced a current account surplus of 2.4 per cent of GNP in 2001. However, the economic upturn in the following year took the current account back into deficit (see Figure 2.6) (Central Bank of the Turkish Republic, 2004).

On the capital account, foreign direct investment in recent years has been disappointing, at less than 1 per cent of GDP (AAF Canada, 2004). The Foreign Direct Investment Law 4875 of 2003 is intended to improve the climate and incentives for this type of investment in the coming years (WTO, 2003b). FDI over the period 1999-2002 was about USD 6.4 billion, with Italy responsible for nearly 31 per cent, followed by the Netherlands (15.1 per cent), the USA (12.4 per cent) and the UK (10.1 per cent). About 12 per cent of this investment went to the food, beverage and tobacco processing sectors (AAF Canada, 2004).

Trade in agricultural products[1]

Table 8.3 shows a marked long-run decline in agriculture's share of exports. This illustrates the unbalanced sectoral growth typical of a developing economy, and reflects the stronger export performance of the manufacturing sector and the higher domestic food demand from an expanding, more prosperous population, rather than to a decline in agriculture. During the period covered by Table 8.3, exported manufactures increased from about 68 per cent to nearly 84 per cent of total exports. This is overwhelmingly due to a strong increase in exports of machinery and transport equipment, and to a much lesser extent to higher export of consumer goods and semi-manufactures (Undersecretariat of Foreign Trade, 2004).

Table 8.3. Agricultural products as a percentage of total merchandise trade

	1990	1995	2000	2001	2002	2003[P]
Total exports, EUR bn	10.18	16.54	30.07	34.99	38.13	41.61
Agricultural products % of total exports of goods						
Agricultural products	25.5	21.1	13.9	13.9	11.2	11.2
of which: food	*22.5*	*19.6*	*12.7*	*12.8*	*10.2*	*10.0*
Total imports, EUR bn	17.51	27.30	59.01	46.22	54.52	60.83
Agricultural products % of total imports of goods						
Agricultural products	n.a.	n.a.	3.6	3.1	3.1	4.2

Source: Undersecretariat of Foreign Trade, 2004

[1] Several definitions of "agricultural products" are used in this chapter. Chapters 01-24 (excluding fish and seafood) of the Combined Nomenclature (CN) is the usual definition adopted by the European Community for its trade statistics. Some additional products are included in the definition of agricultural products used within the WTO's Uruguay Round Agreement on Agriculture (see note to Table 8.5). The definition used in FAOSTAT includes some of the additional WTO products, but omits a few of the more processed items that are included in the regular EU definition.

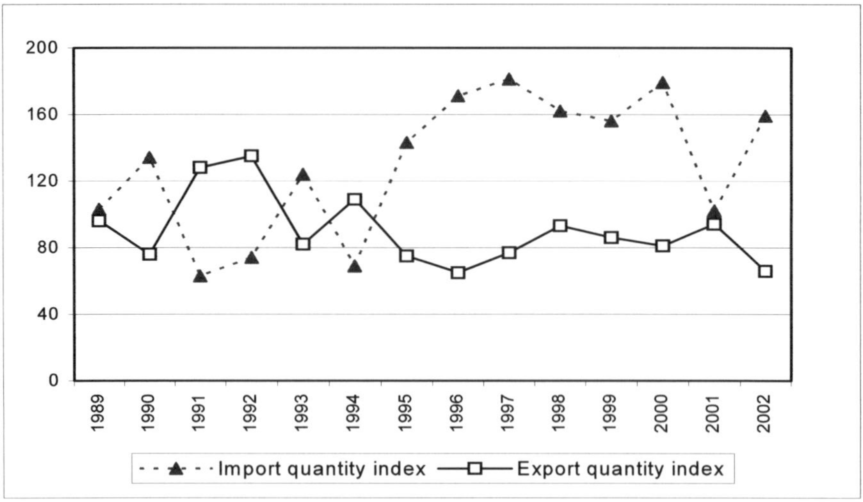

Figure 8.1. Agricultural trade: quantity indices (1989-91=100)
Source: FAO, 2004

Figure 8.1 shows the movement in Turkey's agricultural trade totals, valued at constant prices of the base period. In the early years trade volumes were volatile around a stable level. After the mid-1990s, import quantities increased until the 2001 crisis whilst export quantities remained slightly below the base-period level.

Turkey's terms of trade for agricultural products improved in the mid-1990s (Figure 8.2), but by the early 2000s, most of this improvement had disappeared.

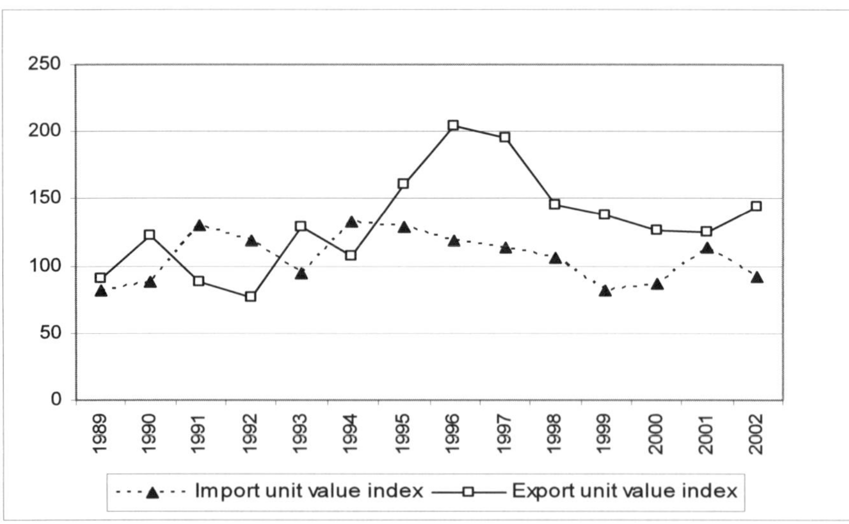

Figure 8.2. Agricultural trade: unit value indices (1989-91=100)
Source: FAO, 2004

Table 8.4 gives a breakdown of Turkey's total agricultural trade flows for selected years during the period 1990 to 2002. The product categories shown accounted for 76 per cent of the value of agricultural imports and 83 per cent of the value of agricultural exports in 2002. Throughout the period, fruit and vegetables comprised over 50 per cent of Turkey's agricultural exports (56 per cent in 2002), with citrus, tomatoes, dried fruit and nuts the most important individual categories.[2]

Table 8.4. Turkey's agricultural trade: Selected product categories and selected years

Product category	1990	1995	2000	2001	2002
	Value of imports (EUR mn)				
Total agricultural products	1776.4	2754.9	3498.3	2703.6	3247.6
Meat	11.3	59.0	1.3	0.4	0.1
Dairy and eggs	14.3	26.5	38.6	23.7	38.5
Cereals[1]	448.4	359.7	458.6	234.7	429.2
Oilseeds	25.7	156.3	252.7	149.2	236.2
Sugar	233.0	147.2	1.0	0.6	1.0
Fruit, vegetables	42.9	66.9	208.2	131.3	147.5
Animal and vegetable oils[2]	230.2	481.0	402.7	357.1	432.2
Olive oil	2.4	0.1	2.7	0.0	3.8
Tobacco	264.8	122.3	379.7	316.2	219.9
Hides and skins	45.1	131.2	155.9	165.4	226.6
Textile fibres	182.7	383.8	822.5	629.4	601.8
Feeding stuffs[3]	27.0	66.4	209.0	132.0	142.5
	Value of exports (EUR mn)				
Total agricultural products	2450.1	3288.1	3920.0	4570.9	3676.8
Meat	16.7	13.2	13.5	20.0	14.7
Dairy and eggs	9.5	16.0	19.8	38.7	26.4
Cereals[1]	58.5	337.1	440.1	366.0	290.0
Oilseeds	5.8	13.9	24.7	32.9	27.1
Sugar	0.1	1.3	139.9	247.6	33.5
Fruit, vegetables	1363.4	1666.0	1967.9	2392.4	2041.1
Animal and vegetable oils[2]	104.6	156.6	74.1	49.6	44.7
Olive oil	3.7	92.2	32.8	151.5	46.0
Tobacco	347.4	291.6	532.1	486.1	388.2
Hides and skins	0.2	0.9	19.3	19.1	18.0
Textile fibres	153.3	40.6	87.5	101.4	87.9
Feeding stuffs[3]	4.5	7.7	10.6	28.2	11.4

[1] Includes cereal preparations. [2] Excluding olive oil. [3] Byproducts of cereals and other agricultural products.

Source: FAO, 2004

Agricultural imports are more diversified. Non-food agricultural commodities (tobacco, hides and skins, and textile fibres) made up one-third of total agricultural imports by value in 2002. Sugar imports peaked in 1996 and had fallen to low levels by the end of the decade. There is relatively little trade in meat, dairy and eggs in either direction. The extent to which these trade flows represent underlying comparative advantage rather than the outcome of trade policies is examined below.

[2] Turkey has been the world's third largest exporter of fruit and vegetables, after the USA and EU-15, for many years.

Turkey's agricultural trade with the European Union

On 1 January 1996, a customs union between the European Union and Turkey came into force. Agricultural products remained outside the customs union, although both the EU and Turkey awarded significant trade preferences to each other's agricultural products (see Section 8.3).

Table 8.5. Bilateral trade (EU – Turkey), EUR million

Product category	EU exports to Turkey		EU imports from Turkey	
	2001	2002	2001	2002
Live animals	14.1	10.0	3.3	2.4
Meat, edible offal (02) and meat preparations (016)	1.4	1.6	1.1	1.5
Dairy produce; eggs; natural honey	11.6	29.1	5.6	23.7
Other products of animal origin	2.4	2.4	37.7	37.7
Live plants and floricultural products	9.9	10.6	14.9	20.3
Edible vegetables, plants, roots and tubers	2.8	7.6	156.5	165.0
Edible fruit and nuts; citrus and melon peel	4.3	6.6	1006.5	863.9
Coffee, tea, maté and spices	4.2	3.4	18.9	19.6
Cereals	22.9	51.1	43.4	32.4
Products of the milling industry; malt; starches	2.9	4.4	12.5	10.4
Oilseeds and oleaginous fruits	27.3	33.9	42.3	40.7
Lac; gums, resins, other vegetable saps and extracts	10.6	13.9	0.2	0.3
Vegetable plaiting materials etc	0.2	0.2	13.5	7.6
Animal or vegetable fats and oils	47.3	58.2	87.0	41.8
Sugars and sugar confectionery	7.7	8.4	23.9	27.7
Cocoa and cocoa preparations	26.3	34.2	14.4	10.1
Preparations of cereals, flour or starch	26.8	27.5	21.0	22.8
Preparations of vegetables, fruit or nuts	6.5	7.1	416.8	394.8
Miscellaneous edible preparations	53.7	46.7	24.2	29.2
Beverages, spirits and vinegar	58.6	64.5	27.5	33.0
Residues and waste from the food industries	21.4	30.4	1.6	2.8
Tobacco and manufactured tobacco substitutes	49.7	70.4	133.2	128.7
TOTAL agricultural products[1]	412.5	522.4	2106.0	1916.3
Other agricultural products included in the Uruguay Round[2]	362.5	437.7	91.2	83.5
TOTAL – AGRICULTURAL PRODUCTS	775.1	960.1	2197.2	1999.8
TOTAL - ALL PRODUCTS	20264.9	24210.5	20231.2	22018.6

[1] Chapters 01 to 24 of the Combined Nomenclature (excluding fish (03) and fish/seafood from other chapters).
[2] Includes fibres for textiles, hides and skins, artificial sweeteners etc.
Source: European Commission, 2003

Table 8.5 shows the trade flows between Turkey and the EU in 2001 and 2002. Turkey had a strong positive balance on agricultural trade with the EU, amounting to EUR 1.4 and EUR 1.0 billion in 2001 and 2002, respectively, for the broader (WTO) definition of agricultural trade[3]. About three-quarters of Turkey's

[3] For comparison, the EU has had a strong positive balance with the USA on agricultural trade in recent years (EUR 3.9 billion in 2002), whereas Turkey has a persistent agricultural trade deficit with the USA, amounting to EUR 425 million in 2002. The value of Turkey's agricultural imports

agricultural exports to the EU in these years consisted of fresh, dried or processed fruit, vegetables and nuts. By contrast, EU agricultural exports to Turkey were much smaller, and more diversified.

8.2 Trade policy

Institutional aspects

The Undersecretariat of the Prime Ministry for Foreign Trade is responsible for Turkey's trade policies, in consultation with other ministries and external bodies (WTO, 2003a). In recent years, a large amount of legislation relating to trade has been passed, necessitated by commitments under the WTO Uruguay Round Agreement and the EU-Turkey Customs Union (see WTO, 2003b). Among other things, this legislation relates to anti-dumping and countervailing measures, sanitary and phytosanitary regulations, changes in intellectual property legislation in the direction of the EU's legislation on intellectual property rights and the WTO TRIPS Agreement, and adoption of the EU's General System of Preferences (2002)[4] and the EU's clothing and textile restraints.

The implementation of trade measures for agricultural products is carried out by MARA. MARA conducts checks of certain imported agricultural products and inputs according to risk evaluation protocols. MARA is also responsible for issuing the "control certificate" required for imported agricultural products. To obtain a control certificate, a number of documents are required[5], which must be obtained from the designated authority in the exporter's country, and accompanied by a Turkish translation. The control certificate is of fixed duration (4–12 months depending on the product).

In addition, importers need an inspection certificate from the Turkish Standards Institution (TSE) in order to import goods for which compulsory standards relating to human, plant or animal health, safety and the environment apply. The TSE is an independent institution and a full member of the International Organisation for Standardisation (ISO), which is responsible for implementing standards for domestically produced and imported products. Turkey has 26,941 standards, of which 1,264 are obligatory. For agricultural products, as defined by ISIC, there are 1,378 standards of which 11 are compulsory and of which about 14 per cent are equivalent to international standards; for the ISIC category "food technology", there are 1,409 standards, of which 104 are compulsory and of which 29 per cent are equivalent to international standards.

Turkey has been a signatory to the GATT since 1951 and a WTO member since its inception in 1995. Within the WTO, Turkey has developing country status, which means that – compared with developed countries - it qualified for a

---

from the USA and EU-15 are roughly the same, the main imports from the USA being tobacco and tobacco products, cereals and oilseeds (EU Commission, 2003).

[4] These preferences do not cover agricultural products (UNCTAD, 2003).

[5] A *pro forma* invoice, a health certificate, a certificate of analysis, a list of contents of the product, a pedigree certificate and a radiation analysis report (WTO, 2003a).

more gradual and less stringent programme of liberalisation measures for agricultural trade under the Uruguay Round Agreement on Agriculture (URAA), starting in 1995. On entering the EU, Turkey would lose this status. It is not clear whether Turkey's WTO Schedules, to be merged with those of the EU, would be adjusted to those of a developed country before the moment of entry.

Since the early 1990s, Turkey has entered into a number of preferential agreements with trading partners. Like the customs union with the EU, these arrangements do not cover agricultural trade. In 1992, a Free Trade Agreement (FTA) with the EFTA countries came into force. This was followed by FTA agreements with Israel in 1997, with 10 of the EU candidate countries between 1998 and 2000, and most recently with Macedonia, Bosnia and Herzegovina, Croatia and Morocco. FTA negotiations continue with a number of other countries (WTO, 2003b).

Turkey is a founder member of the Economic Cooperation Organisation (ECO), formed with Iran and Pakistan in 1985, and expanded in 1992 with the addition of 7 new (mainly southern CIS) members. Turkey's exports to ECO members represented 2.8 per cent of total exports in 2002 (over EUR 1 billion) and imports from ECO members were 3 per cent of total imports (over EUR 1.5 billion). A major aim of the ECO is liberalisation of trade between members. To this end, the ECO Trade Agreement (ECOTA) was signed between Afghanistan, Iran, Tajikistan and Turkey in July 2003. Under this agreement, tariffs will fall over the following 8 years to a maximum of 15 per cent in the highest tariff bracket. As well as tariff reductions, ECOTA regulates other aspects of foreign trade between members.

Turkey is a party to the Barcelona process (Garcia Alvarez-Coque, 2002), involving the EU and 12 Mediterranean countries, which aims to establish the Euro-Mediterranean Free Trade Area by 2010.[6]

Agricultural trade policy measures and WTO commitments

The mainstay of Turkey's trade policy is the tariff. As well as *ad valorem* tariffs, Turkey also uses specific, mixed, compound and formula duties. Following the Uruguay Round, about 46 per cent of all tariff lines (36 per cent of all non-agricultural lines) were bound. From 2005, final bindings for non-agricultural products will range from zero to 102 per cent. Many applied tariffs are currently well below bound levels.

For industrial goods and the industrial component of processed agricultural goods, Turkey applies the EU common external tariff under the EU-Turkey customs union agreement[7]. Agriculture remains outside the customs union. According to the Turkish government (WTO, 2003b), average weighted tariffs on non-agricultural goods imported from third countries fell in 1996 following

[6] Other international economic commitments: Turkey is a founder member of the Black Sea Economic Cooperation Organisation (with 10 other regional nations including Russia, 1992), a member of the Group of 20 (G-20) (set up on the initiative of the G-7 in 1999), and has been for many years a member of the Organisation for Economic Co-operation and Development (OECD).

[7] In 2001, supplementary tariffs above CET levels for certain "sensitive" industrial products were removed.

the introduction of the customs union from about 15 to 5.6 per cent, and further to 4.4 per cent in 2003[8]. Under the EU-Turkey customs union, Turkey applies the same rules of origin as the EU regarding third country trade. These rules of origin differ between countries with and without preferential trade arrangements.

Agricultural trade reveals a very different tariff structure. For agricultural products, all duties were bound under the URAA, and Turkey undertook to reduce them by an average of 24 per cent (and a minimum of 10 per cent for each tariff line) over the period 1995-2004. However, various aspects of the tariff structure remain complicated, with *ad valorem*, specific and mixed tariff regimes. There is scope for rationalisation and greater transparency.[9] There is a strong tendency towards positive tariff escalation (i.e., tariffs increasing with the degree of processing) in the agriculture and food tariff structure. This is shown in Figure 8.3, which depicts average tariffs per product category at the ISIC 2-digit level.

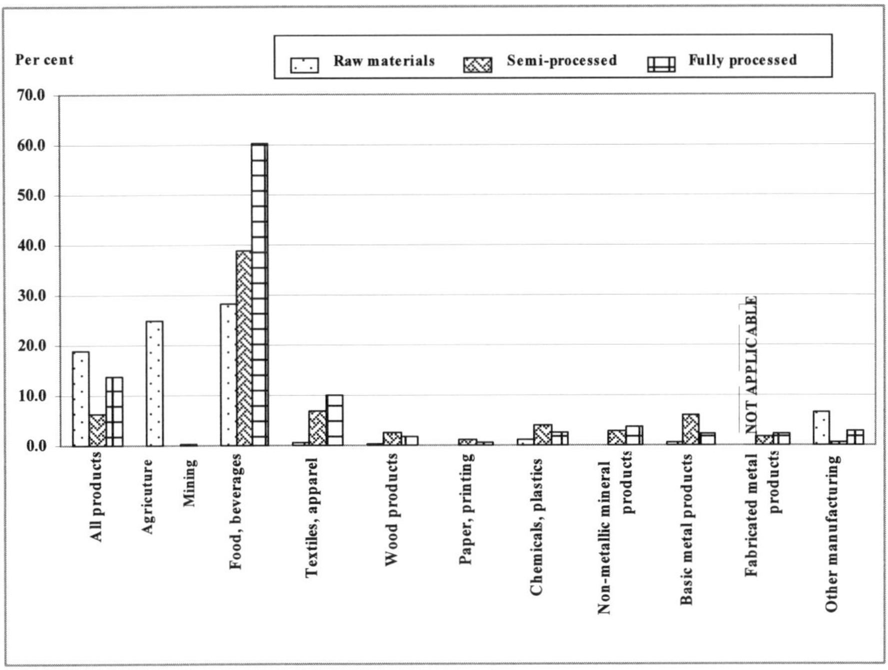

Figure 8.3. Tariff escalation by ISIC 2-digit category, 2003
Source: WTO 2003a: 38

Behind these averages, the variation in tariffs is considerable. In 2003, the highest bound tariff for an agricultural product (ISIC 111) was 147.4 per cent,

[8] A low *weighted* tariff average can hide the existence of very high individual tariffs for goods whose imported volume is (perhaps as a consequence of the tariff) very small.
[9] For example, "some 1.5% of tariff lines carry non-*ad valorem* tariffs while they are bound at *ad valorem* values, risking broken bindings" (WTO, 2003a: viii).

for food products (ISIC 311) 227.5 per cent (for certain meat products), for other food products and animal feeds (ISIC 312) 145 per cent, for beverages 70 per cent and for tobacco 76 per cent (WTO, 2003a:101).

For some agricultural products (cereals, beef, dairy), Turkey's tariff bindings are higher than those of the EU. Therefore, given current differentials, Turkey's adoption of the EU's common external tariff (CET) would lead to a lowering of these tariffs. However, Grethe (2004) lists 40 agricultural products or product categories for which EU tariff bindings appear[10] to be more than 5 percentage points above those of Turkey. If Turkey adopts the EU's CET for agricultural trade in the near future, raising tariffs on these products could provoke claims for compensation from other WTO members if Turkey has been importing these products from third countries. Among products for which Turkey's tariffs are lower than those of the EU are tomatoes (fresh or chilled), olive oil, sugar, animal feed and various animal feed ingredients. The first two products are important exports for Turkey, but imports have been minimal for many years. Turkey has not been a big importer of sugar since the mid-1990s, and the EU was then a main supplier. Turkey has a steady flow of imports of animal feeds from outside the EU; however, the tariff difference appears to be quite small. Therefore, in these circumstances, the risk of major compensation claims appears to be small. However, third countries may attempt to build up an import position in Turkey in the coming years in order to increase the case for a future compensation claim.

In Turkey's URAA Schedules of commitments, Turkey's domestic support was declared either as qualifying for the "green box" (i.e. "minimally trade distorting support") or as *de minimis* support (i.e., product-specific support amounting to less than 10 per cent of the value of that output in the base period). Neither of these categories of support are bound in the WTO. Instead of using the standard methodology of the Aggregate Measure of Support (AMS) for calculating the value of domestic support for its URAA Schedules, Turkey calculated an Equivalent Measure of Support (EMS, see WTO, 1994, Annex 4), which is to be used only when the calculation of an AMS is not practicable. Grethe (2004: 85-88) has queried the way Turkey's EMS calculations were performed, suggesting that it yielded artificially low estimates of the relevant domestic support. Moreover, using the AMS methodology, Grethe estimated support for key commodities (wheat, barley, maize, sugar) and showed that for each of these commodities domestic support as a share of output value, in both the base period *and* the URAA implementation period, was well above the 10 per cent *de minimis* threshold.

Several observations should be made on this issue. First, although Turkey's declared figures for domestic support could have been challenged by other WTO members prior to the final signature of the Uruguay Round Agreement, they were not. Second, the fact that Turkey did not declare any AMS support under the

[10] After conversion of specific tariffs to *ad valorem* equivalents, using "best guess" assumptions about world market prices.

URAA means that it has no AMS allowance that could be "merged" with that of the EU should it become a member. Moreover, Turkey's *de minimis* allowances would vanish. In theory, this situation could be considered a handicap for a prospective EU entrant with a large, supported agricultural sector. Third, however, with the major shift in the CAP away from AMS-eligible payments to payments that are green-box-eligible (at least under current green box definitions), the EU's AMS binding has not acted, and currently does not act, as a constraint on EU agricultural policy makers. Fourth, if Turkey adopts the CAP, it is unclear what the implications of the new blue box rules might be for Turkey's cotton production, which exceeds the joint production of EU-25. Finally, some of the WTO rules regarding "permissible" types and levels of agricultural support are still under discussion within the Doha Development Round negotiations. It is difficult to predict what rules will apply in 2015 and beyond. In short, there are various technicalities regarding domestic support that would have to be clarified during membership negotiations and within the WTO.

Alternative estimates of Turkey's domestic support are provided by the OECD. The OECD's percentage Producer Support Estimate (PSE)[11] for Turkey is 20 and 26 per cent (provisional estimate) for 2002 and 2003 respectively (OECD, 2004).[12] Other things being equal, the switch from output support and input subsidies to direct income support (DIS) introduced under ARIP (see chapter 7) would affect the PSE *only* to the extent that it changed the *aggregate* transfers to producers. Thus, as a larger proportion of support is shifted out of border measures and market prices towards minimally-distorting direct payments, the PSE diverges increasingly from the AMS[13].

The PSE calculation at sector level does not cover all products (for example, most fruit and vegetables are omitted), and so may give a biased view of the *average* rate of support over all commodities. However, even with these gaps in product coverage, the OECD's Total Support Estimate (PSE + other transfers from consumers and taxpayers as a result of agricultural policy) for Turkish agriculture in 2003 amounted to about 4.4 per cent of GDP (OECD, 2004). Thus, despite the absence of a declared AMS, Turkish agriculture can be characterised as heavily supported; and the continuing existence of a number of high tariffs indicates a considerable degree of border protection for the products in question.

Turkey declared export subsidies for 44 products at the HS (4-digit) level in its URAA Schedules. Bindings on export subsidy expenditure have been reduced by 24 per cent, and the subsidised volumes by 14 per cent, over 10 years from 1995. There is very extensive overlap in the coverage of products eligible for export subsidies specified by the EU and by Turkey in their

[11] Transfers to agricultural producers from consumers and taxpayers due to agricultural policy, expressed as a percentage of total production value.

[12] Note that, unlike the WTO Aggregate Measure of Support (AMS), the PSE includes all direct payments to producers, whether or not they satisfy "green box" or *de minimis* criteria and hence are AMS-exempt.

[13] On the question of under what URAA heading DIS payments will be notified to the WTO, the Turkish authorities have indicated that this will be decided on the completion of ARIP in 2005 (WTO, 2004b).

respective schedules (Grethe, 2004: 82)[14]. As Grethe (2004) points out, however, the aggregate orders of magnitude are very different: the final (2004) level of Turkey's total export subsidy bindings is USD 95 million (about 4 per cent of annual agricultural exports), whereas for the EU-15 it is USD 7 billion (20 per cent of annual agricultural export value).

Turkey pays export subsidies on eligible commodities at a rate between 10 and 20 per cent of export values, and on export shares ranging from 30 to 100 per cent depending on the product. Export subsidies available for agricultural (primary or processed) products in 2003 are shown in Table 8.6.

Table 8.6. Export subsidies by product, 2003

Product	Rate (in USD/ton)	Share of export quantity eligible for subsidy, %
Cut flowers (fresh)	285	57
Vegetables, frozen (excl. potatoes)	106	38
Vegetables (dehydrated)	370	33
Fruit (frozen)	92	45
Preserves, pastes	55	85
Homogenised fruit preparations	53	69
Fruit juices (concentrated)	168	31
Olive oil	200	100
Prepared or preserved fish	210	100
Poultry meat (excl. edible offals)	199	28
Eggs	US$7/1,000 pieces	40
Chocolate and other food preparations containing chocolate	110	60
Biscuits, waffles	110	30
Macaroni vermicelli	73	40
Potatoes	20	n.a
Onion (dried)	17	n.a

Source: WTO, 2003a:89

Grethe (2004:83) identifies a number of highly protected commodities, for which Turkey has no export subsidy bindings and for which no export subsidies are notified to the WTO, but for which "implicit export subsidies" have regularly been paid via the commercial losses of state-owned companies[15]. The commodities concerned in the past include tea, sugar, tobacco and barley. Grethe calculates the amounts of implicit export subsidy in 2000 at EUR 8 million (barley), EUR 225 million (sugar), EUR 10 million (tea) and EUR 100 million (tobacco)[16]. As he points out, since Turkey has no WTO export subsidy

[14] Notable omissions from Turkey's list are skim milk powder, sugar, pig meat and wine. The export subsidy for citrus was discontinued recently.

[15] Underwriting of trading losses will no longer be permitted under the Doha Round Agreement.

[16] These amounts, expressed as a percentage of the value (at world market prices) of Turkey's exports of these commodities (obtained from the FAO database), are roughly 35% (barley), 160% (sugar, tea), 18% (tobacco).

allowances for these products to "merge" with those of the EU, they would be problematic for an enlarged EU *if* the EU is already using its full subsidy binding to deal with current surpluses. Sugar was singled out as the commodity most likely to create difficulties. The decision of WTO member countries to eliminate export subsidies by the end of the Doha Development Round implementation period (WTO, 2004d) would exacerbate the problem that will arise for the EU if it gains another member with a structural surplus in sugar. The upcoming reform of the EU sugar regime should provide a more favourable context within which to deal with this concern.

Turkey within the WTO

Turkey's conduct within the WTO presents some interesting contrasts. We have already referred to Turkey's willingness and promptness to adopt the legislation necessary for compliance with its commitments under the GATT and the URA. At the same time, we have also reported evidence that the domestic support for agriculture notified in Turkey's URAA schedules was not measured according to WTO guidelines. In the last WTO Trade Policy Review (WTO, 2003a) and the discussions among WTO members that followed (WTO, 2004b, 2004c), Turkey was congratulated on what has been achieved in terms of domestic policy reform and legislative changes, but concerns were also expressed about specific issues, mainly focussing on agricultural trade.

In the last four years, WTO members have raised concerns about issues such as implicit export subsidies for wheat, domestic support for sugarbeet and import permits for rice (WTO, 2003a). Some trading partners have complained that the control certificate system was being used for various agricultural products in a discriminatory way, akin to an import license, in order to impose delays or quantity limits. For example, in the most high profile case, Ecuador lodged a formal complaint in 2001 to the effect that the way control certificates were being issued for bananas amounted to a *de facto* quantity restriction and was timed to protect the internal market at the moment of the domestic banana harvest (WTO, 2004a)[17]. Turkey has repeatedly insisted that the control certificate is simply a reference document for SPS purposes, and any malfunctioning of the system has been due to weaknesses in infrastructure and technical capacity (WTO, 2004c).[18]

Turkey has been challenged on various occasions about not properly notifying the WTO when changes to import requirements are made. For example, Canada has complained about frequent changes in sanitary and phytosanitary regulations that were neither officially notified nor communicated bilaterally, including a ban in June 2003 on all live animals and meat because of SARS (despite advice from the WHO that such imports from SARS-affected regions posed no threat to human health) (WTO, 2004c).

A more thorny issue has been Turkey's ongoing import controls for live animals

---

[17] This particular dispute (WT/DS237) was resolved bilaterally at the end of 2002.

[18] For another example of apparent administrative non-tariff barriers, see Section 6.2.

and red meat. Since 1996, MARA has refused to issue control certificates for live animals (except breeding stock) and beef from countries considered to involve a risk of animal disease (specifically, FMD or BSE). This measure has been challenged in the WTO Sanitary and Phytosanitary (SPS) Committee, by the USA in 1998 and by Hungary in 2000. Other countries have raised questions about whether Turkey had followed WTO-agreed procedures (official publication and notification, preparation of a risk assessment etc) in connection with this ban. Although Turkey gave assurances in 1999 that the import ban had been lifted, further complaints have followed and it appears that the situation has not been resolved (see WTO, 2004a) (see also Table 8.4 for evidence of Turkey's very low imports of meat). In defending its position on meat and livestock imports (WTO, 2004c: 28), Turkey has claimed that the uncompetitive position of its livestock sector makes it vulnerable to imports, that livestock production is a small-scale subsistence sector on which the livelihood of many people depend, that the country has lost one-third of its livestock capacity in the last two decades and "is in a state of rebuilding its national livestock herd", and hence its need for protection against "mass importation of meat stemming from a high level of subsidisation". This explanation confirms the view expressed by the USA and others that sanitary regulations are in fact being used by Turkey as a protectionist measure (e.g. Sarigedik, 2004).

In the third of the three SPS-motivated complaints against Turkey (WT/DS256), Hungary challenged Turkey's 2001 ban on pet food from all European countries, which was based on an alleged risk of BSE. Hungary claimed that, in its own particular case, the alleged risk was without a scientific basis since Hungary is a BSE-free country; moreover, Hungary complained that the ban had not been officially published, nor notified to the relevant WTO Committee. The USA and the EU associated themselves with this complaint. This issue is apparently not yet resolved (WTO, 2005:18).

Some commentators have also pointed out the complexities of Turkey's position in the Doha Development Round. On the one hand, as a member of the EU customs union, Turkey has been closely following the EU negotiating position as regards non-agricultural products. On the other hand, for agricultural products, the position of Turkey is close to that of the "G-20" developing countries, who claim that further tariff reductions must be conditional on substantial reductions or the elimination by developed countries of their domestic support and export subsidies (WTO, 2001). Turkey has acknowledged the ambivalence of its position, and has defended it as in keeping with its current situation where its trade with respect to agricultural products is not harmonised with that of the EU (WTO, 2004c: 8).

8.3 Turkey's agricultural trade arrangements with the EU

On January 1, 1996, Turkey and the European Union formed a customs union. As a result, Turkey adopted the EU's common external tariff (CET) for non-agricultural products, and eliminated import duties and quotas for non-agricultural items of EU and EFTA origin. For products imported from third countries (non-EU and non-EFTA), adoption of the CET meant the average tariff dropped to below 6 per cent. Turkey was expected to adopt the EU's Generalised System of Preferences in 2001. However, the process began only in 2002 and is expected to be completed in 2006.

The customs union has not been extended to agricultural products. However, the EU has been granting trade preferences to Turkey for agricultural products since the 1963 Association Agreement. As part of this process, the EU abolished most *ad valorem* tariffs on agricultural imports from Turkey in 1987, and rates for some specific tariffs have also been reduced. However, there remain high specific tariffs for many of the core CAP products (such as cereals and cereal products, sugar and sugar products, and olive oil), substantial specific duties for many processed products, an entry price system for some fruit and vegetables, seasonal *ad valorem* tariffs for some fruit (four kinds) and vegetables (nine kinds), and high above-quota tariffs on TRQs.

Grethe (2004:58-59) distinguishes four categories of imports from Turkey to the EU: products without any MFN barrier[19], products with an MFN barrier and for which Turkey receives no preferential treatment, products with an MFN barrier that is applied at a lower preferential rate for Turkey, and products with an MFN barrier that is waived completely for Turkish exports to the EU. Grethe calculates that in 2001, the shares of Turkey's exports (by value) in each of these categories were 7, 2, 36 and 54 per cent, respectively. Product categories which fall for more than 90 per cent in the last category include meat, dairy and eggs, plants and flowers, cereal products, meat and fish preparations, vegetable and fruit preparations, and tobacco. Among those products for which a reduced level of MFN barrier has been retained are, most notably, 59 per cent of fruit and nut imports, 76 per cent of cereal imports, 87 per cent of fats and oils, 53 per cent of sugar and confectionery and 100 per cent of cereals preparations. The total value of imported Turkish agricultural products in this third category amounted to EUR 770 million.

Although only 38 per cent of Turkey's exports are in the second and third categories (no concession or partial reduction of MFN barrier), it is not clear how much more the EU *would* import from Turkey without the remaining MFN barriers. The fact that in recent years Turkey has not filled many of its EU tariff rate quotas for agricultural products, even when (as in most cases) the in-quota tariff rate is zero, does not necessarily imply that, for products with a prohibitive tariff and no TRQ, tariff barriers have not been a strong deterrent.

Since 1998, Turkey has given preferential market access to many EU agricultural products, but for the most part, preferential concessions have been accompanied by a quota limit. In 1998, Turkey created 39 TRQs for agricultural

[19] MFN=most favoured nation. MFN barriers (e.g. tariffs, entry prices etc) are those facing third countries that do not enjoy any preferential treatment.

imports from the EU, for the most part with zero in-quota tariffs. In many cases, these quotas have been filled and exceeded by EU exporters. For example, average annual imports of butter from the EU for 1998-2001 exceeded the TRQ of 3,000 tons by more than 10 per cent, despite an above-quota tariff of 70-100 per cent. More than twice the amount of the (20 thousand tons) TRQ for rye was imported in the face of an above-quota tariff of 60 per cent (see Grethe, 2004: 63-4, for more details and discussion). In this context, we note that the European Commission (2004: 95) reported receiving complaints from some EU companies concerning technical barriers to trade and infringements of the Customs Union or Association Agreements.

Given that, in theory, the industrial component of processed agricultural products does not face any trade barriers in the EU-Turkey customs union, one might expect a greater increase in processed agricultural product trade (in both directions) after the formation of the customs union in 1996. In fact, such an effect appears to have been quite weak. To the extent that this can be explained, Grethe suggests that it is partly because, in fact, most processed agricultural products are classified in Annex 2 of the Treaty of Rome as agricultural products, and therefore have remained outside the customs union.

In response to questioning by the WTO Trade Policy Review Body (WTO, 2004c:10), Turkey admitted that export performance since the formation of the customs union has fallen short of expectations, but claimed the contraction in external markets (due *inter alia* to the Asian and Russian financial crises) as the main reason. This does not shed light on why the share of exports specifically to the EU has not increased. By contrast, the explanation given of why Turkey's imports from the EU have increased faster than exports to the EU since the formation of the customs union (namely, heavy demand for capital goods and intermediate inputs in a rapidly growing industrial sector) is highly plausible and augurs well for Turkey's strength and competitiveness in manufacturing in the longer term.

Grethe (2004) used a simulation model to estimate the effects of extending the customs union with the EU to include all agricultural products. In his simulation, harmonisation is simulated to take place in 2006. His study assumes that the Agenda 2000 reforms have been fully implemented in the EU with consequent price falls, and that, in 2006, all barriers to agricultural trade between Turkey and the EU disappear, Turkey adopts the EU's CET levels for third party trade, and prices in Turkey are aligned with EU prices. The results predict a decline of 5.4 per cent in the total value of Turkish agricultural production (-2.1 per cent in plant production, -13.7 per cent in livestock production). As for net trade flows, Turkey's trade balances deteriorate for virtually all product aggregates: the trade surplus on total plant production falls by nearly 50 per cent (although it remains positive), and a large trade deficit occurs for animal products. Only processed products (a slightly smaller deficit) and fruit (a slightly larger surplus) show an improvement (Grethe, 2004, Table 9.9: 204).

The simulated overall welfare effect for Turkey is positive, as consumers' benefits from lower prices and greater quantities (particularly of animal products) more than offset the welfare losses of agricultural producers. Interestingly, Grethe estimates that *if* the increase in competitive pressure after the removal of trade barriers causes the farmgate-wholesale margin plus the wastage along the chain to fall by just 10 per cent, Turkey's deficit on agricultural trade would become a small surplus, and total welfare gains would more than double (Grethe, 2004, table 9.23). This suggests that there is much that Turkey could do to improve its net trade position even in the current situation, by internal restructuring and raising the efficiency of supply chains. An important question is whether such improvements can only be achieved in response to an external shock, such as facing unprotected competition from EU producers and food manufacturers.

Unfortunately, this model simulates only net trade positions rather than bilateral trade flows in a spatially differentiated framework. Nonetheless, the results are interesting in that they indicate likely directions of changes in Turkey's net trade position and their orders of magnitude, if trade harmonisation occurred under current conditions. The overall conclusion is that admitting agricultural products to the EU-Turkey customs union on a bilateral basis would be to Turkey's net advantage in current conditions but at some cost to its agricultural sector. In the longer term, much would depend on the extent to which Turkish agriculture, particularly the grazing livestock sector, can modernise and restructure in order to withstand competition from EU member countries.

If the scenario simulated by Grethe occurred in 2015 rather than 2006, after some years of population and income growth, demand response to price falls is likely to be greater, particularly for animal products (where price falls would be larger, given the higher protection in this sector). At the same time, whether Turkey became a full member in 2015 or simply aligned its policies and structures more closely to those of the EU over the next 10 years, there would also be significant changes on the supply side.

8.4 Turkey's agricultural trade profile

Turkey's pattern of self-sufficiency in agricultural products can be crudely summarised as over-supply in crop products, and self-sufficiency in animal products (see Table 4.16). At first glance, therefore, it might appear that for most animal products the livestock sector is well matched to the size of the domestic market, whereas in the plant sectors, resources are allocated according to comparative advantage, with fruit and vegetable production outperforming most of the field crops. However, Section 4.4 (Table 4.4) has drawn attention to the much smaller share of livestock output in total production in Turkey compared with the EU. Moreover, the review of trade policy in Sections 8.2 and 8.3 suggests that the internal market balance for livestock products is largely the result of high border protection and import bans, which have constrained consumption to equal what can be produced domestically.

According to FAO (2003), per capita calorie intake in Turkey in 2001 was the highest among the countries of the Economic Cooperation Organisation. Yet in Turkey the share of daily energy intake coming from animal products was the third lowest for the region. Although Turkey's per capita *total* protein consumption was one of the highest in the region, both the share and the absolute level of *animal* protein in the total was below the regional average (see Section 5.3 for more details).

Turkish per capita consumption of meat is estimated at about one-quarter that of EU-15. At the same time, farmgate prices for most animal products (beef, poultry meat, processed dairy products) were in 2003 (post-ARIP) still considerably higher than those in the EU (see Table 4.13). Given the higher farmgate-retail price margins in Turkey, the Turkey-EU price gap was undoubtedly higher at retail level than at farm level. As the average per capita income in Turkey is just 25 per cent of the average in EU-15, it is not surprising that per capita consumption of animal products is so low.

It is clear from the preceding sections of this chapter that these differences between Turkey and the EU in prices and consumption levels for animal products are due to agricultural and trade policies. The market, trade and consumption figures build a picture of a high degree of policy-induced market distortion in relation to the livestock sector, with significant consequences for food consumption patterns and trade outcomes. A combination of structural and cultural reasons is responsible for the weakness of the livestock sector, which is perpetuated in a vicious circle by protectionist agricultural policy measures and low veterinary standards (see Chapter 10). It is inevitable that the opening up of this sector to competition from the EU will have a strong economic and social impact, whose consequences will have to be foreseen and managed by policy makers.

8.5 Conclusions

Turkey has been active in foreign trade for several decades, although it cannot be called an "open economy" insofar as significant trade-distorting policy measures are still used, notably for agricultural products. The EU is Turkey's largest trade partner, for all merchandise trade and for trade in agricultural products, although for agricultural imports the EU and the USA are of similar importance.

Turkey takes part in the international trading system as a member of the relevant international and regional organisations. Turkey is a medium-sized country in agricultural trade terms. It is therefore not surprising that Turkey has had various differences of opinion with its agricultural trading partners and competitors. Turkey's overall performance within the WTO appears to be that of an enthusiastic member, who is generally well prepared to embrace the principles and rules of international trade, but who does not compromise readily on issues that are considered sensitive for domestic policy. As well as high protection for some products, Turkey has had more than its fair share of complaints about violations of

WTO rules. Given the uncompetitiveness of parts of the agricultural sector and the high proportion of the work force still reliant on agriculture for their livelihood, it is inevitable that sensitive trade-related issues show up particularly with respect to agriculture. Turkey's stance in the Doha Development Round negotiations reinforces the impression of a dual perspective on trade policy, which arguably arises from the dual nature of Turkey's economy.

Turkish agriculture can be characterised as receiving substantial support, and the continuing existence of a number of high tariffs indicates a large degree of border protection for the products in question. In particular, the livestock sector is characterised by very substantial policy-induced market distortions, with significant consequences for food consumption patterns and trade outcomes.

Since 1996, Turkey has had a customs union with the EU. As a consequence, Turkey adopted the EU's common external tariff for all non-agricultural goods. However, agricultural trade remains outside the customs union in practice, and takes place on (asymmetric) preferential terms. More than 60 per cent (by value) of the EU's agricultural imports from Turkey face no barriers whatsoever to the EU market, and over 90 per cent have some kind of preferential treatment. Thirty-nine tariff rate quotas have been established for EU exports of agricultural products to Turkey. Exports to the EU have not increased as rapidly as imports from the EU since the formation of the customs union: Turkey's trade deficit has worsened in recent years, and in 2003 stood at over 9 per cent of GNP. According to Turkey's trade performance figures, its customs union with the EU has had no visible effect either on its exports as a percentage of GNP or on the share of its trade with the EU.

Academic analysis (Grethe, 2004) indicates that including agricultural products in the EU-Turkey customs union on a bilateral basis in *current* conditions would worsen Turkey's trading position, although it would lead to a net welfare improvement for Turkey as a whole. This analysis is most useful for identifying to what extent and in which respects Turkey's *current* agricultural trade with the EU diverges from free trade conditions. It is not, however, a prediction of what would happen in 2015, because by that time a number of outside factors will also have changed. However, it is likely that even then small-scale livestock production will feel the greatest impact of trade harmonisation.

References

AAF Canada (Agriculture and Agri-Food Canada) (2004) Turkey. Agri-food trade synopsis. AAF Canada, Ottawa, January 2004.

Central Bank of the Turkish Republic (2004). Turkish Balance of Payments Statistics. http://www.tcmb.gov.tr/odemedenge/menupdf.html (accessed 2 October 2004).

CIHEAM (2000) *Development and agri-food policies in the Mediterranean region.* Extracts from Annual Reports 1998,1999, 2000. CIHEAM, Paris.

European Commission (2003) Agricultural situation in the candidate countries. Country report: Turkey. DG-AGRI, Brussels, November 2003.

European Commission (2004) 2004 Regular report on Turkey's progress towards accession. COM(2004) 656 final. Brussels, 6 October 2004.

FAO (2003) Regional programme for food security in the Near East: Towards sustainable food security and poverty alleviation. Report from a High-level technical workshop, FAO-IDB, Jeddah 8-9/10.2003. www.fao.org/DOCREP/MEETING/007/ad385e/ad385e01.htm (accessed 30 September 2004)

FAO (2004) Agricultural and food trade, FAOSTAT. Last updated: 2 July 2004. FAO, Rome.

Garcia Alvarez-Coque, J.M. (2002) Agricultural trade and the Barcelona process: Is full liberalisation possible? *European Review of Agricultural Economics* 29(3), 399-422.

Grethe, H. (2004) Effects of Including Agricultural Products in the Customs Union between Turkey and the EU: A Partial Equilibrium Analysis for Turkey. CeGE-Schriften, Vol 9, Center for Globalization and Europeanization of the Economy, Georg-August-Universität Göttingen. Peter Lang Verlag, Frankfurt am Main.

OECD (2002) Agricultural Policies in OECD Countries: Monitoring and Evaluation. OECD, Paris.

OECD (2004) Producer and Consumer Support Estimates OECD Database 1986-2003. OECD, Paris.

Sarigedik, Ü. (2004) Turkey: Livestock and products annual report 2004. GAIN Report, Number TU4022. USDA Foreign Agricultural Service, Ankara.

UNCTAD (2003) Generalised system of preferences: Scheme of Turkey. TD/B/GSP/TURKEY1. 28 January 2003.

Undersecretariat of Foreign Trade (2004) Turkey's Foreign Trade, http://www.byegm.gov.tr/references/turkeys-foreign-trade.htm (accessed 23 September 2004).

WTO (1994) Uruguay Round Agreement on Agriculture. WTO, Geneva.

WTO (2001) WTO negotiations on agriculture. Proposal by Turkey. G/AG/NG/W/106. WTO, Geneva, 5 February 2001.

WTO (2003a) Trade policy review: Turkey. Report by the Secretariat. WT/TPR/S/125, WTO, Geneva, 19 November 2003.

WTO (2003b) Trade policy review: Turkey. Report by the Government. WT/TPR/G/125. WTO, Geneva, 19 November 2003.

WTO (2004a) Specific trade concerns. Committee on Sanitary and Phytosanitary Measures. G/SPS/GEN/204/Rev.4. WTO, Geneva, 2 March 2004.

WTO (2004b) Trade Policy Review Body: Trade policy review, Turkey. Minutes of meeting on 8 March 2004. WT/TPR/M/125/Add.2. WTO, Geneva.

WTO (2004c) Trade Policy Review Body: Trade policy review, Turkey. Minutes of meeting on 9 February 2004, and Addendum. WT/TPR/M/125 and WT/TPR/M/125/Add.1. WTO, Geneva.

WTO (2004d) Text of the 'July' package: the General Council's post-Cancún decision. WTO, Geneva, 1 August 2004.

WTO (2005) Update of WTO dispute settlement cases. WT/DS/OV/23, 7 April 2005.

9

Environment and Agriculture

Irene Molina Vílchez

9.1 Introduction

This chapter studies the relationship between agriculture and the environment in Turkey. There is a close relationship between agriculture and the natural environment that sustains it. As an economic activity, agriculture uses natural resources such as land and water to produce goods. The degradation or exhaustion of those resources will have negative consequences for the continuation of production. Environmental targets and criteria are becoming an important part of agricultural policy in the European Union. The environment has started to receive growing attention in Turkey, particularly since the 1990s. In adopting the *acquis communautaire* by 2015, the Turkish agricultural sector would have to align its activities specifically with EU environmental standards. It is important, therefore, to examine the current relation of agriculture and the environment in Turkey.

Section 9.2 describes environmental impacts of Turkish agriculture. Section 9.3 examines the extent to which agri-environmental concerns are dealt with by current policies. Section 9.4 discusses the institutional arrangements for enforcing legislation on agriculture and the environment. Section 9.5 describes the current state and prospects for organic farming. Section 9.6 deals specifically with biodiversity in Turkey and related institutional developments. Finally, the key conclusions are reported in Section 9.7.

9.2 Environmental impacts of agriculture in Turkey

Impacts on water resources

Water consumption

The total water resources[1] that can be economically and technically exploited in Turkey are around 1,600 cubic metres per capita per year. The distribution of water resources is uneven over the year and between regions, with the Western regions (Marmara, Aegean and Central Anatolia) being relatively poor in water (Keskin, 2001). Furthermore, the complicated topography makes it difficult to control water resources in Turkey (Cakmak *et al.*, 2003).

[1] Renewable water resources are rechargeable by the hydrological cycle unless they are overexploited, and comprise groundwater aquifers and surface water like rivers and lakes. Non-renewable water resources are not replenished by nature at all, or not for a very long time (FAO, 2004a).

Pressure on the quantity of water remains moderate in Turkey in comparison with other OECD countries. Turkey used 17 per cent of its total available water resources in 2002, which was higher than the OECD average (11.5 per cent) but lower than in other southern European countries like Italy and Spain who use 32 per cent and 35 per cent of their potential respectively (OECD, 2002).

However, water use in Turkey is increasing. Data from the State Planning Organisation (SPO) show growth of 37 per cent in water extraction over ten years (1990-2000), reaching a total of 42,000 million m^3. Approximately 84 per cent of this came from surface waters and the rest from groundwater.

Approximately three quarters of the total freshwater extracted is used for agriculture. Large-scale dam construction has allowed irrigated areas to expand by two thirds over the last 20 years. Four and a half million hectares are currently irrigated, which constitute 16 per cent of the total agricultural land (DSI, 2004). As regards groundwater, many aquifers are being exploited beyond their sustainable yield, particularly in the Mediterranean region (OECD, 1999; MoE, 2000).

Use of fertilisers and pesticides

Agricultural fertilisers and pesticides are the major source of diffuse water pollution. However, use of these inputs is low in Turkey and seems to be decreasing in total. Figure 9.1 shows the consumption of pesticides for the period 1990-2001.

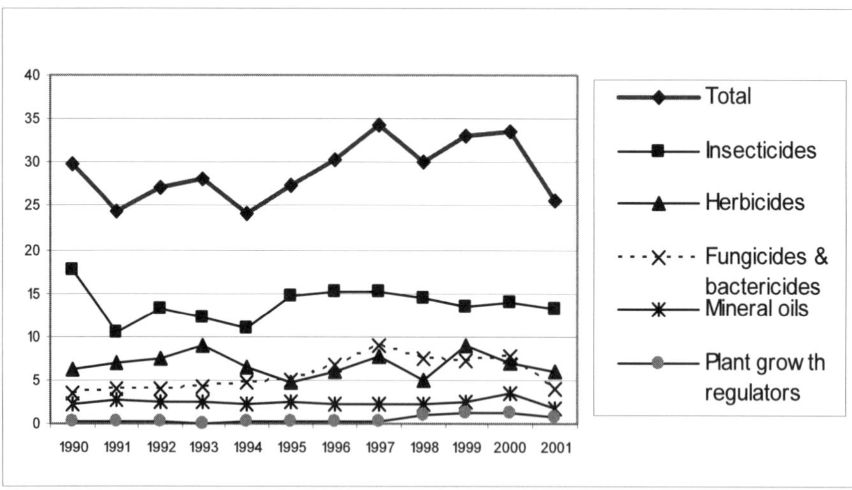

Figure 9.1. Consumption of pesticides (1990-2001), thousand tons active ingredients
Source: FAO, 2004b

An increasing trend in the consumption of pesticides is observed during the first seven years, with a slight decline after 1997, when the phase-out of input subsidies started. Pesticide use averages about 1 kg per ha of arable land, which is below the levels in other southern European countries and far lower than the levels of countries that employ pesticides intensively (OECD, 2003)[2].

According to the European Environmental Agency (Petersen and Hoogeveen, 2004), fertiliser use also remains relatively low in Turkey, at levels similar to central and eastern European countries. In the mid-1990s, nitrogen and phosphate use per hectare of arable land and permanent cropland was among the lowest in OECD countries (OECD, 1999). Figure 9.2 shows the latest figures available on Turkey's fertiliser consumption. Total consumption in 2001-2002 was below recent peak levels. This could be explained by the removal of chemical fertiliser subsidies. Average fertiliser use was 87 kg of nutrients per ha in 2001. Fertiliser use is unevenly distributed across regions, however, with higher levels in the Aegean and Mediterranean regions, the latter using an average of 128 kg per ha.

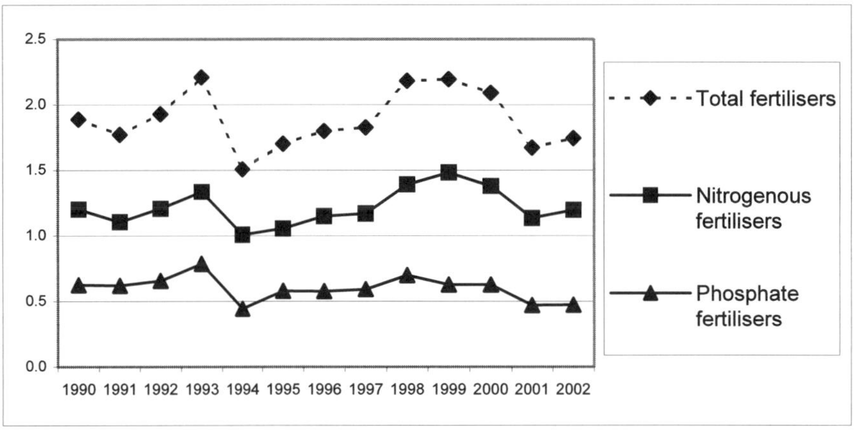

Figure 9.2. Fertiliser consumption in Turkey (1990-2002), thousand tons
Source: FAO, 2004b

Pollution is not exclusive to those areas with high input use. In some irrigation schemes, drained water is re-used or flows to marshes, causing impacts on wildlife (Keskin, 2001). In some zones, run-off, drainage and deep percolated water from irrigated lands contain high levels of fertiliser and pesticide residues. Studies of four river basins in western Turkey showed high levels of pollution from these residues (OECD, 1999).

Impacts on the soil

Like other southern European countries, Turkey is naturally prone to erosion due to its climatic and topographic conditions. The degree of erosion has gradually increased since the 1950s as a result of human activities (Keskin, 2001; MoE,

[2] Usage (in kg per ha of arable land) is 3, 4 and 2 in Greece, Italy and Spain respectively, and 10 and 11 in the Netherlands and Belgium, respectively.

1997). Table 9.1 shows the distribution of categories of soil erosion. A very high proportion of Turkish land (86 per cent) suffers some kind of erosion, with more than half being affected severely or very severely. The problem is especially critical on the extensive slope lands with shallow soil.

Table 9.1. Distribution of soil erosion categories

Degree of erosion	Area (1000 ha)	Share of total area, %	Classification criteria
Slight	5,612	7.2	25% of topsoil eroded
Moderate	15,593	20.1	25-75% of topsoil eroded
Severe	28,335	36.6	Topsoil and 25% of subsoil eroded
Very severe	17,366	22.4	Topsoil and 25-75% of subsoil eroded
TOTAL	66,906	86.4	

Source: Özden et al., 2000

Water erosion affects 15.8 million ha (71 per cent of total agricultural land) and wind erosion affects 0.3 million ha (1.5 per cent of total agricultural land) (Kumuk and Akgüngör, 1995). In addition to problems related to reduced fertility, lost soil estimated at 1 billion tonnes silts up in dams and river deltas, and produces water turbidity, which reduces the amount of dissolved oxygen available to fish and aquatic plants (OECD, 1999).

Inadequate irrigation management is an important contributor to soil damage in Turkey. The problem is especially severe in the Aegean and Marmara regions (OECD, 1999). Poor irrigation management causes salination of large areas and prevents the replenishment of lakes and wetlands. It is estimated that 1.5 million ha of arable land suffer from reduced yield due to salination, and a further 2.8 million from waterlogging (Keskin, 2001). Soil salination is also due to inappropriate nutrient use: farmers sometimes traditionally insist on using the same fertilisers for years, often unnecessarily (Yüksel, 2002). Other factors contributing to the degradation of land include inappropriate land use, excessive grazing, fuel wood use and plant collection. According to the Ministry of Environment (MoE, 2000), 90 per cent of grazing lands are degraded and unproductive mostly due to uncontrolled pasturing. The World Bank (2002) indicates the clearing of forests for maize and hazelnut production as important causes of erosion in the Black Sea region. The use of polluted waters to irrigate agricultural lands is raising concern, especially in western Turkey, which has been experiencing water shortages on a regular basis in recent years (FAO, 1997).

Impacts on climate change

According to data from SIS (1999), agriculture contributed 9 per cent of Turkey's direct greenhouse gas emissions in 1997. Methane emissions constituted the second largest contributor (9.4 per cent of carbon dioxide equivalents). The main source of this gas (60.3 per cent) was livestock enteric fermentation and manure management; an additional 2.2 per cent was generated from rice cultivation and

1.7 per cent from the burning of agricultural residues. Burning these residues also accounted for 3 per cent of emissions of N_2O, 1 per cent of NO_x and 11 per cent of the CO emissions. It has not been possible to find data on specific emissions of carbon dioxide generated from the combustion of fossil fuels in agriculture. However, OECD data on agricultural machinery use (numbers of tractors and combined harvester-threshers) show a sharp rise, from 450 thousand in 1980 to 920 thousand in 2000 (OECD, 2002).

Summary of agriculture's environmental impacts

Overgrazing and irrigation are important agricultural factors affecting the environment in Turkey in an adverse way. Among the various environmental problems facing Turkish agriculture, erosion seems to be the most important. Its main direct impact is the loss of soil fertility, which reduces yield. In a more indirect way, erosion also constitutes a potential future problem since degraded farmland will require more chemical input and more irrigation to sustain productivity (Tanrivermis, 2003). The combined effect of these factors may intensify the chemical content of the soil, incite further water extraction and cause nutrient run-off, therefore increasing the negative externalities of agriculture on the environment. These cause-and-effect interactions highlight the close interrelationship between different environmental problems, and the importance of preventive and integrated management strategies.

9.3 Overview of policies relating to the environmental impacts of agriculture

Environmental policy

Turkey began addressing environmental concerns in the 1970s. Environmental management policies were introduced in the 3rd Five-Year Development Plan (1973-77) (OECD, 1999) and a Prime Ministry Undersecretariat for the Environment was created in 1978. The 1982 Constitution stated the right to live in a healthy, balanced environment and laid upon both the state and the citizens the responsibility for improving the natural environment and preventing environmental pollution. The same year the first Environment Law was adopted. Among its goals was to prevent and eliminate environmental pollution and to improve the management of natural resources and the land. It defined the framework of environmental legislation on the basis of the 'polluter pays' principle, and stated that in all economic activities, every measure should be taken to minimise pollution.

However, it was only in 1991 that environmental issues were considered important enough to be managed by a specific Ministry of Environment (MoE), empowered to implement and enforce environmental policies. The 5th Five-Year Development Plan (1983-1988) introduced the conservation and development of natural resources. Section 9.6 elaborates on the subsequent legislation for the protection of biodiversity. The 6th Five-Year Development Plan (1991-95) brought in the concept of sustainable development and reiterated the goal of integrating environmental concerns into other policies (OECD, 1999).

Laws adopted in subsequent years followed a command-and-control approach for agri-environmental management. For example, cereal stubble burning in rain-fed areas became illegal and a compulsory interval between pesticide application and crop harvesting was established (Tanrivermis, 2003). The National Mobilisation Law for Afforestation, Reforestation and Erosion Control entered into force in 1995. Some of the measures taken by the Government to combat desertification included intensified soil conservation, public awareness-raising and the preparation of drought preparedness and relief schemes (MoE, 1997).

The adoption of international commitments, and in particular the prospect of becoming a member of the European Union, seems to be the driving force behind environmental reforms. The EU Accession Partnership for Turkey establishes a guiding set of short- and medium-term priorities in the field of environment (see Table 9.2.).

Table 9.2. Turkey's Accession Partnership: Selection of priorities for the environment

Short term (to be fulfilled in 2003-2004)	Medium term (to be initiated in 2003-2004 and continued later)
Pursue the development of transboundary water cooperation, in line with the Water Framework Directive and international conventions to which the Community is party.	Set up the administrative structures for the implementation of the EC rural development policy and forest strategy
	Integrate sustainable development principles into the definition and implementation of all other sectoral policies.
Begin to transpose the framework legislation, the international environmental conventions, legislation on nature protection and water quality.	Complete transposition of environmental *acquis* and strengthen the institutional, administrative and monitoring capacity to ensure environmental protection, including data collection.

Source: Council Decision 2003/398/EC of 19 May 2003

The country is making efforts to update its legislation according to the Accession Partnership. A recent step is the Regulation on the Protection of Waters against Pollution caused by Nitrates from Agricultural Sources, adopted in February 2004, which follows the European Nitrates Directive and will be implemented jointly by the Ministry of Agriculture and Rural Affairs (MARA) and the recently merged Ministry of Environment and Forestry (MoEF).

However, despite this increase in regulations, the quality of the agricultural environment still shows much scope for improvement. The latest five-year development plan acknowledges the need for additional policy measures, including economic instruments for environmental management. Farmers in Turkey generally are not encouraged to conserve land and water resources due to the insufficient economic incentives, penalties, training, awareness and participation in the decision-making process (OECD, 2001). The current water pricing system, for instance, does not provide incentives for efficient water use, since payment is related to the area and the crop type, and not to the actual amount of water used. Although supply restrictions exist in the dry season, irrigators tend to waste water when there is enough conveyance capacity. Water is allowed to escape to drainage and night-time irrigation is not used (Ünver and Gupta, 2003). A positive move on this front is the transfer of irrigation management to Water Users' Associations (WUA) (see Section 9.4).

As far as research policy is concerned, projects related to the agricultural environment are limited and are publicly funded only, in the absence of private funds. MARA devoted 4 per cent of its 2001 annual budget to agri-environmental research (Goncagul, 2001).

Agricultural policy

Turkey has only recently started to include environmental concerns in its agricultural and rural development policies. An important element in agricultural policy in the past was the granting of subsidies for inputs such as capital, fertiliser, seeds, pesticides and water (Togan *et al.*, 2003). These subsidies were generally implemented without consideration for the long-term sustainability of soil resources. Subventions for fertilisers, for example, were approved without considering the specific technical requirements of soils and crops (OECD, 1999). Such problems have been addressed by the 7th and 8th Five-Year Development Plans (1996-2000 and 2001-2005). The latter specifically states that one of the main concerns of agricultural policy shall be "to diminish negative impacts of agricultural production to the environment" (SPO, 2001). Moreover, the Plan foresees decreasing support to fertilisers and pesticides and the promotion of organic farming and integrated pest control management.

The Agriculture Reform Implementation Project (ARIP) effectively removed those subsidies, which seems to have led to the decrease of the use of chemical inputs. From 2006, ARIP will include direct payments related to environmental protection schemes (Cakmak, 2004). ARIP also foresees measures for the reforestation of former forestland that was illegally cleared for hazelnut production (World Bank, 2002).

Rural development policy

The Commission reported in 2001 that "rural development programmes as foreseen by the relevant *acquis* do not exist in Turkey". However, a number of 'classical' development projects, aimed at addressing problems of regional poverty and natural resource degradation do exist, mostly funded by international donor organisations (European Commission, 2001: 59). One is the Eastern Anatolia Development Programme (DOKAP), started in 2004, funded by the European Commission and implemented by the MARA, the MoEF, the Directorate of Rural Services and local authorities. It aims at restoring sustainable range, forest and farming activities in the region in order to reduce soil erosion and increase productivity and income.

The Southeastern Anatolia Project (GAP), currently the biggest regional development project underway in Turkey, is causing considerable alterations in the environment. The construction of dams has caused the transformation of steppes into irrigated farmlands and of river beds into artificial lakes, with the subsequent disappearance of the natural living environment of waterborne and steppe species. A major weakness of the project is that it was planned and implemented without formal consideration of environmental management or impacts (SPO, 1998). Cropping patterns and farming practices are changing as a result of the increased irrigated area. For example, increased production of cotton is observed in the GAP region, while in other areas such as the Mediterranean region it is decreasing. Widespread cotton irrigation practices include flooding and in general there is a tendency to over-irrigate, resulting in drainage and salinity problems.

Cultivated soils in the GAP have high lime content and the organic content is poor. Despite a high potassium level, soil fertility is low. Further potential impacts are therefore expected from increased use of fertilisers. Additionally, excess irrigation is expected to cause the soil to compress due to the high clay content (Atabay, 1998). Salinity is expected to build up in the soil, due to the high clay content and the flat gradients, in addition to the high evaporation rates. Changes in the micro-climate are already taking place. The region, historically semi-arid, is now much more humid. It is now common to find aphids and other insect pests that were unknown to the area a few years ago (NWTITC, 2004). It is, therefore, likely that plant protection products will be used. Fertiliser applications are increasing parallel to the intensification of farming practices and monoculture (U. Aksoy, Izmir, November 2004, personal communication). If no appropriate training is given, there is a risk that pesticides and fertilisers will be used in excess as is the case with water.

The GAP Regional Development Administration acknowledges these problems and a number of initiatives have been taken to improve environmental planning and restoration (GAP, 2004). The rural development component of the GAP Regional Development Programme, co-funded by the European Commission, also aims at reducing the environmental hazards due to salination and incorrect use of fertilisers.

9.4 Environmental institutions

Institutional reforms and organisational set-up

Environmental management in Turkey is currently carried out by a large number of agencies; ineffective communication mechanisms hinder their collaboration (Ecotec *et al.*, 2000). For example, water management is performed by a web of ministries and directorates with different mandates, but in many instances with overlapping activities (OECD, 1999; Sezer *et al.*, 2003).

Four developments have occurred since 2001 that have contributed to a more coordinated policy and increased participation of all stakeholders. First, the Ministry of Environment and the Ministry of Forestry were merged in 2003 to form the Ministry of Environment and Forestry (MoEF), in order to increase coordination and effectiveness. Second, an Environment Assembly has been created to provide policy recommendations and support for scientific and technological research. The Assembly includes staff of the MoEF, members from public institutions and agencies, universities, and NGOs. Third, the new Environmental Impact Assessment (EIA) regulation adopted in 2003 is almost fully aligned with the EU Directive on EIA. Finally, the new Law on Associations enacted in the same year allows non-governmental organisations to receive funding, establish partnerships and organise activities without prior authorisation from the government, thereby removing critical constraints on the functioning of these organisations.

Policy implementation, monitoring and enforcement

A study carried out for the European Commission (Ecotec *et al.*, 2000) identified poor implementation and enforcement as important problems for effective environmental policy in Turkey. Municipalities have responsibility for implementing the law. However, smaller municipalities often have major resource problems, which translates into poor implementation of environmental legislation, insufficiently trained and specialised staff, and inability to purchase specialised equipment. Complicated administrative procedures also make it difficult to collect fees. This, combined with the unclear division of roles and responsibilities of different agencies, encourages polluting entities to continue to do so.

Monitoring environmental quality is primarily the responsibility of the MoEF, but this task is performed by different jurisdictions, with little coordination. The Ministry of Health monitors air quality, drinking water and bathing waters, and the General Directorate of State Hydraulic Works (DSI) monitors water quality in rivers, lakes and groundwater. The National Plan for the Adoption of the EU *acquis* recognises the need for the MoEF and MARA to strengthen their organisational infrastructure (offices, laboratories, personnel, etc.) to facilitate effective monitoring and enforcement. Accordingly, in 2001, an Environmental Reference Laboratory was created with specialised staff and equipment. Furthermore, endorsed in 2002, the new Turkish Regulation for Environmental Inspection created an Environmental Inspectorate Department at central level, and delegated inspection responsibilities to other General Directorates within the Ministry, and to the Directorates at provincial level (European Commission, 2001).

Participation, collaboration and partnerships

Mechanisms to involve civil society in policy making have been in place for a long time in the agri-environmental field, such as local environment committees and the Council for the Environment and Forestry (OECD, 1999). The government provides some financial support to non-governmental organisations (MoE, 1997), although during this research we have not found any quantification of that support. The incorporation of the general civil society in policy processes is, however, slow for various reasons. The first is the lack of an effective environmental information system to allow the assessment and reporting of environmental problems. The second is that the majority of the Turkish population perceives environmental problems as less urgent compared to socio-economic problems like unemployment (Magyar Gallup Intézet, 2003). A third factor is the shortage of trained staff and financial resources in non-governmental organisations (Okumus, 2002; Serban, 2002).

Recognising the critical role of NGOs, the Turkish government enacted a new Law on Associations in 2004 to facilitate their continued existence and increase their participation in policy making. For the first time, associations are allowed to receive foreign funding or establish partnerships without prior authorisation. Furthermore, their projects can receive government funding of up to 50 per cent of their budget. Recent regulations also incorporate other mechanisms for involving NGOs and other stakeholders in the policy-making process[3].

Farmer participation in decision-making has increased, thanks to the creation of Water Users' Associations (WUAs). During the 1950s, the DSI encouraged participatory irrigation management in small areas by establishing irrigation groups or WUAs with limited responsibility over the operation and management of irrigation systems. The positive results, which included improved fee collection rates, encouraged the DSI to introduce an accelerated transfer programme in selected areas from 1993. This was coupled with training activities and support from the World Bank, which considered the transfer process as a model for other countries (Cakmak *et al.*, 2004). By the end of 2003, the transfer had been achieved for 1.8 million ha, which represents 40 per cent of the total irrigated area. Positive technical, environmental and financial results are observed, including better use of resources and facilities, increased awareness of the need to save water and energy (Vidal *et al.*, 2001) and higher fee collection rates (Cakmak *et al.*, 2004).

Other forms of public involvement include the participation of rural communities in forest fire prevention. Kurtulmuslu and Yazici (2003) observed that these communities had lower levels of intentional fires compared with national averages.

In the business sector, environmental issues do not appear to receive special attention, mainly because of weak enforcement of legislation and the current business culture that ignores environmental concerns. Although incentives for investment in the environment have been in place since the establishment of the Ministry of Environment in 1991, companies have only recently started to get involved in community environmental protection activities by sponsoring projects executed by non-governmental organisations (MoE, 1997; Serban, 2002).

Financing environmental projects

Turkey's overall environmental expenditure in 2002 was around EUR 278 million, corresponding to 1.44 per cent of GDP. This represented an increase of

---

[3] For instance, the Regulation on Environmental Impact Assessment (EIA) foresees, "when necessary", the participation of representatives from universities, research agencies, vocational chambers, unions and other civil society organisations in the EIA committee, which evaluates the impacts of certain projects on the environment.

3.6 per cent with respect to 2001 (SIS, 2004). In 2002, water and wastewater management absorbed 92 per cent of governmental environmental investment expenditure. In contrast, the protection of biodiversity and landscape received less than 0.1 per cent of the total environmental investment resources (Table 9.3).

Table 9.3. Environmental expenditure of governmental organisations by domain in 2002

Domain	Expenditure EUR 1000	% of total
Protection of ambient air and climate	87	0.03
Water management	135,921	49.0
Wastewater management	119,032	43.0
Solid waste management	79	0.03
Protection of soil and groundwater	590	0.2
Protection of biodiversity and landscape	270	0.1
Other environmental protection activities[1]	22,286	8.0
Total expenditure, of which	278,264	100.0
Current expenditure	113,578	41.0
Investment expenditure	164,687	59.0

[1] Other environmental protection activities includes general environmental management, training etc, and environmental expenditures that could not be divided into environmental domains (such as integrated projects).

Source: SIS, 2004

Typically, public financing of environmental investment comes from the tight government budget. Highly volatile capital and financial markets limit the capacity of commercial banks to extend loans for environmental projects (Sezer *et al.*, 2003). Thus, funding for environmental projects is mostly provided by international development agencies, donors and international organisations.

The majority of these projects (other than infrastructure) are scattered and of small scale. Especially important is the United Nations Development Programme (UNDP) support to the MoEF for the integration of environmental concerns into development policies and for the promotion of dialogues among the public sector, NGOs and other societal actors. The UNDP/GEF (Global Environment Facility) Small Grants programme has supported approximately 90 projects since 1993. These projects have a budget of EUR 10-40 thousand, and are implemented by NGOs (U. Aksoy, Izmir, November 2004, personal communication).

9.5 Organic agriculture and relevant institutions

Organic farming involves producing crops and livestock in ways that favour cultural, biological and mechanical management methods rather than the use of off-farm inputs such as synthetic fertilisers and pesticides. Emphasis is placed on improving environmental protection, and for livestock production, on animal welfare (European Commission, 2000).

Market for organic products

Driven by external demand, Turkey started producing organic products in the mid-1980s. At first, production and export of organic products adhered to the standards of the importing country. Following EC Council Regulation 2092/91 in 1991, Turkey issued its own legislation in 1994, adopting the EU definition of organic agriculture and taking over some parts of the EU regulation with the aim of increasing agricultural exports and providing income to marginal farmers.

Data on the number of farmers and the area devoted to organic farming vary significantly between sources. According to the MARA, organic agriculture was practiced on 103,190 hectares by 13,044 farmers in 2003 (Süngü, 2004; Aksoy and Can, 2004).[4] Organic production reached 292 thousand tonnes in 2003 (Süngü, 2004). Most organic production is of plant origin. Nearly 100 crops are grown organically. According to MARA figures for 2001, after processing the number of products rises to 150. The recent increase in processed organic foods not only adds value, but also reduces transport costs and diversifies the market (Aksoy and Can, 2004). Organic livestock production is still very limited and is expected to reach market shelves in 2005 (Süngü, 2004).

Table 9.4. Exports of major organic agricultural products of Turkey in 2003

Products	Tons	1000 EUR	Main export destinations[5]
Raisins	5677	6238	DE, NL, UK, CH, FR, DK
Dried figs	2027	4567	DE, CH, FR, UK
Hazelnuts	1403	4515	NL, UK, DE, CH, ES, USA
Dried apricots	1688	4185	DE, UK, USA, FR
Apple juice	2528	2701	NL, IT
Frozen fruits	1212	1753	IT, DE, CH, AT
Cotton	865	1216	TH, BG
Pine kernels	70	1071	CH
Lentils	1447	906	DE, UK, IT
Chickpeas	1167	734	UK, IT, NL
Frozen vegetables	841	507	DE, NL, BE, USA
Anise, fennel & coriander seeds	229	400	DE
Honey	109	261	DE
Pistachios	32	234	DE
Olive oil	54	154	USA, JP
Canned cherries	88	129	DE, NL
Tomato paste	134	126	NL
Others	1512	2953	n.a.
Total	21083	32649	

Source: Export Promotion Centre of Turkey, IGEME, 2004

[4] In the current data collection system, if several crops are grown in the same field, the field and the farmer are recorded more than once, causing double-counting. The MARA is currently updating its data collection system (Süngü, 2004).

[5] AT: Austria; BE :Belgium; BG: Bulgaria; CH: Switzerland; DE: Germany; DK: Denmark; ES: Spain; FR: France; IT: Italy; JP: Japan; NL: Netherlands; TH: Thailand.

The majority of organic production is sold in foreign markets, primarily in Europe. Exports have been steadily growing, doubling in euro value and more than doubling in volume between 1998 and 2003. According to the Export Promotion Centre of Turkey (IGEME, 2004), Turkish organic products were sold in 37 countries in 2003. Table 9.4 gives an overview of the most important exports in 2003 and the destination markets. The great majority are dry or processed food or non-food commodities with a long shelf-life (Özkan, 2002). With regard to trade data, there is no separate Harmonised Commodity Description and Coding System for organic products in Turkey. Therefore, these products are also included in the general trade statistics for agricultural products. Export figures for organic products are collected by the Exporter Unions and then transmitted to the Aegean Exporters Union.

The domestic market is still only a small niche (Nicely, 2001). Its development started in the late 1990s with some health food and specialised shops, and some products are now being introduced through supermarkets. In 2003, a total of 15.3 thousand tons were certified and sold on the internal market in 2003 (E. Süngü, Ankara, October 2004, personal communication). In general Turkish consumers lack awareness of what a certified organic product is, and the great majority of Turkish consumers remain price-sensitive (FAO and MARA, 2004). Both aspects have been confirmed in various consumer surveys (Kenanoğlu and Karahan, 2002).

Organic products can be 70-200 per cent more expensive than conventional ones. Premiums of 35-50 per cent are usually not enough to cover laboratory testing costs (FAO and MARA, 2004) and the quantities marketed do not permit economies of scale in distribution. Furthermore, many consumers perceive Turkish food products as already very healthy, as their country has always been an agrarian country that traditionally produces most of its own food. Therefore, the appeal of certified organic products is not immediately apparent to them (Nicely, 2001: 4). Low trust in the authenticity of labels is also a barrier and increases the need for effective marketing strategies.

Institutional set-up for organic farming

The organic industry has developed almost exclusively without government support. From the beginning, exporters had to look for farmers who would be willing to produce organically and organic farming developed on the basis of contract farming (Nicely, 2001). According to Aksoy and Can (2004), this brought advantages, such as the transfer of know-how between the contracting parties. On the other hand, because only companies exporting abroad were involved in organic farming, and these types of contract did not exist in local markets, there has been a limited range of cultivated products.

The 1994 law assigned to MARA the responsibility of overseeing the cultivation of organic crops. The Ministry created a specific Organic Agriculture Committee (OAC), which is the main decision-making body. It implements the regulation, authorises certification bodies, inspects these organisations and coordinates all other activities to improve and foster organic agriculture (Özkan, 2002). The OAC is composed of representatives of the various Directorates of MARA. The representation of stakeholders remains guaranteed in 3 additional committees: The National Orientation Committee, the National Trade Committee, and the National Research Committee (Aksoy and Can, 2004). In support of the work of the Ministry, the Turkish Association of Organic Agriculture (ETO), a non-governmental organisation, provides policy input and contributes to advancing technical improvement in the organic industry. ETO members include producers, exporters, academics and consumers.

The new Law on Organic Agriculture currently being discussed in the Parliament provides for penalties and administrative fines to be applied in the case of non-compliance by operators, monitoring and certification bodies. With regard to the institutional infrastructure, Organic Farming Units have been established at the Provincial Agricultural Directorates. The aim is to provide these teams with specialised staff to collect data, implement standards and perform inspections and certifications of companies (Aksoy and Can, 2004).

There are currently seven certification bodies established in Turkey, five of them branches of European registered companies and two Turkish. A deterrent for new farmers entering the organic sector is the high fees of these bodies compared with fees in other countries. Exporters tend to overcome this by grouping themselves into a so-called 'project' within the same region, which can comprise up to 100 farmers, thereby reducing the cost per farmer. But this prevents a farmer from selling his products as organic to alternative buyers (Kenanoğlu and Karahan, 2002). Another relevant aspect is that organic farming does not receive specific subsidies from the government. Only recently did the Agricultural Bank reduce the yearly interest on loans for the production of organic products and inputs used in organic agriculture (Aksoy and Can, 2004).

Prospects

Turkey has considerable potential for the production of organic products given the diversity of microclimates and the abundance of plant varieties adapted to the local conditions (Kenanoğlu and Karahan, 2002). Bearing in mind that the sector developed initially without governmental assistance, the expansion in the 1990s has been significant. The MARA is currently making great efforts to build up domestic awareness and trust in the organic sector. In addition to further adapting national legislation to the European organic farming regulations, training and research activities are being implemented in cooperation with ETO. In 2003, a vocational school was created to provide the sector with skilled human resources (IGEME, 2004) and in 2004 workshops were organised with FAO support, bringing stakeholders together for discussion (U. Aksoy, Izmir, November 2004, personal communication). Research is being carried out by TÜBITAK, the State Planning Organisation, and by university research funds.

The dissemination and development of organic agriculture is also believed to be an important tool for increasing the incomes of the small family farms that are dominant in Turkish agriculture (Kenanoğlu and Karahan, 2002).

9.6 Biodiversity

Biological diversity

Turkey has a privileged position at the junction of three major phyto-geographical regions: the Irano-Turanian, (present in Central, South and East Anatolia), the Euro-Siberian (around the Black Sea Coast and surrounding mountains) and the Mediterranean (Mediterranean coast and mountains) (MoE, 2000; Karagöz, 2003). Its biological richness is further enhanced by the topographic variety of the country, with altitudes from 0 to 5000 metres, its geological and geomorphologic structure, and the presence of seven climatic regions with several micro-climatic areas. Additionally, different types of farming systems contribute to the existence of different agro-ecosystems (Tan, 2001).

Environmental and ecosystem diversity

The main ecosystems in Turkey are the Forest, the Steppe and the Aquatic ecosystems, including wetlands (Box 9.1). From the economic point of view, the Steppe ecosystem is the most important, due to the large number of food crops that have originated from native species. It is generally accepted that most steppe vegetation in Turkey is the result of long-term human activities.

Since the 1950s, agricultural mechanisation, industrialisation and tourism have increased the pressure on biological diversity, resulting in the reduction and fragmentation of natural habitats (Karagöz, 2003). The total size of grassland areas has been declining since the 1930s due to their conversion into arable land to satisfy the food needs of the growing population. This trend seems to have ended in recent years because almost all potential areas for cultivation have already been converted (Tan, 2001: 3).

Species diversity

Table 9.5. Number and status of species

	Total known species 1992-2000	Number of threatened species 2002[1]	% of species threatened 2002
Higher plants	8,650	3	0.03
Mammals	116	17	14.60
Breeding birds	278	11	3.90
Reptiles	133	12	9.00
Amphibians	23	3	13.00
Fish	162	29	17.90
Invertebrates	80,000	13	0.01

[1] Includes all full species categorised at the global level as critically endangered, endangered or vulnerable. Subspecies, introduced species, species whose status is insufficiently known, and species whose status has not been assessed are excluded.
Source: EarthTrends, 2003; IUCN, 2003

Turkey hosts three quarters of the plant species that can be found in the whole of Europe. According to different estimates, there are between 8,700 and 10,500 different species of flora. One third of these species are indigenous. They also include wild relatives of food crops originating in Turkey such as wheat, chickpea, lentil, apple, fig, pear and pistachio. Turkey is also very rich in wildlife, with over 80 thousand species, most of them invertebrates (Goncagul, 2001). The International Union for Nature Conservation (IUCN) estimates that 85 animal species are in danger of extinction, 12 of which are in a critical situation. Three species of Turkish flora are also threatened (see Table 9.5).

Genetic biodiversity

Various native livestock breeds are present and in general they are better adjusted to their environment (Goncagul, 2001). Although a nation-wide census has not yet been carried out, it is estimated that there are 20 indigenous cattle breeds, 17 breeds of sheep and 5 breeds of goat. Vegetal crop species with a high level of genetic variety are still grown in transitional zones and mountainous areas where farms are small and modern farming is difficult. These breeds are kept owing to cultural and traditional preferences and demands (Tan, 2001).

Influence of agriculture on biodiversity

Habitat destruction is the most serious cause of the loss of flora and fauna. Over-harvesting and pollution cause moderate losses. In addition, fauna losses result from forest fires, and flora losses from urbanisation (MoE, 1997).

Agriculture contributes positively to ecological diversity through the cultivation of plants, including cereals and fruit trees, but it also has negative impacts on biodiversity through species' increased sensitivity and vulnerability to pests and diseases. In Turkey, as in other Mediterranean countries, land abandonment is less of a problem for biodiversity than the intensification of agriculture. Clearing land for fields, overgrazing, forest fires, the construction of dams and pesticide use are mentioned in the literature as some of the factors responsible for the major direct and indirect impacts of agriculture on Turkey's flora and fauna (Ozturk *et al.*, 2002). It has been confirmed that six indigenous plant species that were collected in the 19th century are now extinct. Of these extinctions, two occurred as a result of flooding behind the Keban High Dam water reservoir and four are known to be extinct because of overgrazing and habitat destruction (MoE, 2000). Low-yielding native livestock breeds are also progressively being replaced by foreign high-yielding breeds through artificial insemination and by importing animals that are made available to the farmers. Similarly, during the last two decades native fruit and vegetable species are being replaced by similar imported species that are more expensive in the market but which cause a decrease in the biodiversity of agricultural products. Decreased demand for native species also causes land abandonment in some areas (Yüksel, 2003).

Changes in land use patterns, especially conversion of forests, ploughing up grazing land and wetland drainage to create agricultural areas also have negative consequences for the habitats of fauna and flora (Tan, 2001; MoE, 2000).

Institutional framework

Interest in protecting natural assets is by no means new in Turkey. The concept of endangered species first appeared in legislation in 1937, and in 1956 the first national park, Yozgat Çamliği, was created (MoE, 2000:7). Turkey's Constitution (Article 63) states that the Government should protect its natural values and resources, as well as support and promote conservation efforts. Following the objectives laid out in the Fifth Five-Year Development Plan (1983-1988) on the conservation and development of natural resources, the National Parks Law and Hunting Law of 1983 made very specific references to endangered, vulnerable and rare species of flora and fauna, unique ecosystems and categories of protected areas that had been internationally recognised.

In the last decade, Turkey has multiplied its efforts to include biodiversity protection in its legislative framework, by becoming signatory to relevant international conventions and adopting national legislation.

The signature of the Convention on Biological Diversity in 1992 gave fresh impetus to the recognition of biodiversity and related issues. To apply the Convention, the National Biodiversity Strategy and Action Plan was formulated in 2002 in cooperation with the relevant stakeholders under the coordination of the State Planning Organisation (SPO) and with financial support from the World Bank. In 1994, the country adopted other additional international agreements such as the Bern Convention on the conservation of European Wildlife and Natural Habitats, the Ramsar Convention on Wetlands of International Importance and the UN Convention to Combat Desertification. More recently, Turkey ratified the Council of Europe's European Landscape Convention, which came into force in March 2004.

National legislation arising from Turkey's international commitments includes special conservation measures regarding the collection, cultivation and export of native bulbs, in line with the Convention on International Trade of Endangered Species of Flora and Fauna (CITES), the Animal Breeding Law which determines the registration of new breeds and the preservation of farm animal genetic resources and the Law on Reforestation and Soil Erosion Control. Regarding cross-sector legislation, the Environmental Impact Assessment Regulation requires development investments to evaluate potential impacts on habitats and species.

Economic instruments for the protection of nature are rare in Turkey. One exception is the Law on Grazing and Common Grazing Land of 1998, which introduced a charge for grazing beyond the common pastureland in order to reduce the pressures on steppe ecosystems from overgrazing (Yüksel, 2002).

Protected areas in Turkey are divided into four types: national parks, nature parks, nature reserve areas and natural monuments. Between 2003 and 2004, the protected area increased from 3.8 per cent of the total area to 4.1 per cent (OECD, 2003 and 2004). The MoEF, MARA, and the Ministry of Culture and Tourism are

Box 9.1 Wetlands in Turkey

Turkey hosts 250 wetland areas, covering an area of 1.35 million ha. Two important bird migration routes pass through the country, and given the aridity of the territory, these areas are of particular importance for migrating birds. Wetlands also regulate the hydrological balance and increase water quality through the retention of sediments and toxic materials. Furthermore, they make a significant contribution to the economy by means of fishing, hunting, reed cutting and tourist activities (MoE, 2000).

Drainage works to fight malaria from the 1950s to the mid-1990s converted many wetlands to agricultural use. In the last four decades, more than one third of Turkish wetlands (approximately 700,000 ha) have been lost (Guclu and Karahan, 2004). Some of the remaining wetlands are damaged due to feeding water being stored in dams, deterioration of water quality due to agricultural and industrial pollution and unsustainable hunting and harvesting of wild plants and tubers. Out of 132 Important Bird Areas (IBAs) in Turkey, 50 are significantly threatened by agricultural intensification and expansion (Birdlife International, 2004a: 34).

After signing the Ramsar Convention in 1994, the Turkish Government cancelled projects requiring the drainage of wetlands, and projects likely to damage the ecological characteristics of wetlands were revised. Nine sites were designated as Wetlands of International Importance, covering a surface of 159,300 hectares. Agriculture and livestock grazing, among other activities, are present in six of these sites. The Regulation on Conservation of Wetlands adopted in 2002 established the "National Wetlands Committee" in order to increase coordination and collaboration among the various institutional organisations, public and private. The Regulation prohibited drainage and sand extraction in areas larger than 8 ha of wetlands. Drainage of areas smaller than 8 ha requires special permission from the MoEF.

Despite these legislative advances, wetland conservation problems persist due to the strong economic interest in these areas and lack of coordination between administrations. One example is the controversial project to construct an international harbour in the Gediz Delta, one of Turkey's most important wetlands (Birdlife International, 2004b). Nature conservation NGOs have tried to fill these gaps in order to overcome the *de facto* lack of protection. The private sector is starting to assist these initiatives by sponsoring NGOs' projects (MoE, 2002).

The "2003-2008 National Wetlands Strategy Plan for Turkey", prepared by the MoEF in cooperation with relevant stakeholders, including academia, has two main goals. Firstly, the conservation, management, restoration and rehabilitation of biodiversity, including the sustainable use of those of international importance. Secondly, promoting cooperation and coordination among institutions, and ensuring that the necessary financial resources are mobilised for the appropriate implementation of the Strategy.

Specifically in relation to agriculture, the Strategy foresees the adoption of administrative and legal measures to prevent *in situ* pollution and to control excessive agricultural inputs, to encourage organic methods in agricultural areas around the wetlands, to compile, evaluate and disseminate advice, methods and best practice for the wise use of wetlands, to ensure local communities' involvement in the management of wetlands and to promote public participation in the decision-making processes.

responsible for decision-making in the field of biological and genetic conservation of resources. Within the limits established by the specific legislation, each of these organisations has power to declare conservation areas and management plans or principles for the use of those areas. The MoEF is responsible for the overall coordination of these efforts. 'Specially Protected Areas' (SPAs), which are areas of international ecological importance created in line with national legislation and international conventions on nature conservation, are the responsibility of a special body, the Authority for the Management of SPAs.

The fact that these areas are legally protected does not mean that they are safe from degradation. National parks, for example, are used as recreational sites, often to the detriment of their ecological values. Moreover, not all protected areas are managed appropriately. As occurs in other policy areas, regulation and management of sensitive natural ecosystems and species suffer from problems of overlap and lack of coordination between different organisations. Furthermore, insufficient staff and resources often compromise the correct implementation of legislative measures and biodiversity projects (Hüsnü, 1999).

The EU is currently funding a project to increase Turkish administrative capability to identify, monitor and protect habitats and species according to the Habitats and Birds Directives (Representation of the European Commission to Turkey, 2004). NGOs also play a key role in raising awareness and implementing specific conservation projects in these areas.

9.7 Conclusions

The main environmental impacts of agriculture in Turkey involve the degradation of water and soil due to uninformed farming practices, namely the overuse of water and chemicals. Erosion is a naturally occurring process that is exacerbated by human activities. The use of fertilisers and pesticides seems to be following a slight decreasing trend, enhanced by the removal of subsidies. However, the expansion of irrigated areas may stimulate excessive use of water, leading to more nutrient run-off and salination, and a rise in the use of artificial inputs.

In the last 10 years, environmental policy in Turkey has made a quantum leap in terms of new legislative instruments. The implementation of global and regional conventions, participation in international environmental fora and the goal of joining the European Union have been major driving forces behind these reforms. According to the literature, there are (or have been) projects aiming at the improvement of the environment. However, there is a lack of information about their actual implementation and it is difficult to know whether they have been effective. The literature suggests that institutions dealing with agri-environmental issues seem to be poorly coordinated and there seems to be a lack of effective implementation at local level. Regulations are the main policy tool and there are few economic instruments. Farmers therefore appear to have few incentives or insufficient knowledge in order to implement environmentally friendly practices.

Civil society is gradually becoming more involved in policy making. Non-governmental organisations have an important role to play in increasing environmental awareness and public participation, and in advancing governmental policy. Recent regulatory developments facilitate the registration and financing of non-governmental organisations and their projects. Public opinion gives low priority to the environment.

Organic farming has developed rapidly since the mid 1980s. Production is driven by export demand. The sector could expand further if domestic demand grows in line with income and environmental awareness. Both the government and non-governmental actors are making efforts to develop the organic domestic market.

Turkey is very rich in biodiversity. Some species and habitats are, however, at risk due to various kinds of environmental impact, which include agricultural intensification, land abandonment and the construction of large infrastructure projects.

References

Aksoy, U. and Can, H.Z. (2004) Country report: Turkey. In: *CIHEAM Report on Organic Agriculture in the Mediterranean Area.* CIHEAM (in press).

Atabay, S. (1998) Sustainable development and environmental management – Southeast Anatolia Project example. *38th Congress of the European Regional Science Association,* 28 August-1 September 1998, Vienna, Austria.

Birdlife International (2004a) *State of the World's Birds 2004: Indicators for our Changing World.* Birdlife International, Cambridge, UK.

Birdlife International (2004b) Continuing the fight to save the Gediz Delta - one of Turkey's most important wetlands. Press release 02.09.2004. Birdlife International, Cambridge, UK.

Cakmak, E. (2004) Structural change and market opening in agriculture. *CEPS EU-Turkey Working Papers,* No. 10, September 2004. Centre for European Policy Studies, Brussels.

Cakmak, E., Cetýk, M., Tuzun, A.M., Baysan, N., Karaca Býlgen, G., Terzi, M., Kanber, R. and Yazar, S. (2003) Participatory irrigation management and water cultural heritage, Wasamed thematic network. www.gap.gov.tr/turkish/tarim/wsm2.pdf (accessed 18 august 2004).

Cakmak, E., Erdogan, F.C. and Cetik, M. (2004) The performance of irrigation associations in Turkey after the transfer. Paper presented at the *Wasamed Workshop on Irrigation System Performance*, Hammamet, Tunisia, 25-28 June 2004.

DSI (General Directorate of State Hydraulic Works) (2004) Land and water resources of Turkey. http://www.dsi.gov.tr/land.htm (accessed 7 August 2004).

Earthtrends (2003) Biodiversity and protected areas – Turkey. Earthtrends Country profiles, http://www.earthtrends.org/pdf_library/country_profiles/Bio_cou_792.pdf (accessed 20 September 2004).

Ecotec, IEEP and FEI (2000) Administrative capacity for the implementation and enforcement of EU environmental policy in the 13 Candidate Countries. Final Report for DG Environment. Ecotec, Brussels and Birmingham.

European Commission (2000) Organic farming in Europe: Guide to Community rules. Office for Official Publications of the European Communities, Brussels.

European Commission (2001) 2001 Regular report on Turkey's progress towards accession. SEC(2001) 1756, Brussels.

FAO (1997) AQUASTAT: Turkey - country profile. FAO, Rome.

FAO (2004a) Glossary. http://www.fao.org/glossary (accessed 21 November 2004).

FAO (2004b) FAOSTAT Database. FAO, Rome.

FAO and MARA (2004) Workshop report on the development of organic agriculture in Turkey and alignment of related Turkish legislation, 21-22 January 2004, Ankara.

GAP (2004) SouthEastern Anatolia project website. http://www.gap.gov.tr (accessed 20

September 2004).

Goncagul, T. (2001) Farm animal diversity in Turkey. Paper presented to the *OECD Expert Meeting on Agri-Biodiversity Indicators,* 5-8 November 2001, Zürich, Switzerland.

Guclu, K. and Karahan, F. (2004) A review: the history of conservation programs and development of the national parks concept in Turkey. *Biodiversity and Conservation* 13 (7), 1373-1390.

Hüsnü Can Baser, K. (1999) Country profile: Turkey. In: Heywood V.H. and Skoula M. (eds.) *Wild Food and Non-Food Plants: Information Networking*. CIHEAM-IAMC, Chania, Greece, pp. 303-322.

IGEME (Export Promotion Center of Turkey) (2004) Organic agricultural products of Turkey. http://www.igeme.org.tr/introeng.htm (accessed 15 September 2004).

IUCN (2003) 2003 IUCN Red List of Threatened Species. http://www.redlist.org (accessed 4 September 2004).

Karagöz, A. (2003) Turkey. *Country Pasture/Forage Resource Profiles Database*. FAO Crop and Grassland Service, FAO, Rome.

Kenanoğlu, Z. and Karahan, Ö. (2002) Policy implementations for organic agriculture in Turkey. *British Food Journal*, 104 (3-5), 300-318.

Keskin, S. (2001) Gateway to land and water information: Country report Turkey. FAO, Rome.

Kumuk, T. and Akgüngör, S. (1995) The role of public extension in introducing environmentally-friendly farming methods in Turkey. *Journal of Agricultural Extension*, 1995(4), 65-74.

Kurtulmuslu, M. and Yazici, E. (2003) Management of forest fires through the involvement of local communities in Turkey. In: FAO (ed.) *Community-Based Fire Management: Case Studies from China, the Gambia, Honduras, India, the Lao People's Democratic Republic and Turkey*. RAP Publication 2003/08, FAO, Bangkok.

Magyar Gallup Intézet (2003) Public opinion in the Candidate Countries: Full report. *Eurobarometer* 2003(4), European Commission, Brussels.

MoE (Ministry of Environment) (1997) Implementation of Agenda 21: Review of progress made since the United Nations Conference on Environment and Development, 1992. Information presented to the United Nations Commission on Sustainable Development. Fifth Session, 7-25 April 1997, New York. http://www.un.org/esa/earthsummit/turky-cp.htm (accessed 8 December 2004).

MoE (Ministry of Environment) (2000) First national report on biological diversity in Turkey. MoE, Ankara.

MoE (Ministry of Environment) (2002) National planning tool for the implementation of the Ramsar Convention on Wetlands: Turkey national report. RAMSAR, Gland, Switzerland.

Nicely, R. (2001) Turkey organic products organic food report. GAIN Report TU1029. USDA Foreign Agricultural Service, Ankara,

NWTITC - North West Texas International Trade Council (2004). The GAP irrigation in Turkey. http://www.nwtitc.org/gapturkey.htm (accessed 6 October 2004).

OECD (1999) Environmental Performance Reviews - Turkey. OECD, Paris.

OECD (2001) Environmental Indicators for Agriculture. OECD, Paris.

OECD (2002) Environmental Data Compendium. OECD, Paris.

OCED (2003) OECD in figures: Statistics on the member countries. *OECD Observer* 2003 supplement 1. OECD, Paris.

OECD (2004) OECD in figures: Statistics on the member countries. OECD, Paris, July 2004.

Okumus, K. (ed.) (2002) *Turkey's environment: A review and evaluation of Turkey's environment and its stakeholders*. The Regional Environment Center for Central and Eastern Europe, Szentendre, Hungary.

Özden, M., Dursun, H. and Sevinç, A.N. (2000) The land resources of Turkey and activities of general directorate of rural services. Paper presented at the *International*

Symposium on Desertification, Konya, Turkey, 2000.

Özkan, M. (2002) Organic agriculture and national legislation in Turkey. In: *Organic Agriculture: Sustainability, Markets and Policies*. Proceedings of the OECD Workshop on Organic Agriculture, 23-26 September 2002, Washington DC, pp. 289-294.

Ozturk, M., Celik, A., Yaerci, C., Aksoy, A. and Feoli, E. (2002) An overview of plant diversity, land use and degradation in the Mediterranean region of Turkey. *Environmental Management and Health* 13(5), 442-449.

Petersen, J. and Hoogeveen, Y. (2004) Agriculture and the environment in the EC accession countries: Implications of applying the EU common agricultural policy. *Environmental Issue Report,* 37. European Environment Agency, Copenhagen, 28 April 2004.

Representation of the European Commission to Turkey (2004) EU supports the nature protection in Turkey. Press release, Representation of the European Commission to Turkey, Ankara, 4 May 2004.

Serban, S.D. (2002) REC Extension to Turkey: Feasibility and work plan. The Regional Environment Center for Central and Eastern Europe: Szentendre, Hungary.

Sezer, S., Kocasoy, G. and Aruoba, C. (2003) How vital is the "lack of funding" in effective environmental management in Turkey? *Waste Management* 5, 455-461.

SIS (1999) Greenhouse gas emissions of Turkey in 1990-1997. Environmental Statistics Bulletins, 30 May 1999. SIS, Ankara.

SIS (2004) Environmental employment and expenditure inventory: Governmental organizations survey results, 2002. SIS News bulletin 26 October 2004, SIS, Ankara.

SPO (1998) National environmental action plan. SPO, Ankara, May 1998.

SPO (2001) Long-term strategy and eighth five-year development plan 2001–2005. SPO, Ankara.

Süngü, E. (2004) Data collecting and evaluation of the organic agriculture system in Turkey. *Proceedings of the 1st EISfOM Seminar*, April 2004, Berlin, pp. 87-90.

Tan, A. (2001) Agricultural plant diversity in Turkey. *OECD Expert Meeting on Agri-Biodiversity Indicators*, 5-8 November 2001, Zürich, Switzerland.

Tanrivermis, H. (2003) Agricultural land use change and sustainable use of land resources in the Mediterranean region of Turkey. *Journal of Arid Environments* 54, 553–564.

Togan, S., Bayaner, A. and Nash, J. (2003) Analysis of the impact on agricultural markets and incomes of EU enlargement to Turkey. Paper presented at *Turkey: Towards EU accession*, Bilkent University, Ankara, 10-11 May 2003.

Ünver, O. and Gupta, R.K. (2003) Water pricing: Issues and options in Turkey. *International Journal of Water Resources Development* 19(2), 311-330.

Vidal, A., Comeau, A., Plusquellec, H. and Gadelle, F. (2001) Water conservation: a priority in the Mediterranean Region. Secretariat of the International Programme for Technology and Research in Irrigation and Drainage (IPTRID). FAO, Rome.

World Bank (2002) Turkey - Agricultural Reform Implementation Project (ARIP) Vol. 1. Report No: PID9405, World Bank, 12 November 2002.

Yüksel, E. (2002) Background study on the link between agriculture and environment in accession countries. National report for Turkey. Institute for European Environmental Policy, London, May 2002.

Yüksel, E. (2003) Decrease in soil biodiversity via land use pattern changes due to trade (import) in Turkey: Proposed indicators. *OECD Expert Meeting in Soil Erosion and Soil Biodiversity indicators*, 25-28 March 2003, Rome, Italy.

10

Animal and Plant Health in Turkey

Alison Burrell

10.1 Animal health

Turkey's animal sector

It is widely accepted that grazing livestock numbers (both cattle and small ruminants) have been declining (see Table 4.10), although the reliability of statistics on Turkey's livestock sector is somewhat uncertain and livestock population figures are controversial (Sarigedik, 2003). The main reason for uncertainty about livestock statistics is the decentralised and fragmented nature of Turkey's animal production and marketing systems, which makes accurate recording problematic. Most livestock production in Turkey takes place in small-scale, family-operated units. Over two-thirds of cattle units have 1-4 animals, and about a quarter have 5-19 cattle. In the early 2000s, 2 per cent of cattle units were larger enterprises (with up to 1000 cattle but no calf production). The structure of the sheep and goat sector is even more fragmented. Traditionally, keeping a few livestock has been considered a form of insurance by Turkish smallholder farmers rather than a main enterprise.

Most livestock are grazed on public lands, which leads to over-stocking and (in conjunction with high animal feed prices) to poor animal nutrition. Pastures have become degraded over the years. This has been working against attempts to improve cattle productivity through importing pregnant pure-bred cattle for cross-breeding with local breeds (Karagöz, 2003a). Sedentary transhumance (whereby villagers take their herds to the plateaux during the summer months for better grazing) is still practised in some areas, although fully nomadic herding has virtually disappeared. By contrast, in the poultry sector restructuring, with consequent productivity increases, has been massive. In the mid-1990s, many Turkish poultry operations were still characterised as small (5,000-10,000 bird places) although larger, integrated operations were becoming more frequent. This process has continued so that current observers now describe the sector as consisting mainly of larger, integrated companies. Production costs in modern poultry units are thought to be competitive with imported supplies.

It is also difficult to find an authoritative figure on the number of slaughterhouses. FAO (undated) estimates the number of slaughterhouses at 1200, of which 1000 are owned by municipalities. In the view of FAO, "Although there are large-scale export plants in which the hygienic conditions are of adequate standards, many of the small- or medium-scale slaughter and meat processing plants do not meet the desired sanitary conditions. New legislative arrangements have been made in Turkey in order to improve technical and hygienic conditions in the red and poultry meat plants to affect better control. Thus, what is needed for the improvement of overall sanitary conditions is well-trained personnel that will help implement the new legislation. In addition to that, the number of skilled meat inspectors need to be increased and their inspection procedures should be updated to the modern day". Günes (1998) put the number of slaughterhouses in the red meat marketing chain at approximately 900, with a capacity to handle 50-55 per cent of slaughterings. Although this percentage may have risen since the mid-1990s, it is clear that a significant share of annual production still bypasses conventional recording and monitoring systems. These details give a very imprecise picture of slaughtering facilities, whilst suggesting that veterinary inspection and monitoring of animals at this point in the chain is likely to be inadequate, and far from comprehensive.

Identification and registration of bovine animals started in 1991 but seem to have stopped in the mid-1990s. Harmonisation with the EU *acquis* began in 2001 (preparation of legislation, setting up of a computerised database, start of ear-tagging and registration of all bovines). By August 2004 about 9.5 million cattle had been ear-tagged and registered, and 1.5 million animal holdings had been registered (an increase of 300 thousand over the previous 12 months) (European Commission, 2004).

Statistics on notifiable animal diseases in Turkey are available from the World Animal Health Organisation (OIE, 2004)[1]. The following summary relies heavily on that source. However, these figures cover reported cases only. Inevitably, accurate reporting is difficult given the considerable share of production that has no contact with the veterinary services and falls outside 'regular' marketing channels.

Turkey faces some major challenges with respect to animal health. The first challenge arises from the small-scale structure of the sector, which impedes appropriate monitoring and control. Second, Turkey is prone to recurrent outbreaks of animal diseases that have been virtually eradicated in western and northern Europe. Even after several years without a reported outbreak, the existence of disease reservoirs in more isolated rural areas cannot be ruled out. A third difficulty is related to Turkey's geographical situation. Unless effective border controls for livestock movements can be set up on the eastern and south eastern borders, it will be difficult to protect domestic livestock populations from imported infections. A fourth challenge has been identified at institutional level. It relates to the efficiency with which the veterinary services are organised, the extent of the political commitment to pursue effective control and eradication, and the

[1] A total of 26 animal diseases are notifiable under Turkish law (Official Journal No 24900, 18/10/2002): African horse sickness, American foulbrood, anthrax, bluetongue, bovine brucellosis, bovine tuberculosis, BSE, caprine and ovine brucellosis, dourine, equine encephalomyelitis, equine infectious anaemia, foot and mouth disease, fowl plague, feline spongiform encephalitis, glanders, infectious haematopoietic necrosis, Newcastle disease, *peste des petits ruminants*, rabies, rinderpest, salmonella gallinarum, salmonella pullorum, scrapie, sheep and goat pox, varroosis, vesicular stomatitis.

availability of resources to do so.

The following sections cover the most infectious and economically significant animal diseases (List A diseases), less infectious diseases that also have serious economic consequences (List B diseases) and diseases that can be passed from animals to humans (zoonoses), followed by a discussion of the legislative framework and the resources available for adequate animal health control.

List A diseases[2]

The World Animal Health Organisation (OIE) classifies 15 animal diseases as List A diseases. Seven of these diseases have never been reported in Turkey. Of the remaining eight diseases, Turkey now has official rinderpest-free status, whilst African horse sickness has not been observed since 1961, and highly pathogenic avian influenza has not been reported for many years. By contrast, foot and mouth disease (FMD), *peste des petits ruminants* (PPR) and sheep and goat pox have occurred in virtually every year since 1996, and the last outbreaks of bluetongue and Newcastle disease were in 2000 and 2001 respectively.

The EU has disease-free status for most of these diseases. Individual countries have invested heavily in recent years in maintaining this status or restoring it after a disease outbreak (Burrell, 2002).

Foot and mouth disease (FMD)

Foot and mouth disease is one of the most infectious and economically damaging animal diseases. It affects all cloven-hoofed animals, and has devastating potential for rapid spread with high morbidity rates.

FMD has been endemic in Asiatic Turkey for many decades. Turkey has had outbreaks of FMD in every year since 1996, usually in both the cattle, and the sheep and goat, populations. The worst recent year for FMD in cattle was 2000, when 3705 cases were reported in 110 herds. In 2003, the figures were 776 cases and 46 herds. Officially, control measures in use include surveillance, internal movement controls, border precautions, monitoring, vaccination and zoning. Stamping out on a limited scale was used only in 1999. In 2003, a record number of bovines (10.7 million) were vaccinated according to an established biannual vaccination programme (Figure 10.1). According to the official statistics, 2000 and 2001 were also years of high FMD infection in sheep and goats, with over 440 cases reported each year. Control measures for FMD in sheep and goats are the same as those for cattle. Vaccination of sheep and goats in the period 2000-2003 appears to have been carried out at a lower level than during 1996-1998.

Until recent years, outbreaks have been due to virus strains O and A. Turkey's first outbreak due to serotype Asia I occurred in 1999, but no outbreak of Asia I has been reported since 2002. Virus strains currently circulating in Turkey can be covered

---

[2] Transmissible diseases that have the potential for very serious and rapid spread, irrespective of national borders, that are of serious socio-economic or public health consequence and that are of major importance in the international trade of animals and animal products.

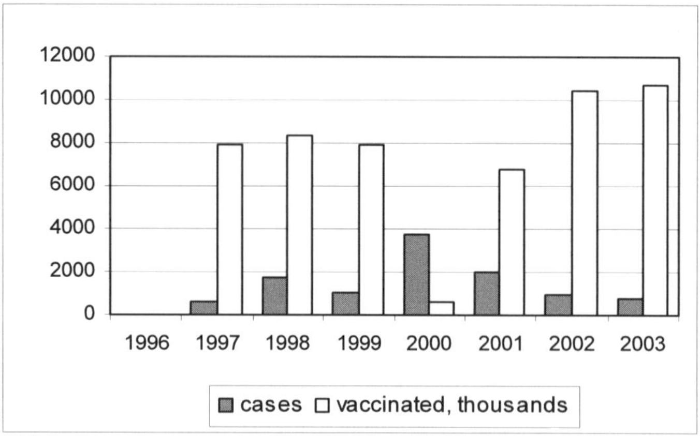

Figure 10.1. Foot and mouth disease, bovines
Source: OIE, 2004

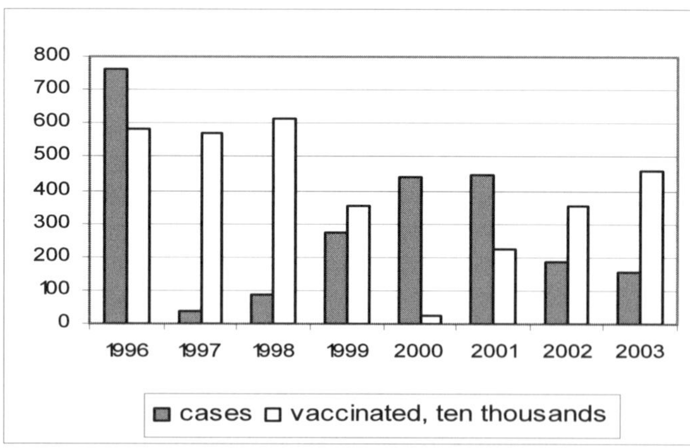

Figure 10.2. Foot and mouth disease, sheep and goats
Source: OIE, 2004

by locally produced vaccines. Active surveillance has been stepped up, particularly in the eastern and south eastern parts of the country. This active surveillance consists of personnel training, encouragement to farmers in these areas to slaughter animals locally rather than transporting them to slaughterhouses in the western (meat-consuming) areas, control of animal movements, investigation of all susceptible cases, and disinfection of vehicles at border inspection points. In 2002, the vaccination programme was stepped up. In Thrace, where there has been no outbreak of FMD since 2001, this involved the vaccination of *all* ruminants with a trivalent vaccine. In other areas, priority has been given to large ruminants on a comprehensive or strategically selective basis. In 2003, random serosurveillance began in Thrace and Anatolia. Regular vaccine testing and improvement is ongoing in collaboration with appropriate research institutes outside Turkey. Until recently, the Şap Institute (FMD Institute) in Ankara was the only government laboratory for vaccine production and diagnosis of FMD in Turkey. New laboratory testing facilities in Bornova, Izmir, are expected to be fully operational in 2004. Contingency plans for an outbreak in south eastern Turkey have been set up (EUFMD, 2003).

Peste des petits ruminants (PPR)

This viral disease was first identified in West Africa in the mid-twentieth century, but has since spread to the Horn of Africa and northwards to the Middle East, adjacent former Central Asian Republics and as far as South Asia. No cases of PPR have been reported in Europe. Goats are the most susceptible species, with sheep somewhat more resistant. Among susceptible populations, mortality is 50-80 per cent, whereas in chronic form it involves a number of symptoms that seriously affect the animal's health and productivity. Infected cattle produce antibodies but show no visible sign of infection.

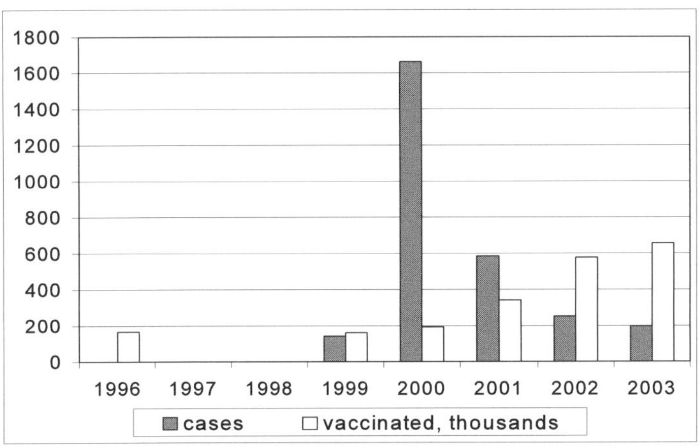

Figure 10.3. *Peste des petits ruminants*, sheep and goats
Source: OIE, 2004

PPR became a notifiable disease in Turkey in 1997 and the first case was reported two years later. The greatest incidence so far was in 2000, with 1662 cases in 43 outbreaks, of which a quarter resulted in death and another 1000 were slaughtered as part of the control measures. Vaccination has increased fourfold since 1999, with over 650 thousand susceptible animals vaccinated in 2003. Vaccine production began in Turkey in 2002. The 2002 World Animal Health report also cites strict control measures such as quarantine, control of animal movements and disinfection.

Sheep and goat pox

These viral infections are spread by direct and indirect contact (e.g. contaminated equipment), and may also be spread by insects. In endemic areas, 70-90 per cent of animals may become infected, with mortality rates of 5-10 per cent. However, mortality rates approach 100 per cent in imported animals with no immunity. Sheep and goat pox are endemic in most of Africa, Asia and the Middle East. No cases have been reported in western or eastern Europe for many years.[3]

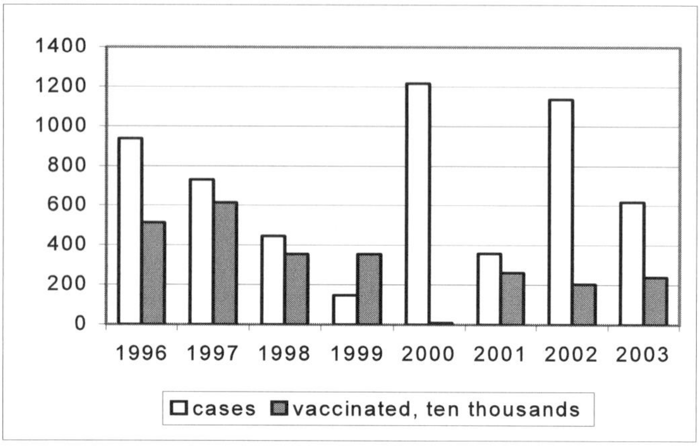

Figure 10.4. Sheep and goat pox
Source: OIE, 2004

High numbers of infected animals were reported in Turkey in 2000 and 2002 (1216 and 1137 respectively). In other years since 1996, the number of cases reported has fluctuated under 1000. Most outbreaks occur in the centre and east of the country. Preventive vaccination (about 2.4 million animals in 2003) is targeted on areas with a high sheep population, and border areas adjacent to countries where the disease occurs. Border controls are also in force. Quarantine and movement controls are applied when an outbreak occurs.

Bluetongue

In the late 1990s bluetongue (a viral disease affecting cattle, small ruminants and deer among other species) appeared in a number of eastern Mediterranean countries (Cyprus 1997, Greece 1998, Tunisia and Bulgaria 1999). It is usually spread by biting midges and possibly other insects, which can travel long distances with wind assistance. Maternal transmission is possible, but otherwise it cannot be passed as a contagious disease between animals. In July 1999, Turkey had its first outbreak for 20 years (FAS, 2002) although no cases were reported by Turkey to the World

[3] The most recent cases were in Bulgaria (1996), Cyprus (1989), France (1964), Czech Republic (1964), Spain (1968), Portugal (1970).

Animal Health Organisation until 2000. In the same year, it moved further west in Mediterranean areas (reaching Algeria, Sicily, Sardinia and Corsica in 2000). In the spring of 2002, it was reported in central Italy and more recently in the Balearic Islands and Portugal. These outbreaks have been due to more than one serotype. It is feared that with global climatic changes, the disease could take hold in southern European countries.

In 1999, Turkey had 15 outbreaks among the sheep population, with 172 cases and resulting in 33 deaths. 732 thousand sheep were vaccinated. Two more outbreaks (involving 22 infected sheep) were reported the following year, and fewer than 6 thousand animals were vaccinated. Since then no further cases have been reported but vaccination has continued on a limited scale in sheep, and more recently in cattle. For the last few years, in parallel with vaccination, Turkey has operated programmes to control the insect vector, as well as surveillance and monitoring, and border and internal movement controls.

Newcastle disease and highly pathogenic avian influenza (HPAI)

According to the 2002 World Animal Health Organisation report, preventive annual vaccination, strict sanitary measures and training of poultry producers have eliminated Newcastle disease from industrial poultry production in Turkey, although not from smallholder village flocks. There was a single outbreak in 2001. In that year, over 27.5 million birds were vaccinated. No further occurrences or mass vaccination have been reported since then. No cases of HPAI have been recorded for many years.

Despite this relatively favourable official record in the poultry sector, a report for the EU Commission (European Commission, 2000) on poultry production in Turkey noted that "doubts remain with regard to the animal health situation in relation to avian influenza and Newcastle disease, in particular because the control of these diseases is on the basis of clinical suspicion only and no monitoring programme is in place" (p.29). Moreover, the officially designated laboratory for these diseases was in fact not able to characterise the viruses. A follow-up report (European Commission, 2003b) found some improvement in monitoring systems and in preparedness for an outbreak.

Rinderpest

The last rinderpest outbreak in Turkey was in 1996, when 24 infected cattle were reported. Since then, preventive measures have consisted of border controls, movement controls within the country, monitoring and surveillance. In 1998, Thrace was declared provisionally free of rinderpest, and vaccination ceased there. Vaccination stopped in the rest of the country in January 1999. A serosurvey carried out in 2001 found no positive antibodies. Since 2002 Turkey has been recognised by the OIE as free of rinderpest, but, along with other countries that have only recently obtained rinderpest-free status, it must confirm each year by letter that the criteria on which this status depends are still met.

List B diseases

The World Animal Health Organisation's List B contains over 90 transmissible diseases that are considered to be of lower socio-economic and/or public health threat than those on List A, largely because of a slower rate of spread, although these diseases may still be significant in the international trade of animals and animal products. This section reports on those with most relevance to Turkey.

Anthrax

Anthrax is a bacterial infection that produces toxins which can be fatal in animals and humans. Anthrax spores can lie dormant in the soil for many years. Because of this, once anthrax becomes endemic in an area, it can be impossible to eradicate but may be well controlled by annual vaccination in high-risk locations.

Anthrax is recurrent in cattle, sheep and goat populations in Turkey. Recent years have also seen a few equine cases. The worst year since the mid-1990s for bovine anthrax was 1999, when 485 cases (in 31 outbreaks) were recorded. According to the statistics, that year was unusual in that a minority of the cases resulted in death. More typically, the infection has been virtually 100 per cent fatal. Vaccination has been performed in most years on about 600 thousand cattle. The incidence of anthrax among sheep and goats has climbed since the late 1990s to 120 cases in 2003, whereas the number of sheep and goats vaccinated has fallen somewhat to around 440 thousand per year in the last few years.

Current control measures for both cattle and sheep/goats consist of monitoring, internal movement controls and vaccination. Vaccination is mainly targeted on susceptible animals in areas where the infection has previously occurred. Animals for export to neighbouring countries are vaccinated before departure.

Bovine tuberculosis

Bovine tuberculosis is an infectious bacterial disease of cattle that involves the development of tubercles in any organ of the body. Apart from cattle, human beings and many other mammals may contract it. Various wild animals can act as natural reservoirs, which makes eradication difficult.

In Turkey, the number of reported bovines cases has been increasing since the late 1990s. In 2002, 64 cases (2 per cent of the animals tested) were reported to be infected. In 2003, 175 cases in 45 herds were reported (the number tested is unknown). Most infected animals are slaughtered.

Brucellosis

Brucellosis is a bacterial infection that can cause abortions, delayed breeding, loss of milk productivity and other clinical symptoms. Most commonly affected are small ruminants, cattle and pigs. Other animals, including humans, may be infected but do not transmit the disease. A national control and eradication programme for bovine brucellosis has been running since 1984. Some measures (serosurvey, vaccination) have also been taken for sheep. Bovine brucellosis was removed from Turkey's list of notifiable diseases in 2003.

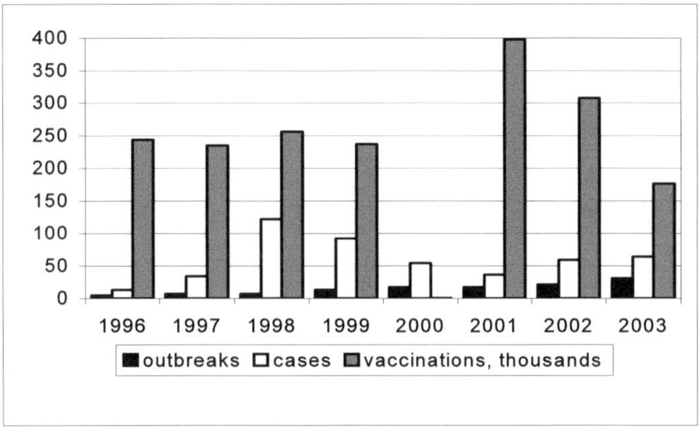

Figure 10.5. Brucellosis
Source: OIE, 2004

There are usually fewer than 100 reported cases each year (1998 was an exception, with 122 cases in 7 outbreaks). However, a local expert has written recently that "...much of the cattle population suffers from brucellosis" (Sarigedik, 2004: 4). Control measures include surveillance, stamping out, internal movement controls and vaccination. Annual vaccinations were between 175 and 400 thousand over the period 2001-2003.[4]

Rabies

Reported cases of rabies among dogs peaked in 2000 (with 200 cases) and fell steeply thereafter, with just 56 cases recorded in 2003. During the same period (2001-2003), well over 200 thousand dogs per year were vaccinated against rabies. These figures for canine vaccination are about 45 per cent higher than those of the late 1990s. A worrying development is the higher level of bovine cases reported in 2002 (122 cases) and 2003 (54 cases). This has been accompanied by a huge increase in anti-rabies vaccination for bovines. By contrast, the number of cases among sheep and goats has remained steady at 20 or less per year. As one might expect, each recorded outbreak among livestock generally consists of just one case and is presumably due to a bite from an infected dog, cat or wild animal. Somewhat paradoxically, the numbers of recorded cases among these latter species have either been falling (dogs) or have remained stable at very low levels (cats and wild animals), whilst infections in grazing livestock have been increasing. This suggests that in many of these latter cases it has been impossible to establish

[4] In EU-15, some countries (France, Germany) have a small number of cases in some years. Spain slaughters some thousands of cattle each year under its brucellosis control programme, but does not report specific numbers of infected animals to the World Animal Health Organisation.

the cause of infection or to test the carrier. If this is so, there may well have been an undetected increase in the wildlife reservoir of carriers.

For all susceptible species (including horses, for which a small number of cases are recorded each year), programmes of monitoring and surveillance, internal movement restrictions and targeted vaccination are implemented. The 2002 World Animal Health report (OIE, 2002) describes vaccination of dogs as the cornerstone of canine rabies control in Turkey. At the same time, it notes that the increasing number of dogs, particularly of free-roaming, abandoned and poorly supervised owned dogs, constitutes a major obstacle to rabies control and indicates the need for new strategies that, at the time of that report, were in preparation.

BSE

So far, no case of BSE has been reported in Turkey either in the domestic herd or in imported cattle. Since 1996, all cattle imported for breeding purposes have been closely monitored. Prior to that date, there were very few imports of live cattle from the UK, but considerable numbers from other sources classified as BSE-risk areas. However, most of these cattle were not older than 24 months old and went straight to slaughter or for fattening.

In addition, no meat and bone meal (MBM) was imported from the UK during the entire period 1980-2000, although there were significant imports of MBM from other BSE-risk countries. The local authorities maintain that these imported MBM supplies, as well as domestically produced meals, were used largely for poultry and pet food.

BSE-prevention measures in place include a ban on imports from countries where BSE has occurred[5], a ban on the inclusion of ruminant material in animal feed, classification of BSE as a notifiable disease (since 1997), the setting up of a BSE Monitoring Committee, introduction of a system for surveillance and monitoring of BSE (in 2001)[6], training of key veterinary personnel in BSE diagnosis, and the addition (in 2001) of scrapie and feline spongiform encephalopathy to the list of notifiable diseases.

Nevertheless, an independent assessment of the BSE risk in Turkey, completed in June 2002, classified the level of Turkey's BSE risk as level III, after taking into account the risk of having imported BSE either in live cattle or infected MBM, together with conditions prevailing internally that could have permitted imported BSE to enter the domestic herd (European Commission, 2002a). Level III risk means that "it is likely but not confirmed that one or several domestic cattle are (clinically or pre-clinically) infected with the BSE agent" (p.16). The internal conditions evaluated included measures to prevent contamination of cattle feed by BSE agents, rendering procedures, and the non-removal of SRM (Specific

---

[5] Since 1996, Turkey has banned virtually all imports of live animals (except for breeding purposes) and red meat, regardless of the BSE risk of the exporting country (see Chapter 8).

[6] Up to June 2003, about 1.5 thousand BSE tests had been carried out (European Commission, 2003a).

Risk Material[7]) from carcasses and BSE surveillance. In particular, it considered that the surveillance system set up in 2001 was not yet adequate for detecting low-level BSE incidence. The report concludes that unless stricter measures are adopted concerning SRM removal and surveillance, the existing level of risk will increase, even in the absence of an 'external challenge' (such as an imported infected animal or material).

Zoonoses in humans

Anthrax

Human infection occurs when spores enter skin lesions, or by inhalation. Infection through eating undercooked contaminated meat is considered rare. In Turkey, recorded cases in humans peaked at nearly 700 in 1997, but were down to about half that level in 2002-2003.[8]

Rabies

Generally, fewer than 5 human cases of rabies are reported per year, the exception being 1999 with 7 reported cases.

Brucellosis

In recent years, human cases of brucellosis in Turkey reported to the World Animal Health Organisation peaked at about 17.6 thousand in 2002. The average levels in the first three years of this decade seem to be significantly higher than in the later part of the 1990s.[9] However, Karabay *et al.* (2004: 13) concluded, on the basis of a survey carried out in 2003 in the Bolu region of Turkey, that the official statistics from the Ministry of Health on human brucellosis seriously under-estimate its prevalence in that region, and that "the recording system in our country is not very efficient". They report that, rather than the 15 thousand or so cases per year reported in recent years, the true figure (including unreported and subclinical cases) is believed to be at least 50-100 thousand (see also Atmaca *et al.*, 2004).

Leishmaniosis

Leishmaniosis is a parasitic infection causing a variety of syndromes due primarily to at least 16 species and subspecies of Leishmania. Wild rodents, wild and domestic carnivores and humans can be reservoirs of leishmaniosis. Dogs are usually affected by only a few species of the parasite, but humans are susceptible to all Leishmania. Humans may experience different levels of infection, ranging from asymptomatic infections to those with high mortality. Three distinct forms are

---

[7] Includes tissues such as the brain, eyes, spinal cord and nerves attached to the spinal cord (of older cattle) and the intestine of cattle of all ages (Regulation (EC) 999/2001, Annex V).

[8] For comparison, human cases in the member states of EU-15 are extremely rare.

[9] For comparison, human cases in the Netherlands, France, Germany and Italy in 2002 were 2, 26, 35 and 101 respectively.

typically described: visceral (VL), cutaneous (CL) and mucocutaneous (MCL). The vectors of these diseases are phlebotomine sandflies. The most common cycle is between dogs and humans, although rodents may also be involved in some areas.

Statistics for the incidence in Turkey are too patchy to allow the identification of recent trends. However, in 2002 and 2003, over 2.5 thousand human cases were reported, which suggests an increase in incidence relative to the late 1990s. Turkey has not reported any cases of Leishmania in animals, so the specific source of these infections is not known[10].

Salmonellosis

This class of bacterial intestinal infections can be present in various species, including poultry, pigs, dairy cattle, sheep and humans. Human infection is usually by faecal contamination of undercooked meat. Turkey has not reported any cases in animals to the World Health Organisation in recent years. However, salmonellosis is a notifiable disease in Turkey.

The number of reported human cases of salmonellosis has fallen steadily from 1997 (33.3 thousand) to 2003 (20.9 thousand)[11]. Detailed control programmes are in operation for poultry: health certificates are issued to hatcheries and breeding stocks that are salmonella free. When infection is found, the entire flock is slaughtered. In 2000, a report for the EU Commission (European Commission, 2000) concluded that the veterinary services in Turkey were "not in a position to give valid guarantees concerning the supervision of poultry meat production and animal health" (p.27). The 2002 Annual Report of the World Animal Health Organisation, however, reported improved levels of technical and sanitary conditions in hatcheries and breeding poultry farms in Turkey, as a result of new regulations brought in at the end of the 1990s (OIE, 2002). In mid-2003, there was a reorganisation of the veterinary services. A follow-up EU mission in September 2003 (European Commission, 2003b) was guardedly optimistic that the new system would lead to substantive improvements, due to better training and the development of specific procedures for supervision of production units.[12]

No information is available about possible salmonellosis control measures for other livestock species.

Current situation, prospects and conclusions

State veterinary services are the responsibility of the Animal Health Section of the

---

[10] In 1995, a group of Turkish and US medical scientists expressed concern about the scope for the spread of disease associated with the GAP project. When new irrigation schemes are introduced into previously dry areas, disease frequently follows due to changing microclimatic and other environmental conditions. An increase in leishmaniasis due to more favourable conditions for the phlebotomine sandflies endemic in that region was one of their specific concerns (Aksoy *et al.*, 1995).

[11] For comparison, human cases in the Netherlands, France, Germany and Italy in 2002 were 1.47, 7.46, 72.38 and 12.57 thousand, respectively.

[12] The same report notes that "animal welfare has improved at the time of slaughter, even though the situation is still not satisfactory" (p.16).

General Directorate of Protection and Control (GDPC) within the Ministry for Agriculture and Rural Affairs (MARA). The GDPC has a very wide remit, which includes preparing plans for combating animal diseases, establishing and enforcing internal and border controls with penalties for transgression, preparing and disseminating guidelines for disease control and biosecurity, overseeing their implementation and monitoring publicly owned slaughterhouses.

The World Animal Health Organisation's 2002 Report states that since Turkey's admission as an official candidate for EU membership in 1999, the veterinary services have been working towards "harmonisation with EU legislation and upgrading enforcement capacity, in particular with regard to laboratory testing and inspection arrangements". Legislation relating to animal disease control is contained in Turkish Law 3285 (1986, and subsequent amendments). This law gives a key role to the veterinary authorities. For notifiable diseases, the scheduled disease has to be immediately notified to the veterinary authorities, which undertake or supervise the necessary measures such as outbreak investigation, taking of specimens for typing at the Şap Institute, ordering movement restrictions to prevent further spread, quarantine, disinfection, compensation (destruction and stamping out), ring vaccination, immediately informing neighbouring districts and provinces etc (EUFMD, 1999).

According to Law 3285 (Article 108), the Animal Health Control Commission has to meet as soon as possible after a FMD outbreak occurs; the measures to be taken are also spelled out. An independent report for the FAO on FMD control in Turkey concluded, after reviewing 40 years of efforts to control FMD that "Despite the national efforts of Turkey and the technical and financial assistance provided by international organisations, attempts made to control FMD in Turkey over many years have met with only limited success" (Garland, 2001). Apart from some success in the geographically isolated region of Thrace, "elsewhere throughout the Anatolian peninsula the disease remains endemic and new types and strains of virus continue to gain access, principally from the east" (p.80). This report considered that the legislative framework for the control of animal diseases was already "by and large adequate" (Garland, 2001). However, it drew attention to the fact that prosecutions for infringing biosecurity regulations are few "and existing penalties do not seem to be a deterrent". Moreover, legislation such as the authorisation to slaughter FMD-infected animals with compensation is hardly applied due to financial constraints. According to this 2000 report, there were no new strategies or recommendations available for FMD control that had not already been made in the past. What was needed was political commitment "...both from Turkey and from the international organisations...in order to achieve the medium-term objective of control and to move towards the longer-term objective of eradication" (p.80).

Since that report was drafted, some positive developments have occurred. The vaccination and surveillance programmes for FMD have been intensified. In 2001, changes to the relevant articles of Law 3285 increased the penalties for illegal cross-border animal movements and smuggling, and also improved control measures for internal animal movements. It is thought that these legislative changes have had a concrete impact. As documented above, vaccination levels for most of the other more threatening diseases are being maintained or increasing. The modernisation of the poultry sector and the upgrading of poultry inspections appear to have brought real improvements in this sector.

Moreover, the Turkish government has set out the steps necessary for harmonising Turkey's legislation and administrative procedures relating to animal health with those of the *acquis* (for the latest action plan, see NPAA, 2003). Under the heading "Alignment to the veterinary *acquis*", the following separate areas for action are identified: adoption of the veterinary framework law and setting up of the necessary administrative structures, border inspection posts, veterinary information system, control and monitoring of animal diseases and contingency plans, animal welfare, veterinary public health and feedingstuffs. Some new legislation had already been enacted in mid-2003 when this document was produced, but the greater part of the legislative changes were scheduled for 2004 or 2005. Some training schemes already began in 2003, and were scheduled to continue into 2005 and beyond.

The total cost of this programme (no part of which extends beyond 2006) is estimated at EUR 44.373 million, of which over EUR 33 million relates to investment in infrastructure (laboratories, veterinary information system and upgrading of border inspection posts), while the remainder relates to adoption of the *acquis* and its implementation (training, etc). About two-thirds of the investment costs and about 10 per cent of the implementation costs are to be borne by the EU, so that the allocation of the total budget for harmonising with the EU's *acquis,* up to the end of 2006, falls more or less equally on the Turkish national budget (EUR 20.8 million) and EU resources (EUR 23.5 million).

Despite this detailed action plan and the greater availability of resources, it seems clear, given the current situation, that effective implementation of the EU *acquis* with respect to animal health and conformity with EU standards, and the effective upgrading of health in Turkey's livestock populations, is a huge challenge and can only be fully met in the longer term. It is reasonable to expect that disease-free status from currently endemic notifiable diseases will have to be achieved region by region within Turkey. If so, as well as strong border controls, it may be necessary to envisage controls between zones within the country that will have to be maintained over the medium term. Much will depend on the political commitment of the Turkish authorities and the rigour with which regulations are implemented at farm, market and border levels. As western European countries have learnt to their enormous cost in recent years, biosecurity in relation to highly infectious animal diseases is only as good as the weakest point in the whole system.

10.2 Plant health

Overview of agricultural and horticultural crops

Over 85 per cent of Turkey's total cultivated land (26-27 million hectares) is used for field crops and fallow. Fruit trees account for about 5 per cent. The remainder is devoted to vegetables, olive trees and viticulture. Vegetable production involves more than 1 million hectares, including 230 thousand hectares of tomatoes and over 100 thousand hectares of melons. Treenuts are grown on nearly half a million hectares, of which more than two-thirds are devoted to hazelnuts. Due to its great variety in geomorphology, topography and climate, Turkey is very rich in indigenous plant species (estimated at approximately 10,000); a number of the world's well known commercial plant species have their origins in this area (Karagöz, 2003b).

In 2001, Öztürk *et al.* reported that 245 insects, 85 diseases and 75 weeds of an economically damaging nature have been reported in cultivated crops in Turkey. Phytosanitary chemicals (1917 registered chemicals, made from 361 active ingredients) were available to producers. About 35 thousand tons of these products were used annually, costing about USD 250 million[13]. According to Öztürk *et al.* (2001), it has been estimated that without the use of any chemical plant protection, 65 per cent of crop output could be lost.

In what follows, information will be given, by category of crop, about various pests and diseases that occur in the arable and horticultural sectors, and the methods adopted by Turkish farmers to prevent or control them. Our literature search has not uncovered any comprehensive studies on this topic for Turkey. Therefore, the following selective overview may not reflect either the relative importance of the diseases covered or the relative priority given to research and extension efforts for improving the situation.

Arable crops

Cereal yields have been stable for some years. Diseases are considered to be among the most important yield-limiting factors in the Turkish wheat and barley crops. The pathogens of most significance are rusts (leaf, stem and/or stripe rust), common bunt and loose smut, as well as root rot (USDA, 1998). Extensive research is being carried out, in collaboration with CIMMYT and ICARDA, on the development of more resistant strains.

Sunn pest is one of the major insect pests of wheat and barley in Turkey. Three of the 14 known species of the pest are prevalent in Turkey. Sunn pest was first reported in 1930, but infestation and damage became economically important only in the late 1980s. Economic losses are due to lower yields *and* lower quality: with more than 3 per cent damage, grain is fit only for use as animal feed, and although with lower rates of damage wheat may be milled, it has to be treated with flour improvers to strengthen its bread-making qualities (Kinaci and Kinaci, 2004).

Chemical sunn pest control is used, which needs to be carefully timed according to the developmental stage of the sunn pest insects but also so as to protect their natural enemies. Until recently, application has usually been by aerial spraying, which

[13] This figure has not been converted to euros, as the precise year of the estimate has not been reported.

is expensive and can have negative environmental effects due to drift. However, in the last few years, there has been a shift to ground control (the Aegean, Thrace and Marmara areas very recently switched completely to ground control) (Gül *et al.*, 2003). In 2003, 56 per cent of Turkey's sprayed area was treated using ground spraying.

The aim of current management policy is to reduce sunn pest infestation below the economic threshold, without disturbing the ecological balance. From the early 1980s onwards, 500 thousand hectares or more were sprayed in most years and this figure increased during the 1990s. Total area sprayed was nearly 1.5 and 1.9 million hectares in 2002 and 2003 respectively. The total cost of the campaign in 2002 was EUR 16.3 billion, of which more than half was labour cost and about one-sixth was chemical cost (Gül *et al.*, 2003).

Clearly, the active cooperation of farmers is necessary to support such a campaign. Gül *et al.* report the results of a survey carried out in the Thrace and Konya (eastern Mediterranean) areas. The farmers surveyed in both regions were of a similar age, and had similar educational levels. Moreover, their ability to recognise sunn pest was also comparable (97 per cent in Thrace, 89 per cent in Konya). However, 82 per cent of the Konya farmers reported that, when they saw sunn pest on their crop, they did nothing (as against 53 per cent in Thrace). Moreover, whereas 38 per cent of the Thrace farmers informed the authorities when sunn pest was observed, this was the reaction of only 6 per cent of farmers in Konya. Both groups of farmers, however, expressed strong willingness to undergo training on how to combat sunn pest. These results show that successful disease control requires not only technical solutions, but farmers who are aware of the action required and convinced that their participation can be effective. For the longer term, in order to reduce the need for chemical use against this pest, researchers stress the need for more research into the potential for biological control and the development of resistant varieties.

Maize is the third most important cereal crop in Turkey, and is mainly grown in the Black Sea, Marmara, Aegean and Mediterranean regions (Kornosor, 1999). Traditionally, maize was grown only as a first crop, but since the mid-1980s, it has been grown as a second and even a third crop in the Aegean and Mediterranean regions, where higher yields can be obtained. As a result, maize pest populations have increased and so too has research into the biology, population development, and the natural predators of the major pests. Harmful and beneficial species have been inventorised in the maize agroecosystem, which has a rich insect fauna in different maize-growing regions. A variety of pests, some specific to certain regions only, are encountered and typically some pests are more dangerous in first-crop maize and others in the second or third crops. *Ostrinia nubilalis Hbn.* and *Sesamia nonogrioides Lef.* are considered to be the most damaging second-crop pests. Given the high occurrence of natural predators of *O. nubilalis* in a few regions, which removes the need for chemical controls in those areas, studies are underway to develop biological control programmes for other regions that exploit these naturally occurring predators (Kornosor, 1999).

For cotton growing, adverse effects of diseases and insects are closely related to climatic conditions and infestation levels. High humidity and temperature create a favourable environment for plant diseases and insects. Therefore, 4 or 5 pest applications per crop at different growth stages are required, particularly in the Çukurova (east Mediterranean) region. For example, one species of thrips (*Thrips tabaci*) is considered a major pest of cotton seedlings in Turkey (Atakan *et al.*, 1996), whilst another thrips species (*Frankliniella intonsa*), which attacks many other flower-bearing plants besides cotton, is a danger to late season cotton in Turkey. *Frankliniella* reached economic damage thresholds in the Mediterranean region in the late 1990s (Atakan and Özgür, undated). The cotton bollworm is an important cotton pest worldwide. By the mid-1990s, this insect had developed resistance to several groups of conventional insecticides such as pyrethroids, organophosphates and carbamates in Turkey and elsewhere (Xiao, 1997). It is therefore not surprising that the world's first serious attempt at organic cotton production began in Turkey in the late 1980s.

Several fungal diseases have in the past led to production damage in rice growing: blast, brown leaf spot, Bakanea and foot rot. Blast is the most significant, which at times has caused crop losses of 25-75 per cent (Surek, 1997). There was a blast epidemic in 1995, due apparently to excessive precipitation and poor training of farmers in preventive techniques (Beşer, 2001; Surek, 1997). This epidemic was followed by another in 1997. According to Surek (1997), the problem could be solved through a policy of replacing local varieties with imported resistant varieties. Although an Italian variety, Rocca, which is resistant to bakane, has been introduced in regions prone to this infection, Beşer indicates that its cultivars are not resistant and so benomyl is still needed to control it. Rice crops are also prone to various insect attacks. These information sources suggests that disease problems have not yet been solved for rice, and that the underlying cause may be the conditions for rice-growing in Turkey, or at least in some of Turkey's producing regions, which are somewhat marginal in terms of temperature changes over the day and night, precipitation levels and so on.

Thanks to Turkey's diverse agroclimatic conditions, potatoes can be grown somewhere in the country most of the year (Rhoades *et al.*, 2000). Farmers in the Mediterranean and Aegean regions may plant several crops per year. Along the Mediterranean coast, winter planting occurs from late November to January and several months later in the Aegean Region, with harvest about four months after planting. Autumn planting takes place in August, with harvest in November or December. At higher elevations (500-2,000 m.), only one crop is planted (in late spring, with harvest in September/October).

Fungal diseases are a serious problem for potatoes in Turkey, especially in the lower warmer plains. The most widespread of these is verticillium wilt, but fusarium dry rot and wilt, early blight and black scurf are also common. Late blight is especially severe in the Black Sea area. Various viruses are commonly found in the plains and sometimes in the highlands. The most serious bacterial disease is common scab, especially in areas of alkaline soil. Major insect pests are potato tuber moth and aphids. The 'imported' Colorado potato beetle is now found in most parts of the country; the economic threshold has been passed for this pest, so chemical treatments are used. Cutworms and wire-worms can also do significant damage.

Pulses, sugarbeet and tobacco are also prone to a number of diseases and pests, for which chemical treatments are undertaken.

Horticultural crops

Protected crops

Turkey's climate permits the cultivation of most temperate and subtropical horticultural crops. Over the last three decades, the areas around large cities and along the coast have seen a shift from lower value field crops to higher value greenhouse horticulture. Since greenhouses are usually not heated because of the high cost, most of the greenhouse and tunnel capacity is located on the Mediterranean coast where climatic conditions are milder. The combination of low temperatures, high relative humidity (especially under plastic houses), intensive cultivation, high plant density and repeated planting of the same crop causes condensation of humidity in greenhouses and favours the development of bacterial and fungal diseases that result in severe crop losses. Thus, Turkish growers are characterised as preferring less heating, but using more chemicals. Viral diseases usually appear in autumn production. Leaf miners, white flies and red spider mites are the most important pests.

Öztürk *et al.* (2001) report that Turkey had 42.6 thousand hectares of protected area (glass or plastic greenhouses, high or low tunnels). About half this capacity is under low (temporary) plastic tunnels, whereas 11 per cent consists of glass greenhouses. Tomatoes are the main protected crop, followed by cucumbers, peppers and eggplants. About three-quarters of cut flower production takes place under plastic.

In greenhouses, soil-borne pathogens may cause severe problems. This situation mainly results from not adopting crop rotation and from the inappropriate use of growing techniques. In particular, intensive use of greenhouse soil for 10-11 months of the year leads to an increase of nematodes. Fumigants are often the method of choice due to their relative low cost and broad spectrum of activity. Horticultural producers have routinely used the ozone layer-depleting substance methyl bromide (MeBr) as a fumigant to control soil-borne pathogens and nematodes due to its broad spectrum, and its quick and deep penetration into the soil. However, the Turkish government has undertaken to phase out MeBr use by 2008. Recently, soil solarization techniques are being adopted as an alternative to MeBr for controlling soil-borne pathogens. In 2000, a World Bank-supported project began trials of MeBr alternatives in the eastern Mediterranean region, and early results look promising.

Tomato production is approximately 8.3 million tons a year in Turkey, most of which are grown as protected crops. Tomatoes have serious disease problems particularly when grown under cover. The fungal diseases of tomato grown under cover in the Çukurova region are foliar diseases such as powdery mildew, downy mildew, grey mould, early blight, leaf mould and *Sclerotinia sclerotiorum* that attacks root, stem and branches. The main soil-borne fungal diseases are damping off and root rots diseases, fusarium wilt, and Sclerotinia rots. An integrated disease management approach is being used to control tomato plant diseases. There are also many pathogens, which are known to cause tomato pith necrosis. For the last five years these pathogens have caused major damage in tomato greenhouses, especially in the Eastern Mediterranean region. Symptoms depend on the pathogen species but generally involve wilting, flocking on the stem, and browning and softening of the pith. The pathogens enter the plant from pruning wounds and spread rapidly in greenhouses. Recently, therapeutic antimicrobial compounds derived from some medicinal plants have been researched as a means of controlling plant pathogens (IOBC, 2000).

Verticillium wilts and rot-knot nematodes are soil-borne diseases that cause serious problems in vegetable, field, tree and ornamental crops. For example, in eggplants these conditions can cause yield losses of 30-60 per cent. There are also harmful insects in greenhouses, such as whitefly, spider mites, aphids and leaf miner. Generally, chemicals are used against the leaf miner, but they are not effective enough and residue problems arise, especially on vegetables. Recently, the leaf miner population in protected vegetable production has been controlled using integrated pest management (IPM) programmes.

In the Eastern Mediterranean area, the carmine spider mite has become one of the main pests in both greenhouses and open fields. Due to its ability to develop resistance to a wide range of pesticides, biological control programmes have been initiated in greenhouses. So far, however, very few bio-control agents have been registered in Turkey. These agents have a narrow spectrum activity, so they are used to control specific pests.

In the relatively new but fast growing areas of ornamentals, and aromatic and medicinal plants, research on plant health and protection, as well as on pest and disease resistance, is in its infancy, and little information is available (Kuden, 1998).

Tree crops

This subsection gives a non-exhaustive description of some of the plant diseases that are present in this economically very important production sector. Over 75 virus and virus-like diseases have been identified in citrus worldwide, but only 15 of them have been reported in Turkey (Yilmaz, 1999). Detection methods involve observation of symptoms in the field and, since some viruses do not cause visible symptoms, other laboratory-based diagnostic tools such as biological indexing and serology are used. Yilmaz suggests that if more advanced detection techniques and a detailed survey were carried out in all citrus growing areas, other virus diseases might be detected. The main concern is the introduction of graft-transmissible viruses in propagation materials from abroad, which has occurred in the past. Among other things, Yilmaz recommends stronger quarantine measures and cooperation with other Mediterranean countries on quarantine regulation, legislation for mandatory certification of citrus, and the setting up of an advanced laboratory for early disease detection.

The most damaging diseases for satsuma mandarins (a major export crop for the Aegean region) are green and blue mould caused by *Penicillium digitatum, P.italicum* and sour rot. Postharvest *Penicillium* decays induce substantial crop losses during the storage, transport and even the marketing of the satsuma mandarin (Kinay *et al.*, 2002). A dangerous pest for the citrus crop, the citrus leaf miner, was first noted in mid-1994 and just one month later acquired the status of an economic pest warranting immediate pesticide applications. It is now found in every citrus orchard except those on the Black Sea coast and all varieties are susceptible. Economic losses due to the citrus leaf miner are not known in monetary terms, but the fact that a large-scale control campaign is being carried out signals its status as an economic pest. Pesticides are recommended only for young trees and nursery seedlings. Recently, great attention has also been given to developing effective biological control of citrus leaf miner (Kuden, 1998).

Plum pox potyvirus (PPV), the causal agent of Sharka disease, is considered the most devastating viral disease of stone fruit crops because of damage to fruit quality, premature dropping of fruit, its wide host range in cultivated and wild Prunus species, its rapid natural spread, and the rapid decline and death of infected trees when jointly infected with other viruses. Almond, apricot, nectarine, peach, plum and cherry are all hosts. PPV was long known to be present in Marmara and Central Anatolia, and surveys have continued in these regions over the last 20 years. Recent surveys have found that the eastern Mediterranean region of Turkey is free of PPV. In particular, the Malatya province, which is the most important apricot production area, has been surveyed and no PPV infection on the apricot trees was found (Sertkaya *et al.*, 2003). Biological control of capnodis in stone fruit has also been heavily researched (Kuden, 1998).

Nearly 150 insect and mite species have been identified on hazelnuts, although few of them cause significant economic losses (Tuncer and Ecevit, 1997). The European chestnut is a native species of Turkey that grows in coastal regions. Recent statistics refer to the presence of about 2.5 million chestnut trees, with an annual production of 61 thousand tons of chestnuts. Chestnut blight was first reported in 1967 and is present now in all growing areas, threatening the economic cultivation of chestnuts (Celiker and Onogur, 2002).

Current situation, prospects and conclusions

New agricultural quarantine regulations (2002/61 and 2002/62) were put into force for the management and control of potato wart disease, potato cyst eelworm, potato ring rot and potato brown rot. These regulations are in compliance with EU legislation. Also, the potato testing capacity of the Ankara, Izmir and Adana Plant Protection Institutes and Nigde Potato Research Institute laboratory has been upgraded in line with EU standards. As of 2002, plant quarantine offices are linked to a database where information and statistics are recorded. Turkish legislation for the establishment of a national system of plant variety protection has been completed and published in the Official Gazette 15.01.2004/25347; over 60 new plant varieties were registered in 2004.

In June 2002, a new regulation on agricultural quarantine was adopted. The regulation sets out rules and procedures on importation and transit of plants, plant products and other substances in order to protect Turkey from the contamination of plants and plant products by harmful organisms. The importation of all kinds of soil, weed, natural fertiliser, leaves, stems and hay is prohibited, other than the products listed in a 'Special Requirements' Annex. Plant and plant products can only be imported through specified ports of entry. Plants and plant products to be imported in Turkey have to be issued with a 'Plant Health Certificate' granted by the relevant body of the exporting country. Applications for import permits are dealt with by MARA (FAOLEX, 2004).

A further regulation (11 July 2004) provides for the setting up and training of a quarantine inspectorate with responsibility for preventing and controlling pests and diseases due to importation of plants, plant products, industrial and forestry products, for compliance with the quarantine legislation of buyer countries on exportation, and for conducting necessary supervision and monitoring during importation, exportation, re-exportation and transit movements. Their main duties are issuing Plant Health Certificates for export products in accordance with the respective legislation of the importing country, advising respective customs offices whether the imported products conform with Turkish quarantine legislation, and checking packages and transport vehicles for transit movements (FAOLEX, 2004).

Some time will be needed before the upgrading of the plant quarantine system makes its impact, but it is clear that it is needed urgently as new diseases are continuing to manifest themselves. Among the diseases reported for the first time in 2002-03, in Turkey as a whole or in hitherto unaffected regions of Turkey, are bacterial canker in apricots, a type of bacterial rot in lettuce, bacterial spot in pepper in the eastern Mediterranean region, crown gall disease on rose cultivars and apricot, and bacterial leaf necrosis on melon. In each case, the disease is thought to have been introduced by contaminated seeds or transplants, imported from abroad or from another region of the country (Plant Pathology, 2003).

In the Turkish government's National Plan for Adopting the Acquis (NPAA), detailed steps for the adoption of EU phytosanitary legislation and standards are set out (see NPAA, 2003) under the following headings: legislation on harmful organisms including plant passport practices and border inspection posts, plant protection products (pesticides), seed and seedling quality, and a schedule of necessary institutional change (involving the Ministry of Agricultural and Rural Affairs and the Ministry of Environment and Forestry). The timetable is spread over the period 2004-2006; the projected budget cost is EUR 19.7 million (most of which goes to MARA), of which the national budget bears about EUR 3.3 million and the EU contributes the remainder. In 2003, European Commission (2003a) reported that 'transposition and implementation' of the phytosanitary *acquis* was in general at an early stage. The report singled out several recent developments: four potato communiqués in line with the *acquis* on the control of ring rot disease, cyst nematodes, wart disease and brown rot disease, which came into force in September 2002, and an action plan to fight aflatoxin in hazelnuts, pistachio and dried figs, in which about 90 food inspectors would be trained on mycotoxins in foodstuffs, sampling methods for some of these toxins and on basic principles of HACCP (Hazard Analysis and Critical Control Points) legislation and inspection. The Commission's 2004 Regular Report (European Commission, 2004: 98) suggests that this initiative is still not operational, and that aflatoxin contamination of hazelnuts and dried figs is still "excessive", requiring strict monitoring.

An EU-funded 'twinning' project on phytosanitary issues kicked off in 2004, with a budget of just over EUR 1 million, with the objective of institution-building through long-term partnerships between administrations. This programme, which follows similar projects with the 12 recent accession candidate countries, involves the long- or short-term secondment of experts to the Turkish administration, in order to develop the capacity for fulfilling the phytosanitary *acquis* (Representation of the European Commission to Turkey, 2004).

On the international level, Turkey is a party to the International Plant Protection Convention (IPPC), the international standard-setting body for plant protection, which is recognised under the WTO-SPS agreement as providing the framework within which countries take action and cooperate to control pests and prevent their spread across national boundaries. Turkey is a member of the European and Mediterranean Plant Protection Organisation (EPPO), an intergovernmental plant protection organisation for that region. Furthermore, Turkey is also well integrated into international plant research networks via FAO, ICARDA, CIMMYT and other more crop-specific international bodies.

10.3 Conclusions

Two main conclusions emerge from this overview. First, much institutional change (new legislation and procedures, expertise formation, training and improved awareness of all participants in the growing and marketing chains) is still needed before Turkey can effectively implement the veterinary and phytosanitary *acquis* of the EU. Second, and this holds particularly for the animal sector, some further time will be needed *after* the measures required by the *acquis* are in place, before the health of animals and plant populations, and the risks associated with the most important diseases, converge to the levels reached by EU-15, or indeed EU-25.

It is not realistic to imagine that legislative and administrative changes alone can bring rapid improvements. Turkish agriculture is hampered by its small-scale and fragmented structure, the low educational level of the farm population, and the fact that for many products, a sizeable proportion of production does not enter conventional supply chains, being either sold informally or consumed by the producing household. This makes monitoring and control more difficult, and arguably reduces motivation for some producers to incur costs to achieve higher standards. And yet, the public-good aspect of plant and animal health is important: as long as reservoirs of disease and pests persist, risks are higher for all producers. Getting the right incentive system (including penalties) for encouraging appropriate biosecurity at every link in the chain has proven to be a difficult task in any country. In Turkey, it is likely that, with respect to both larger-scale commercial production as well as small-scale semi-subsistence producers, governance issues regarding the veterinary *acquis* and biosecurity will not be negligible.

There are, however, important differences between the plant and animal sectors. First, a substantial share of the plant sector is internationally competitive, and its output is actively traded. The phytosanitary standards reached by these products are clearly conformable with those required by international trade. Moreover, any plant disease outbreaks or pest infestations that might occur in the exporting sector will usually result in nothing more serious than local production losses, particular consignments failing to get a Plant Health Certificate from the Turkish quarantine authorities and/or particular shipments of goods being refused entry at export destinations. By contrast, given the highly infectious nature of List A and some List B diseases (a number of which remain endemic in Turkey), the potential trade consequences of an outbreak of one of these diseases are catastrophic, involving a ban on all imports from the source country or region until the disease has been eradicated according to the protocols agreed within the OIE, and international confidence has been restored. Currently, Turkey's ruminant animal sector is virtually closed to trade and it is predicted that animal production would fall if all trade barriers for agricultural products disappear between Turkey and the EU (see Chapter 8). Therefore, the likelihood of Turkey becoming a significant exporter of animal products is small. However, even with Turkey trading as a net importer, it seems that stringent measures for separating the Turkish and EU-25 markets would be needed for some years in order to maintain the confidence of trading partners in the health status of meat and dairy products coming from current EU member states. In the poultry sector, which is largely modernised and *does* export, there is already a good grip on veterinary standards, which are improving towards EU levels.

In the plant sector, there appears to be considerable and diversified research activity, focussing on a number of dimensions including optimal production techniques, varietal breeding for local conditions, biological control regimes and so on. Of course, with so many commercially relevant plant varieties and disease types, there is much work to be done. However, Turkish researchers appear to be well integrated into the international research network, participating actively in international efforts on the plant disease and pest front. As far as animal health is concerned, on the other hand, the most urgent challenge facing Turkey is the eradication of diseases such as FMD, PPR and brucellosis[14]. The science and technology for doing this are already largely known and available. The impediments arise from structural features, institutional inadequacies, geographical characteristics, poor incentives, insufficient trained personnel and low awareness levels. A successful eradication campaign for FMD, for example, will require not only the resources for large-scale vaccination and testing programmes, but also procedures that can work within the existing structure of the livestock sector *and* organisational and enforcement strategies that are adapted to these local conditions.

References

Aksoy, S., Ariturk, S., Armstrong, M., Chang, K., Dörtbudak, Z., Gottlieb, M., Ozcel, M., Richards, F. and Western, K. (1995) The GAP project in Southeastern Turkey: the potential for emergence of diseases, *Emerging Infectious Diseases* 1(2).

Atakan, E. and Özgür, A.F. (undated) Preliminary investigation on damage by Frankliniella intonsa to cotton in the Cukurova region of Turkey. *Proceedings of the 7th International Symposium on Thysanoptera*, pp. 221-224.

Atakan, E., Coll, M. and Rosen, D. (1996) Within-plant distribution of thrips and their predators; effects of cotton variety and developmental stage. *Bulletin of Entomological Research* 86, 641-646.

Atmaca, S., Özokyncy, T., Akpolat, N., Elçy, S., Suay, A. and Arikan, E. (2004) Brucellosis seroprevalence in Southeast Turkey. *Turkish Journal of Medical Sciences* 2004 (4).

Beşer, N. (2001) The new development in rice agronomy and its effects on paddy yield and rice quality in Turkey during the last decade. In: Chataigner, J. (ed.) *The New Development in Rice Agronomy and its Effects on Yield and Quality in Mediterranean Areas.* Proceedings of a Workshop on the New Development in Rice Agronomy and its Effects on Yield and Quality in Mediterranean Areas, 13-15 September 2000, Edirne (Turkey). CIHEAM-IAMM, Montpellier.

Burrell, A. (2002) Animal disease epidemics: Implications for production, policy and trade. *Outlook on Agriculture* 31, 151-160.

Celiker, N.M. and Onogur, E. (2002) Studies on the control of chestnut blight by hypovirulent strains in Turkey. *IOBC/wprs Bulletin* 25(10), 203-206.

EUFMD (1999) Report of the 33rd session of the European Commission for the Control of Foot and Mouth Disease, Rome, 7-9 April 1999. FAO Report.

EUFMD (2003) Report of the 35th session of the European Commission for the Control of Foot and Mouth Disease, Rome, 9-11April 2003. FAO Report AGA:EUFMD/3.

European Commission (2000) Final report of a mission carried out in Turkey from 20 to 24 November 2000 concerning food production: Poultry meat. DG(SANCO)/1203/2000-NR Final.

European Commission (2002a) Final report on the assessment of the geographical BSE risk of Turkey. Report to the SSC of the European Union. 27 June 2002.

European Commission (2003a) 2003 regular report on Turkey's progress towards accession. European Commission, Brussels.

European Commission (2003b) Final report of a follow-up mission carried out in Turkey from 16 to 26 September 2003 in order to assess the conditions of production of fresh poultry meat intended for export to the EU. DG(SANCO)/9130/2003 – MR Final.

European Commission (2004) 2004 regular report on Turkey's progress towards

[14] Our literature search has failed to find information on the situation in Turkey regarding less threatening but still economically important diseases such as mastitis or foot rot.

accession. COM(2004) 656 final. European Commission, Brussels.

FAO (undated). Meat inspection and hygiene, Project TCP/TUR/7821, FAO Turkey Office, available at http://www.un.org.tr/fao/Animals-Livestock.htm, (accessed 30 September 2004).

FAOLEX (2004) FAOLEX Data Base. FAO, Rome.

FAS (2002) Bluetongue: Mediterranean emergence 1997-2002. USDA Foreign Agricultural Service, Washington.

Garland, A. (2001) A review of the foot and mouth disease situation in Turkey during the last decade, including a critical assessment of past national and international control programmes, and with recommendations for future control. Report for the European Commission on Foot and Mouth disease control, Report of the 34th Session of EUFMD. FAO,Rome.

Gül, A., Aw-Hassan, A., Kütük, H. and Canhilal, R. (2003) Sunn pest management policy in Turkey: A sociological analysis of shifting from aerial spraying to ground spraying. Integrated Sunn Pest Management in West Africa, Report of Turkey. ICARDA, Aleppo, Syria, September 2003.

Günes, T. (1998) Red meat marketing systems in Turkey. In: Belhadj, T., Boutonnet, J.P. and Di Giulio, A. (eds.) *Red Meat Industry in the Mediterranean Countries.* CIHEAM-IAMZ, Zaragoza pp. 179-186.

IOBC (International Organisation for Biological Control) (2000) Report from a workshop of WPRS/IOBC Working Group on 'Protected crops, Mediterranean climate', Workshop held in Turkey, April 2000.

Karabay, O., Serin, E., Tamer, A., Gökdogan, F., Alpteker, H., Ozcan, A. and Gündüz, H. (2004) Hepatitis B carriage and Brucella seroprevalence in urban and rural areas of Bolu province of Turkey: A prospective epidemiologic study. *Turkish Journal of Gastroenterology* 15, 11-13.

Karagöz, A. (2003a) Turkey. *Country Pasture/Forage Resource Profiles Database.* FAO Crop and Grassland Service, FAO, Rome.

Karagöz, A. (2003b) Plant genetic resources conservation in Turkey. *Acta Horticulturae* (ISHS) 598, 17-25.

Kinaci, E. and Kinaci. G. (2004) Quality and yield losses due to sunn pest in different wheat types in Turkey. *Field Crops Research* 89, 187-195.

Kinay, P., Yildiz, M. and Droby, S. (2002) Biocontrol of Penicillium decays with epiphytic yeasts on satsuma mandarins. *IOBC/wprs Bulletin* 25(10), 207-210.

Kornosor, S. (1999) Entomological problems of maize in Turkey. Paper presented at the *20th IWGO Meeting*, Adana, Turkey.

Kuden, A. (1998) Horticultural research in Turkey. *Proceedings of a World Conference on Horticultural Research*, Rome, 17-20 June. International Society for Horticultural Science.

NPAA (2003) Turkish national programme for the adoption of the acquis. Chapter IV: Ability to assume the obligations of membership. Section 7: Agriculture. *Turkish Official Gazette,* No 25178, 24/7/2003, pp. 194-311.

OIE (2002) *World Animal Health*, 2002. Country report on Turkey. OIE, Paris, 329-334.

OIE (2004) Handistatus II database. http://www.oie.int/hs2/ (accessed November 2004).

Öztürk, A., Deviren, A. and Özçelik, A. (2001). Methyl bromide in Turkey. http://www.un.org.tr/unido/documents/METHYL%20BROMIDE%20IN%20TURKEY1.doc (accessed 26 September 2004).

Plant Pathology (2003), volume 52, various issues.

Representation of the European Commission to Turkey (2004) EU-funded twinning project on phytosanitary issues to start in Turkey, Press Release 4 May 2004.

Rhoades, R.E., Hijmans, R.J. and Huaccho, L. (2000) Turkey. In: *World Potato Atlas*, International Potato Centre, Lima, Peru.

Sarigedik Ü. (2003) Turkey: Livestock and products annual. *GAIN Report*, Number

TU3019. USDA Foreign Agricultural Service, Ankara.

Sarigedik, Ü. (2004) Turkey: Livestock and products annual report. *GAIN Report*, Number TU4022. USDA Foreign Agricultural Service, Ankara.

Sertkaya, G., Ulubaş, C. and Caglayan, K. (2003) Detection and characterisation of Plum Pox Potyvirus (PPV) by DAS-ELISA and RT-PCR/RFLP Analysis in Turkey. *Turkish Journal of Agriculture and Forestry* 27, 213-220.

Surek (1997) Rice diseases in Turkey. In Chataigner J., (ed.) *Maladies du riz en région méditerranéenne et les possibilités d'amélioration de sa résistance.* CIHEAM-IAMM, Montpellier, pp. 45-47.

Tuncer, C. and Ecevit, O. (1997) Current status of hazelnut pests in Turkey, in *ISHS Acta Horticulturae* 445, Proceedings of the 4th International Symposium on Hazelnuts.

USDA (1998) Items from Turkey. *Annual Wheat Newsletter* 44. USDA, Washington.

Xiao Bin (1997) The preliminary biochemical analyses of resistance of the cotton bollworm in China. *Resistant Crops Newsletter* 6 (2).

Yilmaz, M.A. (1999) Virus and virus-like diseases of citrus in Turkey. *Options Méditerranéennes, Série* B/n° 21, - *Proceedings of the Mediterranean Network on Certification of Citrus*, pp. 85-93. CIHEAM.

11

Consequences for the EU-27 of Enlargement to Turkey

Arie Oskam, Natasha Longworth and Irene Molina Vílchez[1]

11.1 Introduction

The decision to start accession negotiations with Turkey was taken in December 2004 by the Council of the EU-25. If Turkey accedes to the EU in 2015, the decision will be taken by the European Union of 27 member states, and ratified either by their parliaments or by means of a referendum if a country wishes to use such a procedure. It is therefore very relevant to ask what the consequences of EU enlargement to Turkey would be for those countries that are already EU members, or that will be members by 2015[2].

Various geopolitical and strategic considerations, as well as general economic criteria, will have considerable weight in this decision. Here, we concentrate mainly on economic issues arising in relation to the agriculture and food sectors, rural development and structural policy. We do not explore in detail the pros and cons of Turkish accession for sectors other than agriculture and food. Of course, a view has to be taken on the macroeconomic context at the time of accession before we can identify the consequences for agriculture, food, rural development and structural policy.

This chapter considers the main consequences of Turkey's accession for EU-27 regarding agriculture and agricultural policy, the food industry, the likely budget flows and other less tangible, related consequences. It draws on information presented in earlier chapters. The discussion in this chapter and the next requires assumptions about future EU policies, and about exogenous developments in Turkey and the EU. Assumptions regarding exogenous policy trends are given in Section 11.2, whilst specific agricultural policy changes likely to be made in preparation for Turkish accession are discussed in Section 11.3. Sections 11.4 and 11.5 consider consequences relating to the Single Market and to EU agribusiness. Specific quantitative assumptions needed in order to calculate budget payments are set out in Section 11.6, and the decision-making background to these flows,

[1] With contributions from Siemen van Berkum and Alison Burrell.

[2] It is assumed that by 2015 the EU will consist of EU-25 plus Bulgaria and Romania. Croatia is also likely to be an EU member already when Turkey enters, but is not included in our analysis for lack of information.

including structural fund spending, is explained in Section 11.7. Calculations of budget flows are contained in Sections 11.8 and 11.9. These flows are summarised in Section 11.10, and Section 11.11 contains discussion and conclusions.

11.2 Assumptions regarding policy trends

In general, we assume that EU agricultural policy will become increasingly market oriented, following the trend set by the 1992 reform towards lower support prices and decoupled income transfers. This strategy has progressively brought the prices of the main products more in line with world market prices (including the 'southern products' rice, cotton, olive oil and tobacco), although there are still substantial import tariffs. The sugar policy of the EU is assumed to adjust along the lines proposed by the European Commission in July 2004.

It is assumed that the budget costs of Turkey's direct income support will shift to the EU, but on condition that Turkish farmers observe the EU's cross compliance rules associated with decoupled direct income support.

Three developments will influence future EU policies for agriculture, food and rural development. First, the restrictions on the EU budget will impose tighter rules on an enlarging Union. This holds in particular for the Brussels agreement, which limited spending on the first pillar of the CAP to 1 per cent growth in nominal terms over the period 2007-2013 of the Financial Perspective (European Council, 2002: 5). Because there are only limited 'reserves' within the first pillar (for example, release of amounts used for export subsidies), a next step towards funding enlargement would be to reduce either direct income support and/or the budget for rural policy. As a likely response to this budgetary pressure, we assume an additional reduction in nominal terms in these direct income transfers by an average of 2 per cent per year, starting in 2006.

Second, there will be more pressure on the EU market for agricultural products, due to preferential market access arrangements with ACP countries, least developed countries under the EBA agreement, South Africa, the Balkans, Mercosur and so on. This will increase the need for further policy reforms.

Third, a new round of policy revisions will be needed, probably before 2010, to incorporate the Doha Round agreement and to achieve fully decoupled direct income payments. During this period, it is likely that further changes will be made to the CAP to accommodate the entry of AC-2 countries and Turkey.

WTO and trade policy

Full incorporation of the WTO Doha Round agreement (elimination of export subsidies and further tariff cuts) is assumed. Specifically, we assume that import tariff bindings will be 50 per cent lower in 2015 compared to the levels in 2000. For most products, the EU can accommodate these reductions, although for sugar and milk it seems unlikely, even after the proposed policy changes for sugar. Much will depend on exchange rates.

Food policy

The EU's General Food Law (Regulation EC/178/2002) means that, for food safety, policies in all member countries aim at a common standard. By contrast, demand for food quality depends on per capita incomes and, above minimum guaranteed levels, is largely left to market forces. Poorer member states may opt for lower average levels of food quality.

Environmental policy

Taking into account subsidiarity, the tendency to reduce the central administrative burden and the slowdown in economic growth compared to the last decade, we assume that EU environmental pressures on member states will not increase in the period 2005-2015.

11.3 Consequences of Turkey's accession for EU agriculture and agricultural policy

Long-term projections of the impact of harmonised agricultural trade with Turkey for a range of products show only relatively modest effects for EU-25 agriculture, arising from small changes in imports or exports. Turkey is internationally competitive in the fruit and vegetable sector, and has a significant world market share for some individual products. For these products, Turkey is already present to a significant degree on EU markets. Overall, either because of relatively low production efficiency or lack of competitiveness on quality, the effect of Turkey's entry on EU-27 agriculture is expected to be small.

However, Turkey's accession may have consequences for EU commodity regimes, at least for some products that are important for Turkey. These are the typical 'southern' products such as fruit, vegetables, tobacco, cotton, and also sugarbeet. It is likely that, for at least a few of these products, CAP support regimes would be adjusted prior to Turkey's entry with the specific aim of limiting budget expenditure in these areas.

Tobacco

Turkey's entry would increase EU production of tobacco by more than 30 per cent. Within the EU, tobacco consumption is declining. Until now, most EU-grown tobacco has gone for domestic consumption and that would not change fundamentally after Turkey's entry. Giving Turkish farmers the same level of direct payments recently granted to Greek farmers would lead to large budget costs. Therefore, further cuts in direct income support for tobacco can be expected.

Cotton

Turkey is a relatively efficient cotton producer whose output is far greater than that of EU-27. Extending the same premia and direct income support to Turkish cotton producers as are presently applied for Greek producers would significantly increase pressure on the budget. Given the external pressure on developed countries to remove protection for cotton, it is realistic to assume that the EU cotton regime will be further adjusted before 2015.

Olives and olive oil

Turkey is a relatively small producer of olives and olive oil compared to Spain, Greece and Italy. Turkey is a net exporter on an international market where the EU plays an important role. We assume, however, the policy changes already agreed for these commodities will not be changed in the light of Turkey's entry and that full decoupling of direct income support in line with the recent decisions on the 'southern' products will apply.

Sugar

The proposed reform of the EU sugar regime would lead to better market orientation and a phasing out of the quota system. If this proposal is adopted, Turkey would have to adjust its recently established quota system in the same direction as the EU. Turkey's accession would strengthen the case for this reform.

Beef

In this market, there will be considerable opportunities for EU-27 producers to increase their exports to Turkey when existing trade barriers are removed. By accession or even before, a price decrease in Turkey and good export opportunities for EU-27 beef in Turkey are expected.

Dairy

Although Turkey is not a high-cost producer of milk, the dairy chain appears to be quite inefficient (see Chapter 4). This implies either opportunities for export from EU-27 or for FDI by EU-27 companies after Turkish accession.

11.4 Sensitive single market issues

Animal disease status

Turkey's livestock and animal product sector is characterised by an under-developed supply chain for livestock products and high prices, due to low productivity in the domestic livestock sector, high tariffs and (for red meats) a long-standing and contested import ban.

Turkey's accession to the EU would, on the one hand, increase access for EU exports of animal products to Turkey's domestic market. It is likely that, as incomes grow in Turkey and animal product prices fall, this will be a fast-growing segment of the Turkish food market. On the other hand, Turkey's current status with regard to foot and mouth disease, *peste des petits ruminants* and sheep and goat pox means that free flows of meat and dairy products in the opposite direction could threaten the disease-free status that EU member countries have built up over many years. Long-standing disease-free status means that domestic livestock populations have no natural immunity, and once an outbreak occurs, epidemics can be extremely destructive amongst vulnerable animals. By contrast, in countries where these diseases are endemic, subclinical cases can go unnoticed and outbreaks in small herds may not be reported.

The assumption that Turkey's demand for animal products will expand as incomes grow and prices fall means that Turkey is likely to be a deficit area for meat and dairy products, and potential trade flows in these products from Turkey to the EU are likely to be small. However, disease-free status represents both a production asset (higher-yielding herds) and a trading asset (open borders and confidence of trading partners). Thus, there are two possible consequences to be avoided by the EU: the actual import of infected animals or products, and the lowering of confidence in EU exports or an increasing perception of disease risk on the part of trading countries that could affect their willingness to import EU animal products.

New external borders

Turkey has 2627 km of land borders and 8333 km of coastal borders. Only 475 km of the land borders would be internal borders after Turkish accession (206 km with Greece and 269 km with Bulgaria). The remaining 2152 km will be external borders. Turkey also has several international airports and seaports.

Managing Turkish external borders will be a major challenge for the European Union, particularly given the importance of the country as a transit area and the topography of the eastern borders. In order to secure the internal market, future external borders will have to be impermeable to smuggling and to goods that do not meet EU standards. In addition, borders need to be equipped with veterinary and plant border inspection points in compliance with EU legislation. There is currently a high risk of penetration of animal diseases due to the uncontrolled entry of animals across the eastern and south-eastern borders, and to the weak internal control on movements of live animals and animal products.

During the negotiations for the 2004 enlargement, no concessions were made to the NMS on border issues[3] and it is very unlikely that an exception would be made for Turkey. Only the posts complying fully with EU minimum standards upon accession will be approved. If there are insufficient guarantees of the safety of animal and plant products, Turkey risks the use of safeguard measures. Projects are already being carried out in order to improve the capacity of the Turkish administration to carry out border controls and internal monitoring for disease.

An implicit requirement for accession is that applicant countries should solve any pending territorial disputes with other EU members or neighbouring countries (Tocci, 2001). At the moment, most of Turkey's eastern borders are formally closed. It is difficult to foresee how relationships with neighbouring

---

[3] A special regime was agreed however, for the temporary border between Hungary and Romania. (European Commission, 2003).

countries will evolve up to 2015, but the prospect of accession may act as a catalyst for their improvement. Nonetheless, it would be optimistic to assume that in the next 10 years all political conflicts will be solved and Turkey's eastern borders can be controlled according to EU standards.

By 2015, the new European Neighbouring Policy will have been in place for over 8 years. A special provision has been made to include Turkey (even before the start of negotiations) in maritime programmes. At the moment the policy covers all the countries bordering Turkey by land except Iran and Iraq[4] and all non-EU Mediterranean countries. In the Black Sea region, the policy covers Ukraine and Georgia but not Russia. Cooperation with neighbouring countries will be relevant for both food safety and environmental issues. Water will continue to have an important strategic role in the relations with Syria, Iran and Iraq[5].

11.5 Consequences for the EU food industry

The growth in the EU internal market by about 82 million consumers in 2015 could increase the scale of operation of the food industry in the Union, and stimulate more trade and investment flows. Turkey's geographical position is more remote from the high-income consumption centres of the EU than that of the NMS. This is important to those industries where transport costs are significant. On the other hand, distance may become an important factor in companies' decisions on where to locate processing operations. The impact of Turkish accession on trade and FDI flows between the EU-27 and Turkey depends on many country- and company-specific factors (van Berkum, 2002) and is therefore difficult to predict. In this section, we discuss a number of general trends that affect trade and FDI opportunities for the international food industry in Turkey.

Upon accession or perhaps before, tariffs on agricultural products entering Turkey from EU-27 will be abolished. The EU food industry will thus compete on equal terms with the Turkish food industry in Turkish markets. The future competitiveness of food companies in both the EU-27 and Turkey will depend on their response to two important trends.[6] The first is the switch from supply to demand orientation, whereby the focus moves from production to consumer demand. This implies a shift from quantity to quality, both in terms of product and production process. The second trend is from tariffs to standards, which means that trade and competition will depend less on tariffs and prices, but increasingly on standards. These standards refer especially to food safety, environment, animal welfare and labour. An OECD survey of retailers indicates that private standards are becoming more important and are already more demanding than public standards. Both trends indicate that in order to be

---

[4] That is, Armenia, Azerbaijan, Georgia and Syria.

[5] An example of the strategic importance of water in the region is the current cooperation with Israel, which imports Turkish water by tanker.

[6] Following and interpreting Tangermann (2004).

successful in a market where quality, product differentiation and complying with private standards attract a premium, the emphasis is increasingly on vertical coordination in the food chain. Being part of a vertically coordinated chain becomes a major competitive asset.

With respect to these two trends, the food industry based in Western European countries has a large lead on its competitors in the NMS, AC-2 and Turkey. It is better prepared for and already used to the requirements that these two trends imply. West European food industries are generally sourced by well educated, market-oriented farmers, whose output is geared for transformation into high quality food products. Furthermore, many of them are already part of an international network of food chains.

The Turkish food industry may also join with international food chains but it will take time. As noted in Chapter 6, the Turkish food industry needs restructuring and modernisation to improve the quality of products and production processes. Of course, over time - and certainly if Turkey becomes an EU member - its food industry including the primary sector will enjoy greater access to processing technology, production, (post-)harvest and marketing techniques, which will improve output in terms of quantity and quality. However, a lack of capital may be a hindering factor. We expect it will be some years before the Turkish food chain will be able to supply mass markets according to the standards (such as HACCP, EUREPGAP) and quality levels required by most European retailers.

However, producers are a heterogeneous group and individual progress should not be underestimated. For instance, although their number is small, in 2004 some large fruit and vegetable growers in Turkey were complying with EUREPGAP requirements. This is a sector where Turkey has a comparative advantage and consistently has a positive trade balance with EU-15. Overall, however, we expect that the EU food industry will benefit from increased access to Turkish markets rather than lose out to competition from the Turkish food industry on EU markets.

Moreover, foreign (EU and non-EU) companies are more likely to invest directly in local production facilities in Turkey. Political and economic stability, a consistent legal framework and transparent bureaucratic procedures, are crucial factors for attracting FDI, as well as factors that affect production costs more directly (e.g. labour costs, energy costs, taxes, etc.). The 2003 FDI law abolished many technical barriers to FDI. Bureaucratic procedures were reduced and delays were shortened. Several other laws (on social security, work permits for foreign personnel, customs, property rights, etc.) for improving the investment environment were passed in mid-2003 (YASED, 2003). This legislation embodies the principle of equal treatment, whereby foreign and domestic investors acquire the same rights and obligations. It is too soon to see the effect of these changes. However, it is expected that FDI inflows to Turkey will increase substantially in due time. Experience in the NMS indicates that foreign capital inflows took off as soon as a starting date for EU accession negotiations was announced (see e.g. PAIiIZ 2004). This announcement appears to be a signal to the business community that a country has passed a point of no return on the way to accession, which boosts the confidence of business in the country's political and economic stability. There is no reason to assume that this would not happen in Turkey.

11.6 Assumptions underlying budget estimates

In order to calculate budget payments and financial flows, we need to make specific quantitative assumptions about some key variables. Assumptions are also needed regarding economic growth rates in Turkey and EU-27, the lira-euro exchange rate, and EU policies relating to agriculture, food, trade, regional development and the agri-environment. Other aspects of our analysis are purely qualitative, and do not depend on specific numerical assumptions.

Economic growth and inflation

Table 11.1 shows real growth rate projections for Turkey, the EU-15, NMS and AC-2 from various independent studies. There seems to be a consensus that the economies of EU-15 will grow by about 2 per cent per year in the coming years. This is plausible, given their stagnant and ageing populations. The NMS and AC-2 also have stagnant populations, but are assumed to have higher growth rates. This is due to a catching-up process with EU-15, the effect of pre-accession funding, and more generally because of strong targeting on economic growth within the EU framework in new member states where educational levels and female labour participation were already quite well developed. All the NMS have gone through a severe dip in their economic development, but they are all now on a stable growth path (OECD, 2004). Since the economic size of the NMS and AC-2 is relatively small, their growth rates (which we assume to be about twice as high as that of EU-15) raise the growth rate of EU-27 as a whole only to 2.1 per cent.

Table 11.1. Overview of economic growth figures

Publication	Period	Annual growth %			Remarks
		Turkey	EU-15	CEECs	
IMF (2004)	2004-05	6.0	1.9	5.2	Averages of 2 years
OECD (2004)	2003-15	5.2			
Lejour *et al.* (2004)	2003-25	5.6	2.0	2.9	
Quaisser & Reppegather (2004)	2002-13	5.0	2.0	4.0	5% for AC-2 also
Kalshoven & Küçükakin (2004)	2003-14	4.9			Turkey: 5-6.2% over the period after 2015
Basic projections of this report	2003-15	5.2	1.9	3.8 (NMS)	AC-2 countries: 4%

Source: see publications above

For Turkey we assume the same growth rate as the OECD medium growth path (OECD, 2004: 25), that is, annual economic growth of 5.2 per cent per year[7].

[7] OECD assumes annual population growth of 1.65 per cent, which results in per capita growth of

Since international organisations (including OECD, World Bank and IMF) have tended to overestimate Turkey's growth rate, 5.2 per cent may be too high: Turkey's long-term growth rate has been following a declining trend, so this trend would have to be reversed. Moreover, given the large current account deficits, it may be difficult for Turkey to sustain its present growth level (Togan and Ersel, 2005).

On the other hand, there are reasons why economic growth could in fact be higher. First, when the share of the informal economy increases at the expense of the formal economy, as has been happening in recent years, growth estimates based on the formal economy may be biased downwards. Second, preparation for EU accession (partly discounted already in the growth paths projected by the international organisations) will mean more pre-accession funding and a higher rate of FDI. Third, the psychological effect of future EU entry on the confidence of the domestic sector may stimulate an increase in investment. Optimistic scenarios give more weight to these last two reasons (see e.g. Kalshoven and Küçükakin, 2004).

Annual economic growth of 2 per cent is assumed for the EU-27, with some differences between country sub-groups (see Table 11.1 for details). Inflation of 2 per cent for the euro zone is assumed. This implies that by 2015, the price level will be 24.3 per cent higher than that in 2004 (26.8 per cent higher than in 2003).

Population growth

The following average annual rates of population growth are assumed for the period 2004-2015: Turkey (1.29 per cent), EU-15 (0.10 per cent), NMS (-0.20 per cent) and AC-2 (-0.37 per cent).

Exchange rate

The exchange rate between the Turkish lira and the euro plays a key role in determining some important financial magnitudes. These include the future budget contributions of Turkey, price levels for products in Turkey relative to the EU, and the upper limit of absorption capacities for structural funds.

We have found no exchange rate projections for Turkey in the literature so we resort to *ad hoc* reasoning. Turkey has a dual economy with a large informal sector and a recent record of economic instability. Not surprisingly, therefore, the value of the Turkish lira in PPS was about double the market exchange rate in 2003. In other words, the market exchange rate undervalues the lira relative to the euro in terms of relative purchasing power. We assume that during the period 2005-2014, the market exchange rate increases so that it moves towards the PPS exchange rate. A similar development has been observed for Central and Eastern European countries (De Broeck and Sløk, 2001). Our central assumption is that the gap between the market exchange rate and the PPP rate measured in 2003

3.5 per cent per annum. We use a lower population growth rate of about 1.3 per cent (based on UN estimates), which implies per capita income growth of nearly 3.9 per cent.

will reduce by 20 per cent by 2015. Given the 1:2 ratio of these rates in 2003, this assumption implies a 20 per cent appreciation of the lira against the euro in real terms.[8] Full adjustment of the market exchange rate to the PPP rate is unlikely because of high transaction costs and continuing high levels of corruption in Turkey, because the informal economy is unlikely to be fully absorbed in just 10 years and because holding lira may still be seen as relatively risky.

Given the uncertainty about the exchange rate and economic growth, we experiment with two alternative scenarios: a lower level of economic growth and stronger currency appreciation (see Table 11.9). For other scenarios involving higher growth or less appreciation, results can be extrapolated. For the NMS and for EU-15 members outside the euro zone, we assume no change in exchange rates; for the AC-2 countries, a 10 per cent appreciation of the exchange rate against the euro is assumed.

11.7 Financial framework, structural policy and decision-making context

The Common Agricultural Policy, and the structural policies that aim to promote convergence between the most backward (predominantly rural) areas and the more prosperous parts of the Union, give rise to large redistributive budget flows. They are the two most important items in the EU budget. Therefore, a complete picture of the consequences of Turkish accession for the EU, arising from Turkey's agricultural and rural situation, would be incomplete without some discussion of the likely budget implications.

The context for the discussion is, however, fraught with uncertainty, as the budgetary framework for the period 2007-2013 has not yet been agreed by member states, and proposals for reform of the EU's structural policies have also been tabled by the Commission. In the following section, we briefly outline the proposed changes and the underlying issues.

It is striking that, in discussions on EU policies, the link between budget transfers and economic growth has been considered only peripherally. There has been relatively little research on the local or macroeconomic growth impact of redistributive policies such as the CAP or structural aids (see, for example, Psaltopoulos *et al.*, 2004; Doyle *et al.,* 1997). In the case of direct income payments to farmers, the contribution of such transfers to economic development partly depends on the behaviour of farm families. In countries like AC-2 and Turkey, given the present state of agriculture and agricultural incomes, the lack of attractive investment opportunities and ongoing inflation, most of these transfers are likely to end up in consumption. If the tax revenue to fund such payments were also raised within these countries, then the issues (as far as the growth effect is concerned) would be whether the money transferred has a greater multiplier effect within the economy if it is spent by farmers than in its

[8] Our assumption does not imply a prediction of the actual future exchange rate between the lira and the euro, because higher inflation in Turkey than in the EU may cause further adjustment.

next most likely use, and the size of the transaction costs of implementing the transfers. Similarly, for structural policy, whose aim has been to stimulate the convergence of regions that are lagging behind, a minimum expectation would be that those funds are spent in a way that is at least as productive as other competing investments in the local economy.

EU budget payments to member states, however, imply transfers *between* member states, which will influence the economy of both the transferee and the transferor. If the transfer means that purchasing power moves from areas where returns or multiplier effects are lower to areas where these returns are higher, then the average growth of both economies combined should increase. However, when looking at the effects on the economy of the transferring country alone (in this case, EU-27), the question is to what extent the transfer of income out of the transferor's economy lowers it own rate of growth. Moreover, it is not sufficient to look at the first round effects only. For example, the European Commission (2004g:xviii) argues that EU structural spending in the NMS is likely to benefit growth in the rest of the Union because of these countries' high marginal propensity to import, and the high share of their imports (particularly in the case of investment goods) coming from the rest of the EU. Hard evidence on such questions is more difficult to find.

The budget process 2007-2013

This section provides information on the Financial Perspective for 2007-2013, as proposed by the European Commission in February 2004. The annual changes in the main budget items set out in Table 11.2 show the changing priorities reflected in the financial proposal.

The Financial Perspective assumes that the own resources ceiling is maintained at 1.24 per cent of the Gross National Income (GNI) and appropriations for payments remain at 1.14 per cent of the EU GNI. This is based on an assumed average growth rate of 2.3 per cent, which breaks down as follows: EU-15 (2.2 per cent), NMS (4.1 per cent), AC-2 (5.6 per cent) (Council of the European Union, 2004). Further sensitivity analysis indicates that with an average 2 per cent annual growth rate, commitments equal 1.3 per cent of GDP (Council of the European Union, 2004).

In the 2007-2013 proposals, growth in commitment appropriations is highest for the expenditure headings 'citizenship' (including 'security and justice') and 'competitiveness'. 'Cohesion' shows a 1.2 per cent annual growth between 2007 and 2013, but makes a big jump between 2006 and 2007 to incorporate the accession of Bulgaria and Romania. Taking the final year of the Commission's Financial Perspective for 2007-2013 as a basis for predicting the shares of the various budget items after 2013, we see that natural resource conservation and management (which includes expenditure on both pillars of the CAP) and structural policy could jointly take 68 per cent of the budget at the moment of Turkey's accession.

Table 11.2. Budget of the proposed Financial Perspective in 2013 (EUR mn, 2004 prices)

Expenditure headings	2013	% of total	Annual average change 2007-13 (%)
Competitiveness (education, training, research, others)	25,825	16	13.50
Cohesion	50,960	32	1.20
Convergence of less developed member states[1]	40,074	25	
Competitiveness outside convergence regions[2]	7,385	5	
Territorial cooperation	2,245	1	
Others (EU Solidarity Fund, administration)	1,256	1	
Conservation and management of natural resources	57,805	36	0.18
Promotion of competitive, market-oriented and sustainable agriculture	42,293	27	-0.50
Supporting and strengthening rural development	12,975	8	1.90
Environment, fisheries, others	2,537	2	
Citizenship, freedom, security, justice	3,620	2	14.20
External policies	15,740	10	5.50
Administration	4,500	3	3.40
Total commitment appropriations	158,450	100	

[1] Including 'phasing-out' regions due to statistical effect.
[2] Including 'phasing-in' regions due to natural growth.
Source: Council of the European Union, 2004

The current budget debate among EU member countries focuses not only on the appropriate size of the specific GNI ceilings for appropriations and spending, but also on the philosophy underlying budgetary policy. The Commission has based its proposal on an estimate of what the Union needs to fulfil its commitments under the Treaties and the political decisions already taken by the Council for the coming period. By contrast, a sizeable minority of member states advocate a 'top-down' procedure whereby an overall ceiling is agreed first, as a function of general economic and political considerations, after which EU spending objectives are then prioritised in order to fit within the budget limit.

These conflicting philosophies have different implications for the approach to be taken regarding the funding of Turkish accession. According to the first philosophy, the starting point would be the assumption that Turkey should have the same kind of opportunities that have been available to other acceding countries in the past, extended perhaps to accommodate Turkey's less developed situation. The budgetary resources would then have to be found, if necessary by increasing the appropriations ceilings after 2013. By contrast, according to the top-down approach, the EU could be expected to modify significantly both its agricultural and structural policies, or to accept the crowding out of other budget items, rather than allow the cost of Turkish accession to impose too heavy a burden, economically or politically, on the existing member states.

Competition between growth strategies and rural/agricultural policies

The EU's ambitious Lisbon Strategy (preparing the transition to a competitive, dynamic and knowledge-based economy) implied pressure to increase the budget for R&D in the EU, in competition with spending on agriculture and rural development. This is reflected in the European Commission's proposed Financial Perspective for 2007-2013 where competitiveness is given top priority (see Table 11.2). Increasing competition, making the labour market more flexible and coping with the ageing population are important items for the EU-27. Research and development is a key element in a growth strategy. Member states and private business are expected to contribute most to R&D expenditure. However, the European Commission has argued for an increase in the EU budget contribution to R&D activities above the current 0.04 per cent of EU GDP.[9] If own resources are fixed, this will mean reducing funds in other sections of the budget. Competition could arise between the higher-return investments in R&D, and agricultural and rural policies. Although both are part of the budget, R&D expenditure tends to be directed to 'rich' regions, where it contributes to higher growth. Concentrating on 'rich' regions would deepen regional disparities and, if the cohesion objective is to be addressed, would increase the need for redistributive policies such as structural and rural policy.[10]

Sapir (2003) argued for a redirection of EU budget spending, involving a reduction in expenditure on rural policy and a cut in the direct income support for agriculture. This would give opportunities to spend more on 'competitiveness'. This was also the line of reasoning taken in the Kok report (Kok, 2003). This advice is indeed partly reflected in the Commission's proposal for the Financial Perspective 2007-2013.

Proposed structural policy reform

The size of structural policy budgets increased as a share of the EU total budget expenditure during the period that the Single European Market was created (European Commission, 2004e). Structural and cohesion policy foster the creation of an integrated European market economy by supporting regions whose economic development is lagging behind the EU average. The basic principles of the current structural and cohesion policy in the EU are: multi-annual programming to guarantee sufficient duration and concentration, broad and effective partnerships

[9] At present Europe devotes 2 per cent in total of its GDP to research, compared with 2.7 per cent in the United States and over 3 per cent in Japan. The EU goal is to increase the GDP share to 3 per cent by 2010, with two thirds of it financed from private sources and the rest by public sources (European Commission, 2004b: 8, 10).

[10] For a relevant study, see Acconcia *et al.* (2003).

of actors at various levels, co-financing based on the complementarity of Community and national aid, and evaluation of performance and quality at various stages in the process (European Commission, 2004f). Widening regional disparities in economic development due to recent enlargements, the need to improve EU growth and competitiveness and the institutional changes associated with the adoption of a Constitutional Treaty have led to the need for a reformed approach to structural and cohesion policy (Bachtler and Wishlade, 2004).

In July 2004, the European Commission (2004d) put forward its legislative proposal for reforming cohesion policy (structural and cohesion funds). The proposal represents a significant simplification of the existing model. The current objectives and financial instruments would be replaced by three objectives and just three financial instruments, compared with the current six. The three financial instruments are the European Regional Development Fund (ERDF), the European Social Fund (ESF) and the Cohesion fund. The proposed three objectives, the related target areas, eligibility and the proposed funding allocations, are shown in Table 11.3. The basic elements of structural aid would be maintained, but the reforms propose a simpler and more strategic approach to programming, and further decentralisation of responsibilities to partnerships within member states.

Under the proposed reform, the contribution of EU structural and cohesion funds to public expenditure must be at least 20 per cent of the total and subject to maximum limits that differ for each priority. These maximum contributions are 85 per cent for the Cohesion Fund, 75 per cent for the ERDF and ESF under the convergence objective (80 per cent for regions within cohesion countries), 50 per cent for the ERDF and ESF under the competitiveness objective and 75 per cent for the ERDF under the cooperation objective (European Commission, 2004f)[11].

Public expenditure is expenditure whose origin is the budget of the State, regional or local authorities, or the structural and cohesion funds themselves.

The proposed criteria for the allocation of funds according to the different priorities to a particular member state are complex and difficult to assess at this stage. As for the distribution of the financial resources among member states, the Commission proposes to apply the method based on objective criteria used at the time of the Berlin Council (1999) for the 'convergence priority' (European Commission, 2004d: 12). The Berlin formula is an allocation method based on the principle that the level of assistance should be related to disparity in regional development, and also account for national prosperity and unemployment levels (Bachtler and Wishlade, 2004: 26).[12] As well as individual allocation methods for each funding priority, a number of overall guidelines and criteria are proposed.

---

[11] Additional rules regarding maximum contribution limits are proposed for outermost regions, regions with natural handicaps, and sparsely populated areas (European Commission, 2004f).

[12] See Bachtler and Wishlade (2004) for more details, and an application for the 2007-2013 period.

Table 11.3. Proposed framework for structural and cohesion funds 2007-2013

Programmes and instruments	Eligibility criteria	Priorities	Allocations (2007-2013)
Objective 1 – convergence			78.5% EUR 264 bn
National and regional programmes (ERDF, ESF)	Regions with per capita GDP in PPS < 75% of EU average	Innovation Environment/risk prevention Accessibility Infrastructure Human resources Administrative capacity	67.34% EUR 177.8 bn
	'Statistical effect': regions with per capita GDP in PPS < 75% of EU-15 and > 75% of new EU average		8.38% EUR 22.14 bn
Cohesion Fund	Member States with per capita GNI in PPS < 90% of EU average	Transport networks Sustainable transport Environment Renewable energy	23.86% EUR 62.99 bn
ERDF	Outermost regions		0.42% EUR 1.1 bn
Objective 2 - regional competitiveness and employment			17.2% EUR 57.9 bn
Regional programmes (ERDF) and national programmes (ESF)	Member states propose a list of regions (NUTS I or II) not eligible for parts 1 and 2 of Objective 1	Innovation Environment/risk prevention Accessibility European employment strategy	83.44% EUR 48.31 bn
	'Phasing in' regions covered by Objective 1 in 2000-2006 but no longer eligible due to natural economic growth		16.56% EUR 9.58 bn
Objective 3 – European territorial cooperation			3.94% EUR 13.2 bn
Cross-border and transnational programmes and networks (ERDF)	Cross-border cooperation: NUTS III regions along internal and certain external land borders and some maritime borders Commission to adopt list	Innovation Environment/risk prevention Accessibility Culture, education	47.73% EUR 6.3 bn
	Transnational cooperation: Commission to adopt list of eligible areas		47.73% EUR 6.3 bn
	Cooperation networks: entire Community territory eligible		4.54% EUR 0.6 bn

Source: European Commission, 2004d and 2004f

The Third Cohesion Report indicates that the cohesion fund should represent a third of the total structural and cohesion financial allocation for New Member

States (European Commission, 2004g). Total annual allocations from the structural and cohesion funds (including some of the rural policy funds) will be subject to a cap of 4 per cent of each member state's GDP (European Commission, 2004d).

The analysis of Bachtler and Wishlade (2004: 29) suggests that in practice the allocation of convergence region funding under Objective 1 for NMS, AC-2 and any other new members is unlikely to be determined by the Berlin formula, since resulting allocations are overridden by other rules, such as the 4 per cent absorption cap, the total budget limits and the split between structural and cohesion funds.

Rural policy

The Commission's proposal is that rural development expenditure for the period 2007-2013 be increased to EUR 88.75 billion at 2004 prices[13]. Of these resources, more than one third will be devoted to regions eligible for the Convergence Objective. The new strategy will review the classification of intermediate Less Favoured Areas (LFAs) according to permanent handicap criteria, such as low soil productivity and poor climatic conditions. The current system for rural policy consists of a number of different instruments that support rural development. These will be simplified and grouped in one single instrument (the European Agricultural Food and Rural Development Fund) under the Common Agricultural Policy. This new instrument will have three objectives: increasing the competitiveness of the agricultural sector through support for restructuring, enhancing the environment and countryside through support for land management, and enhancing the quality of life in rural areas and promoting diversification of economic activities (European Commission, 2004a).

Restructuring and modernisation of Turkish agriculture and the diversification of rural areas will be a priority task requiring significant effort. The Commission stresses the importance of this matter by suggesting the creation of a specific pre-accession rural instrument for the period 2007-2013 (European Commission, 2004c: 33). In the 2004 enlargement, the allocation of rural development funds was 50 per cent higher in per capita terms for the new Member States than for the EU-15[14]. For Turkey, the Commission suggests that "given the differences between Turkey and the CEECs, and the scale of adjustment required in Turkey before accession, existing rural development measures may need to be supplemented and or replaced with new ones" (European Commission, 2004c: 32).

Changes due to the Constitution

We assume that the European Constitution will enter into force as foreseen in

[13] For comparison, EU funding for rural development for the period 2000-2006 accounts for over EUR 50 billion. EUR 33 billion of this comes from the EAGGF Guarantee section and EUR 18 billion from the Guidance section.

[14] EUR 5.1 billion were earmarked for the period 2004-2006, with a wide range of measures being co-financed at a maximum rate of 80 per cent from 2004 (German Federal Foreign Office, 2004).

2006. The Constitution introduces the co-decision procedure for legislation on the Common Agriculture Policy (CAP)[15], giving the European Parliament (EP) a power equal to that of the Council of Ministers. However, for the adoption of instruments for fixing prices, levies, aid and quantitative limitations, the Council would still act alone (without consultation with the EP).

For the first time, the financial framework is introduced into the text of the treaty. For its adoption, unanimity is required in the Council. The budget on CAP market measures is currently a compulsory budgetary line, in which the final decisions are taken by the Council. The abolition of the current budget distinction between compulsory and non-compulsory expenditure will give the EP power over the financing of the CAP.

Assumptions made about policies

Underlying our estimates of budget costs are the assumptions that the proposed structural policy reform is accepted for 2007-2013, and that the rules concerning eligibility and payment rates carry over into the next planning period. Furthermore, we assume that the framework for rural policy laid down for 2007-2013 will remain beyond that date, and that Turkey will use its allocation for rural development initiatives to the full. These assumptions are not predictions of the outcomes of the decision-making process. However, they provide a transparent set of assumptions that will probably be quite close to what is actually decided.

Regarding the budgetary ceilings, it is still unclear what will be decided regarding the 2007-2013 period. The debate on the following financial framework (for 2014 on) will probably be complicated. We make no assumptions, therefore, about the size of the total EU budget for 2015, the hypothesised year of Turkish accession. This means that we do not compare the estimated budget costs of Turkey's accession with an overall budget limit, and do not therefore make predictions about whether the spending ceiling would have to rise above the current 1.14 per cent of GDP, or whether there might be strong pressure to downsize certain policies or switch allocated funds from other priorities.

Inevitably, the budget issue will play a role in EU decisions about Turkey. Given the vigorous and continuing discussions within the EU on the size and the distribution of the budget, on both the income and expenditure sides, the cost of Turkish entry will be closely scrutinised. The European Commission (2004c: 46) has suggested that the budget for agricultural and structural policy could be phased in over a 10-year period after accession. However, as was seen with the last enlargement, it is not enough for the Commission or even the Council to take this viewpoint. Although the provision of direct income payments was originally not intended by either the Commission or the Council, this policy measure entered the negotiation process and was fiercely argued for by the ten accession countries. Given this experience and the difficulty of predicting exactly how payments might be phased in, we use the simpler assumption that

[15] Currently, decisions are taken by the Council by qualified majority with prior consultation in the EP.

Turkey will become eligible for budget payments of the full amount from the year of entry. This approach seems more realistic, at least as a first approximation, given another conclusion of the European Commission (2004c: 47) "... the cost of Turkey's accession will be a matter for the negotiations and will be decided upon on the basis of what the EU's Member States are prepared to accept, as has been the case in previous enlargements".

11.8 Turkish accession and the implementation of agricultural and rural policy

Payments from the EU budget to Turkey under the CAP have two components: (a) CAP first pillar budget, separated into market and price support, and direct income support, and (b) CAP second pillar budget (rural development policy).

Payments under the first pillar of the CAP

Budget payments to Turkey for market and price support will be small, because decoupled direct income support is now the dominant instrument. Quantifying future market and price support is often dropped from this kind of analysis. In this sub-section, we quantify both these items.

Market and price support

Our attempt to quantify the amount of market and price support for 2015 yields two alternative results, which are presented in Table 11.4. The first column uses as a reference the budget cost of market and price support of the EU-15 in 2002. Projections from this baseline assume no supply response, annual yield increases of 1.5 per cent in the EU and 1 per cent in Turkey, and price declines with technological change. Recent and future policy changes have been incorporated for cotton, sugar, tobacco and olive oil. Market and price support is allocated to Turkey in proportion to Turkey's agricultural production relative to that of EU-15.

Since there will be no export subsidies in 2015, the burden of internal market support (intervention, storage, etc) is assumed to increase. Additional internal market support is assumed to be equivalent to 80 per cent of the export subsidy budget (also corrected for a number of policy changes). This is included in the second column of Table 11.4. It should be recalled that Turkey is specialised in a number of products for which market and price support is either non-existent or quite low, such as tobacco and cotton.

The amounts shown in Table 11.4 are small and quite uncertain. We see them as indicative only. They are much smaller than the estimates made by the European Commission, who estimated this item at EUR 1 billion at 2004 prices (European Commission, 2004c:46). The background to the Commission's estimate is not clear. Differences in assumptions about future policy changes between our approach and that of the Commission may explain the difference in estimates. In our calculations of the total budget effect, we adopt the figures in the second column of Table 11.4.

Table 11.4. Market and price support for agricultural products of Turkey in 2015 (EUR million)

	M&P Support	M&P Support (+ 'Exp Refund')
Arable crops	47	63
Tobacco	0	0
Sugarbeet	6	6
Cotton	0	0
Olive oil	15	15
Fruit and vegetables	46	57
Beef	5	19
Sheep meat	0	0
Cows' milk	68	120
Total	187	280
Total (real) prices 2004	150	226

Source: own calculations

Direct income support

The direct income support for Turkey will be of a far greater order of magnitude. We assume that CAP direct income support is paid immediately and in full (no phasing in). This support serves two purposes. It *replaces* the direct income support (DIS) introduced by the Agriculture Reform Implementation Project (ARIP). In addition, it compensates for price reductions incurred due to accession, which will depend very much on the lira-euro exchange rate.

The amounts shown in Table 11.5 are lower than the estimates of the European Commission (2004c) and Grethe (2004). The European Commission (2004c:46) mentions an amount of EUR 5.3 billion in 2004 prices or EUR 6.6 billion at 2015 prices; Grethe (2004: 9) estimates the cost of fully implementing direct income payments at EUR 5.2 billion (at 2004 prices).

There are several reasons for these differences. *First*, our analysis assumes an annual degression of direct income payments by 2 per cent per year in nominal euros, starting in 2006. This assumption explains about EUR 945 million of the difference at 2015 prices. *Second*, we include modulation and a ceiling for farms above 50 hectares. This explains a further EUR 145 million of the difference. *Third*, we assume reform of the sugar market and also lower direct subsidies for cotton and tobacco producers. In general, it is difficult to say whether our yield assumptions are too pessimistic for some traditional agricultural products (cereals, livestock products, etc.). However, a doubling of the annual yield growth from 1 to 2 per cent would only increase the level of direct income support from EUR 4218 to EUR 4439 million at 2015 prices. Thus, the effect is small.

Our analysis appears to go into more detail than that of the European Commission (2004c: 46), although we cannot check this because the Commission has not provided detailed information on its calculations.

Table 11.5. Overview of direct income payments for Turkey in 2015 (EUR, current prices)

Product	Euro/ ton	Ref. yield	DIS/ha	Total ha	DIS (EUR mn)
Common wheat	63.0	2.41	152	8190	971
Durum wheat	63.0	0.96	346	910	246
Barley	63.0	2.46	155	3547	429
Maize	63.0	4.81	303	537	127
Rye	63.0	1.82	114	144	13
Oats	63.0	2.08	131	145	15
Rice	102.0	1.43	146	61	7
Oilseeds	63.0	1.86	117	1381	126
Tobacco	1732.0	0.89	1543	150	136
Sugarbeet			380	349	103
Cotton			1455	543	462
Dry beans	72.5	1.01	73	172	10
Broad beans	72.5	2.03	147	18	2
Chickpeas			181	645	91
Lentils			181	468	66
Hazelnuts			121	563	55
Olive oil	1322.5	0.36	478	615	238
			DIS/head	Total head	
Beef					
Headage payment			186	2021	303
Slaughter premium adults			80	1545	100
Slaughter premium calves			50	953	39
Sheep					
Milking sheep			24	7559	146
Non-milking sheep			28	11339	257
Cows' milk	35.5	1.92	68	4995	275
Total (nominal) (mn EUR in 2015 prices)					4218
Total (real) (mn EUR in 2004 prices)					3392

Source: own calculations

Payments under the second pillar of the CAP

The cost of CAP rural policy measures has been estimated by Grethe (2004) and the European Commission (2004c). The European Commission estimates the cost at about EUR 2.3 billion at 2004 prices. Grethe (2004: 12) provides a detailed formula based on agricultural land, agricultural labour and relative GDP in purchasing power standards to calculate the rural policy funding for Turkey in comparison to the AC-2 and NMS countries. Using this formula, Grethe arrives at a total expenditure of EUR 1.6 billion using the AC-2 as a reference, and EUR 2.5 billion using the NMS as a reference. Applying the same approach, but with our data and for 2015, gives rather similar results. Assuming full adoption from the first year (no phasing in), we calculate total expenditure on rural policy at EUR 1.6 billion.[16] We consider it more appropriate to base the calculations on the AC-2 countries, because they are more similar to Turkey than the NMS. Moreover, the Rural Development Envelope for the AC-2 countries has been determined recently and probably reflects best the new rules of the European Commission.

The actual expenditure on rural development in Turkey depends on its ability to set up the right programmes, which is hard to predict. On the one hand, there is a long lead period to 2015, and the 2004-06 National Development Plan has already tried to incorporate some of the concepts of the EU's rural policy. However, as observed in Chapter 7, a rural development strategy in Turkey is still missing.

11.9 Turkish accession and the implementation of structural policy

Turkey would enter the European Union with a triple handicap. First, its GDP per capita would be lower than that of any member country of EU-27. Second, the income disparity between its own regions is very wide relative to other EU countries. Third, the remoteness of the country, especially the Eastern regions, will act to enhance the disparities, particularly with southern and eastern borders formally closed. Erkut and Özgen (2003) analysed the core-periphery pattern of Europe, focusing on the border regions of south-eastern Europe during the years 1997-2000, and concluded that regional inequalities with the rest of Europe were deepening. The comparison of all regions at European level revealed that the south-eastern European regions bordering with external states seem to be the most depressed areas. Statistics for 1998, 2000 and 2001 confirm that the Turkish regions at NUTS II level with the lowest GDP per capita were those at the eastern border[17]. At the moment of accession, Turkey will still be a country with severe structural deficiencies, for which the EU's cohesion policies are intended. In this section, we examine the consequences of Turkey's entry for the allocation of EU structural expenditure amongst member states.

Methodology

Our objective is to explore the impact of implementing structural policy in Turkey on the eligibility of member states (regions) in the EU-27 for structural aid[18]. We use a relatively simple model that translates EU rules to effects at member state and regional level. We focus on the implications of Turkish accession for those countries that will be eligible in 2015 for Objective 1 funding (structural and cohesion funds).

---

[16] Transfers of modulation funds from CAP pillar I have not been included, but this is a relatively small amount (EUR 38 million).

[17] The group of provinces Bitlis, Hakkari, Mus and Van had the lowest GDP per capita in 2000 (EUR 1061). The provinces Agri, Ardahan, Igdir and Kars scored the lowest GDP per capita with EUR 968 in 1998 and EUR 815 in 2001.

[18] Quaisser and Reppegarther (2004), Terluin and De Kleijn (2004), Togan (2004) and others have suggested partial answers to this question.

Given our assumptions about economic trends and the parameters for structural policy, our procedure involves the following stages:

1. It is assumed that structural and rural development policies remain unchanged when Turkey enters the EU. This might not be realistic, because EU rules could be changed to accommodate the accession of a large and poor member state.
2. Projections of regional per capita income in PPS at NUTS II level are made for the regions in EU-27. Here, it is assumed that regional incomes grow at the same rate as the corresponding national income. Four groups of countries are distinguished according to annual real GDP growth (given in parentheses): EU-15 (1.9 per cent), NMS (3.8 per cent), AC-2 (4.0 per cent) and Turkey (5.2 per cent).
3. EU-27 regions are identified that have projected per capita income levels (in PPS) below 75 per cent of the EU average, calculated without and then with Turkey as a member.
4. EU-27 countries are identified that will be eligible for cohesion funds, calculated both without and then with Turkey as a member.

The eligibility criteria are based on projected income levels for 2015. In practice, eligibility criteria are usually based on data from the last three years available on a date prior to the adoption of a regulation. This date is not yet specified in the proposed regulations (European Commission, 2004d). Clearly, a region's identification as eligible for Objective 1 funding in 2015 depends on the assumed growth path of countries and regions until 2015.

'Statistical effect' of Turkish accession

It is estimated that, after Turkish accession, around 160 billion people will be living in regions with GDP per capita in PPS below 75 per cent of the EU average. Our model predicts that 71 regions will qualify as 'convergence' regions under Objective 1. Only 12 million of these people will live in the EU-15; the majority will be in the NMS (42 million), AC-2 (27 million) and Turkey (79 million; that is, all Turkish NUTS II regions except the region of Kocaeli).

Our model predicts that with Turkey inside the EU in 2015, average EU per capita income measured in PPS decreases from EUR 27,610 to EUR 25,090 in 2004 prices (i.e. from EUR 34,480 to EUR 31,330 at 2015 prices). This reduces the per capita GDP threshold defining eligibility for the convergence regions under Objective 1 of the structural and cohesion funds by EUR 1894 in 2004 prices (EUR 2365 at 2015 prices). This is called the 'statistical effect' of accession. Our model suggests that the total population of regions losing eligibility for convergence funding under Objective 1 is 33 million: 6 million in the NMS and 27 million in the EU-15 (see Table 11.6).

Within the NMS, only the Czech Republic is affected by the statistical effect of Turkish accession: five regions would lose Objective 1 status. Under the current structural fund proposals, regions affected by the statistical effect will receive transitional funding under Objective 1 (see Table 11.3). This funding decreases significantly from the first year of the Financial Perspective and there is no indication that this transitional funding would be maintained in a new Financial Perspective period (Bachtler and Wishlade, 2004).

Table 11.6. Regions affected by the 'statistical effect' of Turkey's accession

Country	Regions (NUTS II level)	Population in 2015 (1000s)
Czech Republic	Stredni Cechy	1108
	Severozápad	1108
	Severovýchod	1465
	Stredni Morava	1216
	Moravskoslezko	1250
Germany	Brandenburg – Nordost	1179
	Mecklenburg-Vorpommern	1771
	Chemnitz	1615
	Magdeburg	1204
	Thüringen	2426
Greece	Peloponnisos	598
	Voreio Aigaio	202
	Kriti	596
Spain	Galicia	2787
	Castilla-la Mancha	1760
	Andalucia	7454
France	Martinique (FR)	411
Italy	Calabria	1955
Portugal	Alentejo	747
United Kingdom	West Wales and The Valleys	1930
Total	20	32783

Source: own calculations.

For comparison, we note that current estimates of the size of the statistical effect for this year's NMS enlargement suggest that 17 or 18 regions, with a total of 18 million people, have lost their Objective 1 status (Bachtler and Wishlade, 2004). The potential statistical effect of Turkish accession appears much larger than that experienced in the NMS enlargement.

Cohesion fund

Countries with per capita GNI in PPS less than 90 per cent of the EU average qualify for cohesion funds. Turkish accession also has potential consequences for EU-27 recipients of cohesion funds, since the 90 per cent of average EU per capita GNI will be lower with Turkey as an EU member. Based on our assumptions regarding growth rates and the use of GDP per capita (rather than GNI) in PPS[19], we estimate that of the EU-15 countries, Greece and Portugal would still be eligible for cohesion funding in 2015. Spain would not be eligible for cohesion funding; in fact, Bachtler and Wishlade (2004) estimate that Spain will already no longer be eligible in the 2007-2013 period. The NMS and AC-2 countries will still be eligible, except for Slovenia, Malta and Cyprus, which are ineligible for cohesion payments with or without Turkish entry.

Our model predicts that the Czech Republic will no longer be below the 90 per cent threshold following Turkish accession. Estimated GDP per capita in the Czech Republic is 89 per cent of the EU-27 average in 2015, but 97 per cent of the EU-28.

11.10 Budgetary consequences

Estimated budget items from the literature

Before presenting our own estimates, we provide an overview of the most relevant budget items and the estimates made of them by other studies. Previously published studies have estimated the budget costs of Turkey's accession in various ways. Often it is not clear which budget items have been taken into account, whether a gross or net figure is given, and whether current rules or expected future rules have been assumed. The most important budget items are: (i) CAP budget (first pillar); (ii) rural policy (second pillar); (iii) structural and cohesion budgets, which also include the budget from the Guidance Fund; and (iv) other budget items. The 'other budget' heading covers items like animal disease actions approved by the veterinary committee, research, administration, justice, etc. This component is often simply estimated as fixed proportion of total expenditure. The gross approach calculates only the costs to the EU budget, whereas the net approach also deducts Turkey's contribution to the EU budget.

Table 11.7 gives an overview of recent relevant studies of budget costs, and shows quite a range of estimates. Differences can be partly explained by the following choices:

1. *Degree of comprehensiveness*. Only Derviş *et al.* (2004) try to provide a full coverage of all budget costs.
2. *Timing*. The budget estimate may relate to the first year of accession, or to a later year if phasing in of certain payments is assumed. Several studies in Table 11.7 assume phasing in.
3. *Estimation methodology*. This ranges from 'best guesses' (e.g. Hughes, 2004) to more structured ways of calculating budgets using details on the situation in Turkey (e.g. Grethe, 2004).

---

[19] Bachtler and Wishlade (2004:27, table 12) show that for nearly all member states, and all those in question here, the difference between GNI and GDP per capita, as a percentage of EU-25, is very small.

Table 11.7. Overview of budget cost estimates (first year of official entrance to the EU)[20]

Publication / author(s)	Year of entry and other assumptions	Results (all amounts in EUR bn, 2004 prices)	Comments
Quaisser & Reppegather (2004: 68)	2013 Three different scenarios with different levels for (1) phasing in direct income support; (2) structural funds as share of GDP; (3) contribution of Turkey to the EU budget	CAP (Pillar 1): 1.4 - 4.5 Structural funds: 3.9 – 11.3 Other: 1.6 Contribution to EU budget: 1.7 – 3.4 Net budget transfer: 5.2 – 14	CAP expenditure based on the per capita TSE (Total Support Estimate – OECD methodology)
Lejour *et al.* (2004)	2025 Structural budget on the basis of 4% absorption capacity	Structural budget: 8 All other budgetary transfers are ignored	Potential growth effect of structural funds (= 4% of GDP) may be 0.7% p.a. (assuming 'appropriate' spending)
Terluin & De Kleijn (2004)	2014 Based on expected budget for a number of products; no expenditure for market intervention; pillar 2 based on Bulgaria and Romania	CAP (Pillar 1): 3.6 – 4.3 (first year: 35%) CAP (Pillar 2): 2.3	Phasing in of direct income payments similar to the NMS is assumed
Hughes (2004: 22)	2015-2017 Agricultural budget similar to Bulgaria and Romania (per hectare) Structural budget phased in at 60% of GDP limit	Agriculture: 1.9 – 4.0 Structural: 8.1 – 13.5 Other: 1.0 – 1.6	First figures for 2015, last ones for 2017
Derviş *et al.* (2004)	2015 Structural budget on the basis of 4% absorption capacity; CAP budget, including rural policy, 20% of value added	Structural budget: 21 CAP budget: 10 Other: 1 Contribution to EU budget: 7 Net budget costs: 26	Presented as an 'upper' limit

[20] Other studies not shown in Table 11.7: Flam (2003), Togan *et al.* (2003), Togan (2004). These studies are not included because they assume an entry year prior to a number of policy and budget changes that are now already operating.

Table 11.7. contd. Overview of budget cost estimates (first year of official entrance to the EU)

Publication / author(s)	Year of entry and other assumptions	Results (all amounts in EUR bn, 2004 prices)	Comments
Grethe (2004)	2014 Based on expected budget for a number of products; no expenditure for market intervention; second pillar based on a formula, where some variables are compared to Bulgaria and Romania	CAP (Pillar 1): 5.2 CAP (Pillar 2): 1.6 - 2.5	If phasing in of direct income support is assumed, then the amounts for CAP (first pillar) are substantially lower in the beginning.
European Commission (2004c)	2015 All amounts in 2004 prices Phasing in of direct income payments over a 10-year period Estimates based on 2025	CAP(Pillar 1): 6 CAP (Pillar 2): 2.3 Structural budget: 5.6 % of GDP Contribution to EU budget: 5.6 % of GNI	CAP first pillar costs include EUR 0.7 bn for market intervention

Source: see publication/authors in table

The most 'thorough' estimate for each budget item has been selected from Table 11.7 and combined to give a synthetic total estimate. Of course, this procedure contains a large subjective element. The results of this exercise are given in Table 11.8, with budget costs shown both in euros at 2004 values, and in nominal 2015 euros (assuming inflation of the euro at 2 per cent per year).

The synthetic estimate in Table 11.8 leads to an annual net budget cost for the EU-27 of EUR 20.6 billion (at 2004 values), which is lower than that of Derviş *et al.* (2004), who estimate a higher cost of CAP measures. Three-quarters of the net synthetic estimate is due to structural and cohesion funds. The figure shown is based on the assumption that structural and cohesion spending will immediately reach the cap of 4 per cent of Turkey's GDP. It is not expected, however, that during the early years after accession such levels of structural aid could be absorbed easily in useful projects and programmes approved by the Commission (Grethe, 2004; European Commission, 2004c).

Table 11.9 presents the results of our own calculations based on the assumptions in Sections 11.2 and 11.6, and the methodology explained in Sections 11.8 and 11.9. As for any enlargement, the administrative burden will increase, as more has to be done with respect to justice, home affairs, research, veterinary costs, fisheries, etc. The *ad hoc* assumption of EUR 2 billion (at 2015 prices) is made for this item. We calculate Turkey's budget contribution at 1.14% of GDP, or EUR 6.8 billion (at 2015 values) under the basic scenario, although actual contributions are based on GNI.

Table 11.8. Synthetic estimate of the main budget costs of Turkey's accession

Item	Study	Estimate (EUR billion) 2004 prices	2015 prices	Comments
CAP first pillar M&P supports	European Commission (2004c: 46)	0.7	0.8	The Commission is assumed to know best and others did not calculate this
CAP, first pillar DIS	Grethe (2004: 9)	5.2	6.5	Under the assumption of full phasing in
CAP second pillar	European Commission (2004c: 46)	2.3	2.9	Similar to Grethe (2004: 13) who obtained EUR 2.2 bn
Structural and cohesion funds	Grethe (2004: 16; based on the European Commission)	15.3	19.0	The budget has been determined at 4% of GDP
Other budget items	Derviş *et al.* (2004: 4)	1.0	1.3	The only study to provide an estimate
Contribution of Turkey to EU budget	Grethe (2004:16; based on the European Commission)	3.8	4.8	Here we have chosen the study that is closest to our central 'projection' with respect to macroeconomic developments
Net budget transfer to Turkey	Net budget transfer to Turkey, derived from the above items	20.6	25.7	Synthetic result

Source: see studies in Table 11.7

Table 11.9. Overview of the results of budget cost calculations for the EU-27 by 2015 (EUR bn)[1]

Scenario	CAP pillar 1	CAP pillar 2	Structural funds[2]	Other items	Turkey's EU budget contribution	Net budget costs for EU-27
1. Central growth projections (Table 11.1)	4.5 (3.6)	2.0 (1.6)	11.9-20.8 (9.5-16.6)	2.0 (1.6)	-6.8 (-5.4)	13.6-22.5 (10.9-18.0)
2. Lower economic growth for Turkey (4%)	4.5 (3.6)	2.0 (1.6)	10.3-18.1 (8.3-14.5)	2.0 (1.6)	-5.9 (-4.7)	12.9-20.7 (10.4-16.6)
3. Higher exchange rate for Turkey (50% increase instead of 20%)	4.5 (3.6)	2.0 (1.6)	14.8-26.0 (11.9-20.8)	2.0 (1.6)	-8.5 (-6.8)	14.8-26.0 (11.9-20.8)

[1] At 2015 prices (figures at 2004 prices in parentheses).
[2] Including cohesion funds.
Source: own calculations

Our estimates imply a lower net budget transfer than the synthetic result derived from the literature. This difference is largely due to *assumed* lower expenditure on structural and cohesion policy, reflecting the view that Turkey's absorption limit is reached before structural spending reaches 4 per cent of its GDP. The lower figure in the range assumes an absorption limit of 2 per cent of GDP, while the higher figure corresponds to a limit of 3.5 per cent. These results show that the proposed structural policy reform and its implications for Turkey are crucial uncertainties from a budget perspective.

Various reports (e.g. Quaisser and Reppegather, 2004; Hughes, 2004; Derviş *et al.*, 2004) simply assume that the budget expenditure ceiling of 4 per cent of Turkey's GDP determines the expenditure on structural policy. Our approach is somewhat different. First, the rural development envelope is part of the 4 per cent, which effectively reduces the ceiling to 3.7 cent. Given the recent proposals for Bulgaria and Romania and for the NMS, it seems unlikely that structural funds can be even used up to this limit of Turkey's GDP. On the basis of Beutel (2002) and the budget proposal of the European Commission for Bulgaria and Romania, we assume an annual budget for structural and cohesion policy equal to 2.0-3.5 per cent of Turkey's GDP during the first years after accession.[21] The lower figure is based on the experience of the NMS. Structural funds for these countries, however, had a much shorter lead time than can be expected for Turkey, which makes it a minimum estimate under the present rules. For comparison, the European Commission has proposed an allocation of structural funds to 2.4 % of GDP in the first accession year for AC-2 countries (European Commission, 2004h).

Moreover, recent experience demonstrates that total expenditure on structural funds will remain under the maximum, certainly for the first years. This leads us to assume a maximum of 3.5 per cent for structural funding.

As already stated, the structural policy rules underlying these calculations are only at the proposal stage for 2007-2013, and even if accepted for that period may not be continued into the following period. Nonetheless, it seems clear that, whatever is decided for structural policy in 2015, structural payments will be a large item in the budget cost of Turkey's accession. As Grethe (2004:16) states, the "costs resulting from the CAP seem to take a backseat" in the financial discussion on Turkish's accession, compared to the cost of structural aid.

Our estimate for CAP (pillar 2) spending is also lower than the figure retained for the synthetic estimate. This is because we base the estimate on the way this item has been handled for AC-2, rather than for the NMS (see Section 11.8).

Table 11.9 shows that lower economic growth for Turkey results in lower net budget costs. This is due to the assumed link between absorption of the structural budget and the level of GDP. A similar but opposite development can be

[21] The financial burden will be even lower because structural programmes are mostly financed over a longer period.

observed in scenario 3, where the Turkish lira is assumed to appreciate so as to close 50 per cent of the gap between GDP in euros and the GDP in purchasing power standards. This leads to no change in CAP spending, but it should be mentioned that such a revaluation would hurt the Turkish agricultural sector severely because of a decrease in agricultural revenue of EUR 8.5 billion (rather than the EUR 4.5 billion fall experienced with the smaller revaluation assumed in scenario 1). This would increase the pressure for direct income compensation, but makes agricultural exports from the EU-27 to Turkey more price competitive.

Under Scenario 1, net budget revenue is about 2.5 to 4 per cent of Turkey's GDP in 2015, which is a substantial injection into the economy. Our estimate of EUR 13.6-22.5 billion at 2015 prices is smaller than EUR 33 billion implicitly given in Derviş *et al.* (2004: 4). Derviş *et al.*, however, present a maximum estimate. The main reason for our lower estimate of total net budget receipts comes from a more detailed analysis of the CAP and the structural budgets.

11.11 Discussion and Conclusions

If the European Union takes the Lisbon Agenda seriously, then a growth strategy as recommended by the Sapir Report (Sapir, 2003) is needed for the EU-15 member states. Even 10 years of high growth in the NMS and Turkey would be insufficient to raise the overall EU growth rate very far, since the greater part of the Union's economic activity takes place within EU-15. To achieve an overall increase in growth, a higher budget for competition, research and development is essential. This need is also reflected in the Commission's budget proposals, where this budget item is planned to grow faster. Making room in the budget for pre-accession funding for Turkey and higher allocations for agricultural income support and structural support in less developed regions will make it less probable that the European Union can achieve this part of the Lisbon Agenda.

Turkey's accession to the EU would not only reduce average EU per capita income by about EUR 3150 (at 2015 prices) but would also bring a country with large differences in institutional performance into the EU. This could divert attention and funding away from a growth strategy for the more developed regions of the EU, whilst also reducing the emphasis given to further improvements in areas of concern like food safety, food quality, environmental quality, veterinary and phytosanitary standards, and animal welfare. Turkey would join the group of countries within the EU that find it difficult to keep the standards observed in the EU-15 and in more advanced parts of the NMS. Moreover, it is hard to predict what the combined effects would be of a new Constitution with a stronger position for the European Parliament and a large new member state like Turkey that differs in many respects from the average of the EU-27.

Macro-economic consequences of Turkey's entry for the EU-27 are still highly controversial. The most common projection in the literature is an increase of 0.0 to 0.3 per cent in economic growth due to accession (Lejour *et al.* 2004:

42; Hughes, 2004: i). It is difficult to know whether these analyses have fully incorporated the effects of budget transfers away from EU-27 to Turkey. If this has not been done, then the net effect of Turkish entry on EU-27 would be even lower than the estimate just given. We note that the estimate of budget transfers to Turkey is EUR 13.6-22.5 billion (2015 prices), whereas a 0.1 per cent increase in economic growth in EU-27 in 2015 equals EUR 16 billion.

Turkish accession would bring in a member state with a relatively high price level for many agricultural products. It would represent a potential export market for EU-27 and a country where foreign direct investment can make a difference in technology and quality. Export opportunities will be most attractive in the animal sector (beef and dairy). But the low level of per capita income and the long distance between typical production regions in the EU and Turkey will be inhibiting factors. For other products (fruit and vegetables, hazelnuts, cotton, etc.) Turkey is an efficient producer that is already competing with the southern EU member states. Direct income payments for products like cotton and tobacco will probably be reduced throughout the Union, because of very high budget costs for the EU if all producers in the enlarged EU (including Turkey) received the same direct income support as at the moment.

Turkey's accession would bring a large increase in the borders of the European Union, and five new neighbouring countries, which arouses some concern (European Commission, 2004c). Establishing these new borders will give rise to additional costs for the EU, which have not been quantified. Maintaining them, once established, would be Turkey's responsibility. For plant and animal health, it will be of crucial importance that external borders are impermeable. This will require good relations between Turkey and bordering countries.

Already in the pre-accession period but even more so after accession, Turkey will be an interesting market for the food industry, both for exports and for foreign direct investment (FDI). At present, the food industry and the retail sector operate at much lower levels regarding food standards and efficiency than in the EU-15. In particular, the transfer of the food chain approach to the Turkish market would give EU companies a competitive advantage in the Turkish food market. The rate of expansion would depend on disposable income growth in Turkey. Such developments would be encouraged by the implementation of the *acquis communautaire* with its uniform standards and procedures.

EU structural policy is highly concentrated in Objective 1 regions of the Union. If Turkey enters the EU, then Turkish regions with a combined population of 79 million inhabitants become eligible for structural funds with a high rate of EU contribution. This would constitute the greater part of the budget transfer flowing from the EU-27 to Turkey. Turkey's accession would reduce the Objective 1 threshold expressed in GDP per capita by EUR 2365 (2015 prices). We estimate that 20 regions in EU-27 will no longer be eligible for structural funds. These regions have nearly 33 million inhabitants.

Many studies have estimated the budgetary consequences of Turkey's accession to the EU. We have summarised the results of seven studies, and have produced a synthetic result that combines the most convincing estimates of each budget item from these studies. Inevitably, there is a subjective element in this selection process. The estimate of net budget cost obtained by this process is EUR 20.6 billion in 2004 prices (EUR 25.7 billion at 2015 values). The corresponding gross budget cost of EUR 30.5 billion means that, if the 1.14 per cent ceiling applies, about 16 per cent of the EU budget would go to Turkey.

Our own analysis comes up with slightly lower figures for net budget costs. Our estimates are between EUR 10.9 and EUR 18.0 billion in 2004 prices (EUR 13.6 and EUR 22.5 billion at 2015 prices), depending on the size of the structural aid transferred to Turkey. The lower amount is based on the assumption that Turkey has an absorption capacity for structural funds in the early years of 2 per cent of its GDP, and the higher limit corresponds to an absorption capacity of 3.5 per cent of GDP. These budget costs will play an important role in the accession negotiations with Turkey, and may lead to adjustments in EU policy. Higher economic growth and more appreciation of the Turkish lira both increase the net budget costs of Turkish accession, unless the budget share for structural and cohesion funds declines.

References

Acconcia, A., Espasa, M., Leonida, L. and Montolio, D. (2003) EU policy, regional growth and convergence across European regions: Evidence from non-parametric and semi-parametric approaches. Workshop: *Ricerca e sviluppo, produttività ed occupazione*. Università della Calabria, Italy. 10 April 2003.

Bachtler, J. and Wishlade, F. (2004) Searching for consensus: The debate on reforming EU cohesion policy. European Policies Research Centre, Paper 55, Glasgow, UK, November 2004.

Beutel, J. (2002) The economic impact of objective 1 interventions for the period 2000 – 2006. Final Report to the DG-Regio of the European Commission. Konstanz, Germany, May 2002.

Council of the European Union (2004) Financial framework for 2007-2013: Analytical report. Report of the Presidency to the European Council. 10219/04, Brussels, 7 June 2004.

de Broeck, M. and Sløk, T. (2001) Interpreting real exchange movements in transition countries. *IMF Working Paper* WP/01/56. International Monetary Fund, Washington DC.

Derviş, K., Gros, D., Öztrak, F., Işık, Y. and Bayar, F. (2004) Turkey and the European budget: Prospects and issues. *CEPS EU-Turkey Working Papers,* No. 6, August 2004. Centre for European Policy Studies, Brussels.

Doyle, C.J., Mitchell, M. and Topp, K. (1997) Effectiveness of farm policies on social and economic development in rural areas. *European Review of Agricultural Economics* 24, 530-546.

Erkut, G. and Özgen, C. (2003) The economic and spatial peripherality of border regions in Southeastern Europe. *43rd European congress of the Regional Science Association*, Jyväskylä, 27th-30th August 2003, Finland.

European Commission (2003) EU enlargement: Questions and answers on food safety issues. MEMO/03/88, Brussels, 5 December 2003.

European Commission (2004a) Proposal for a Council Regulation on support for rural development by the European Agricultural Fund for Rural Development (EAFRD).

COM(2004)490 final, Brussels, 14 July 2004.

European Commission (2004b) Building our common future: Policy Challenges and budgetary means of the Enlarged Union 2007-2013. Communication from the Commission to the Council and the European Parliament. COM(2004) 101 final. Brussels, 10 February 2004.

European Commission (2004c). Issues arising from Turkey's membership perspective. SEC (2004) 1202, Brussels.

European Commission (2004d) Proposal for a Council Regulation laying down general provisions on the European Regional Development Fund, the European Social Fund and the Cohesion Fund, COM(2004)492 final, Brussels, 14 July 2004.

European Commission (2004e) European Union financial report 2003. Office for Official Publications of the European Communities, Luxembourg.

European Commission (2004f) Cohesion policy: The 2004 watershed. DG Regional Policy, Inforegio factsheet, 2004.

European Commission (2004g) A new partnership for cohesion: Convergence, competitiveness and cooperation. The third cohesion report. Office for Official Publications of the European Communities, Luxembourg.

European Commission (2004h) A financial package for the accession negotiations with Bulgaria and Romania. SEC(2004) 160 final, Brussels.

European Council (2002) Brussels European Council, 24 and 25 October 2002: Presidency conclusions. Cover Note, 14702/02. Brussels, 26 November 2002.

Flam, H. (2003) Turkey and the EU: Politics and economics of accession. Seminar Paper, No. 718, Institute for International Economic Studies, Stockholm University.

German Federal Foreign Office (2004) The enlargement of the European Union. http://www.auswaertiges-amt.de/www/en/eu_politik/vertiefung/erweiterung_html (accessed 5 October 2004).

Grethe, H. (2004) Turkey's accession to the EU: What will be the Common Agricultural Policy cost? *Humbolt University Working Paper* 70/2004. Humbolt University, Berlin.

Hughes, K. (2004) Turkey and the European Union: Just another enlargement? A Friends of Europe Working Paper, June 2004, Brussels.

IMF (2004) *World Economic Outlook: The Demographic Transition.* Statistical Appendix. International Monetary Fund, Washington DC, September 2004.

Kalshoven, C. and Küçükakin, S. (2004) Opportunities beyond the Bosphorus: Exploring growth, FDI and trade flows in view of Turkey's EU-accession. Study commissioned by the Dutch ministry of Economic Affairs. ABN-AMRO, Netherlands, May 2004.

Kok, W. (2003) Enlarging the European Union: Achievements and challenges. Report of Wim Kok to the European Commission. European University Institute, Florence, Italy, 26 March 2003.

Lejour, A.M., de Mooij, R.A. and Capel, C.H. (2004) Assessing the economic implications of Turkish accession to the EU. CPB document No 56. CPB Netherlands Bureau for Economic Policy Analysis, The Hague, March 2004.

OECD (2004) *Economic Survey for Turkey 2004.* OECD, Paris.

PAIiIZ (Polish Information and Foreign Investment Agency) (2004) The list of major foreign investors in Poland. PAIiIZ, Warsaw.

Psaltopoulos, D., Thomson, K.J., Efstratoglou, S., Kola, J. and Daouli, A. (2004) Regional social accounting matrices for structural policy analysis in lagging EU rural regions. *European Review of Agricultural Economics* 31, 149-178.

Quaisser, W. and Reppegather, A. (2004) EU-beitrittsreife der Türkei und konsequenzen einer EU-mitgliedschaft. OstEuropa-Institut Working Paper 252. OstEuropa-Institut, Munich.

Sapir, A. (ed.) (2003) An agenda for a growing Europe: Making the EU economic system deliver. Report of an Independent High-Level Study Group. Brussels, July 2003.

Tangermann, S. (2004) Overview of international developments with relevance to agriculture. Paper presented at High-Level Conference, *Europe's Agriculture in a Competing World*, 28-29 October 2004, Maastricht, Netherlands.

Terluin, I. and de Kleijn, T. (2004) Schatting EU-landbouwuitgaven bij toetreding Turkije tot de EU (concept). 7 July 2004, Landbouw Economisch Instituut, The Hague

Tocci, N. (2001) Our future Southeastern Turkish frontiers. *CEPS policy briefs*, No. 6, September 2001. Centre for European Policy Studies, Brussels.

Togan, S. (2004) Turkey: Toward EU accession. *World Economy* 27 (7), 1013-1045.

Togan, S. and Ersel, H. (2005) Macroeconomic policies for EU accession. In: Togan, S. and Hoekman, B. (eds.) *Turkey: Economic Reform and Accession to the European Union*. World Bank, Washington DC.

Togan, S., Bayaner, A. and Nash, J. (2003) Analysis of the impact on agricultural markets and incomes of EU enlargement to Turkey. Paper presented at *Turkey: Towards EU accession*, Bilkent University, May 10-11 2003, Ankara.

van Berkum, S. (2002) Trade and foreign direct investment patterns: The case of Dutch agribusiness. Ponsen and Looijen, Wageningen, Netherlands.

YASED (2003) YASED website. http://www.yased.org.tr.

12

Expected Consequences for Turkey of EU Entry in 2015

Tugrul Temel[1]

12.1 Introduction

This chapter draws on the previous chapters of the book in order to derive the likely consequences for Turkey's agricultural and agrifood sectors of future EU membership. If Turkey becomes a full member of the EU in 2015, it means that Turkey has to comply fully with the EU *acquis communautaire* by that date. Therefore, part of the discussion in this chapter focuses on the implications for Turkey of adopting the *acquis*. We do not explore the expected consequences for Turkey in the years following EU entry in 2015, since this would require making assumptions about the precise entry terms agreed in the negotiations, as well as predictions about many exogenous factors that are hard to foresee so far ahead.

Two groups of issues are explored in this chapter. The first set of issues relates to Turkey's ability to assume the obligations of the *acquis* regarding competition rules, property rights, and free movement of agricultural products by 2015. The second set of issues relates to the implications of membership for agricultural markets in Turkey and their capacity to cope with competitive pressure and market forces implied by the EU single market. The emphasis will be on how the policy changes necessary for membership affect the functioning of agricultural and food markets, given the assumptions in Section 12.2.

This chapter is organised in eight sections. Following the introduction, Section 12.2 presents the key assumptions underlying the discussion. Section 12.3 critically examines the implications of EU membership for Turkish policies that relate to agriculture, food and rural development, international trade, and agri-environmental interactions. Section 12.4 explores the implications for agrifood chain developments. Implications for the institutional framework of Turkish agriculture are discussed in Section 12.5, with an emphasis on the quality and functioning of economy-wide and agrifood-specific institutions. Section 12.6 comments on the budget consequences, already reported in Chapter 11. Section 12.7 discusses the findings of recent selected studies that assess the consequences for Turkey of accession. Section 12.8 concludes the chapter.

[1] With contributions from Siemen van Berkum, Alison Burrell, Irene Molina Vilchez and Arie Oskam.

12.2 Summary of assumptions

Macro-economic assumptions

The assumptions presented in Section 11.2 are applied in this chapter. They include: annual economic growth rates of 5.2 per cent for Turkey and 2 per cent for the EU-27 are assumed, with some differences between country sub-groups (see Table 11.1 for details). Inflation of 2 per cent for the euro zone is assumed. This implies that by 2015, the price level will be 24.3 per cent higher than that in 2004, or 26.8 per cent higher than that in 2003.

The following average annual rates of population growth are assumed for the period 2004-2015: Turkey (1.29 per cent), EU-15 (0.10 per cent), NMS (-0.20 per cent) and AC-2 (-0.37 per cent).

We assume that the gap between the market exchange rate and the PPP rate measured in 2003 will have closed by 20 per cent by 2015. Given that GDP in PPS in 2003 is double the GDP in euro, this implies a 20 per cent appreciation of the lira against the euro. For the NMS and for EU-15 members outside the euro zone, we assume no change in exchange rates; for the AC-2 countries, a 10 per cent appreciation of the exchange rate against the euro is assumed.

It is assumed that the Commission's proposal for the Financial Perspective 2007-2013 is accepted and continues to 2015. The Financial Perspective assumes that the own resources ceiling is maintained at 1.24 per cent of the Gross National Income (GNI) and appropriations for payments remain at 1.14 per cent of the EU GNI.

Assumptions with respect to agricultural policies

Although Turkey is assumed to adopt the *acquis* fully, its precise content will depend on the evolution of EU policies between now and 2015. Therefore, assumptions about future EU policies need to be made. We assume that EU agricultural policy will become increasingly market oriented, moving towards lower support prices and decoupled income transfers (see Section 11.2). This will result in lower market prices. The budget costs of Turkey's direct income support will shift to the EU, on condition that Turkish farmers observe the EU's cross compliance rules associated with decoupled direct income support.

Full adjustment of policies to the WTO Doha Round agreement (elimination of export subsidies and further tariff cuts) is assumed. Specifically, we assume that import tariff bindings will be 50 per cent lower in 2015 compared to the levels in 2000. A new round of policy revisions will be needed, probably before 2010, to incorporate the Doha Round agreement and to achieve fully decoupled direct income payments. During this period, it is likely that further changes will be made to the CAP to accommodate the entry of AC-2 countries and Turkey. Regarding the CAP second pillar, we assume policy rules in line with the proposals of the European Commission for the financial perspective 2007-2013 (see Section 11.6). Furthermore, we assume that Turkey will use all the funds available to it under the second pillar. For structural and cohesion programmes, we assume that the Commission's proposal for structural fund reform for 2007-2013 is adopted, and that these rules remain unchanged for 2015.

With respect to environmental policy in relation to agriculture, we expect a strong influence of EU regulations on Turkey's national policies (Nitrate Directive, Directive for Plant Protection Products, Water Directive, etc.). At the same time, we assume that EU environmental pressures on member states will not increase in the period 2005-2015.

12.3 Consequences for Turkish policies

Implications for agricultural, food, rural and structural policies

Agricultural policy

The basic principles of Turkey's agricultural policy have been laid down by the ARIP programme (2001-2005). Adapting current policies to the EU's agricultural policy *acquis* over the period 2005-2015 is expected to pose no fundamental problems. The impact of accession in 2015 would depend greatly on the lira-euro exchange rate at the moment of accession and on how far the Turkish food chain has already adopted a market-oriented approach without state economic enterprises and enforced cooperatives.

Turkish agricultural policy already incorporates a direct income support system with a flat rate payment per hectare and a cap at 50 hectares per farm, whereas it will take a few more years to fully phase in such a system for EU-27. Given this existing direct income support scheme and the relatively high prices for agricultural products, there is at first sight no strong reason to phase in CAP direct income support gradually, as has been done for the NMS and is agreed for the AC-2. There would be a strong tendency to start CAP direct payments in 2015 at least at the level of direct income support current at the moment of accession. Complications might arise if by 2015 Turkish crop prices have not been aligned with EU prices and income compensation for a price fall is demanded. This would interfere with the flat rate system and would also lead to an upward pressure on direct income support.

By 2015, a large share of the CAP budget will be spent on decoupled direct income support (the Single Farm Payment). By 2006, existing EU members will have converted their area and headage payments to a flat rate per hectare, and will have adopted the cross compliance requirements related to these payments. In Turkey, there are currently no cross compliance rules, which Turkish farmers would have to accept under the CAP.

More difficulties will be experienced in the animal sector. This sector is both declining and inefficient. Producing beef, sheep and milk generates direct income support as long as 'registered' land is involved. At least for beef, a price fall is to be expected. Here, direct income compensation based on the number of cattle might be demanded. But this would not fit into the flat rate per hectare and would open up demands for substantial compensation.

A basic dilemma, already extensively discussed during the negotiations of the NMS, is whether, instead of direct income payments to farmers, these funds should be allocated to restructuring the agricultural sector so that land, labour and capital are used more productively, either inside or outside agriculture. From an economic perspective, supporting the agricultural sector by means of direct income support is not very productive. It keeps labour in agriculture, and it hinders restructuring and farm consolidation because the payment is capitalised into the value of land. For the NMS, however, there was a political decision to introduce the CAP direct income payments, because otherwise there would be a difference in the way farm policy works across EU member states. The fact that farmers in the NMS experienced lower prices before accession and that there was no reason to compensate them for price decreases, was used only as a justification for *phasing in* the direct income support gradually. This justification does not apply for Turkey. Nevertheless, phasing in has been proposed by the European Commission (European Commission, 2004b). This time, budget restrictions are used as the main argument for limiting direct income support for Turkey.

There will be strong political pressure in Turkey to raise the direct income support payment level *before* 2015 as close as possible to the EU-27 level. After entering the EU, only 3 cents of every euro spent on direct income support will be perceived as being paid by Turkey. Moreover, these transfers might be considered helpful for influencing the Turkish agricultural population to vote in favour of entering the EU.[2]

Other products where difficulties may arise when adopting the CAP are sugar and tobacco. Sugar is highly protected in both Turkey and the EU. The EU's reform proposal for sugar involves bringing internal prices more into line with international prices and shifting from a quota system with high prices to lower prices and direct income support (European Commission, 2004c). Specifically, this means a 33 per cent institutional price reduction and a direct income payment that compensates for 60 per cent of the price decrease. Total sugar quota will be reduced by 2.8 million tons or 16 per cent, and total isoglucose quota will be increased by 0.1 million tons. This implies that policy has to adjust in the same direction in Turkey. Policy changes will run parallel with the privatisation of TÜRKŞEKER, a state company that dominates the sugar sector. Tobacco prices in Turkey are also far above EU prices and are likely to adjust downwards, although it is difficult to say how much of the price differential is due to quality differences.

For most other products, either (a) they are not highly concentrated in one particular area or their production value per hectare is relatively low, or (b) prices are not much higher or even lower than in the EU, or (c) in the EU there

---

[2] As was observed in Poland. Polish farmers, however, did not receive the full CAP direct income payment from the beginning. Moreover, their starting price level was lower than that of the EU-15.

is no direct income compensation for the particular product. Under such circumstances, either no direct income payments are required or a slight general increase in the flat rate direct payment can be expected.

Rural policy

The expected decline in certain sectors of Turkish agriculture in a more competitive environment, together with existing rural-urban income differences, means that rural and structural policy measures will be important for generating rural employment and income. New jobs are likely to come from small businesses that provide services to agriculture, industry and other parts of the service sector. Job-creation policies, although present in the National Development Plan (NDP) (Republic of Turkey, 2003: iv), have not been used so far in Turkey.[3] Moreover, the NDP lacks useful approaches and instruments for these problems, in particular for rural areas. It will therefore be crucial for the EU and Turkey to develop a joint strategy that allows spending for rural development to be adapted to the needs of the Turkish situation. Although the menu approach of Regulation 1257/99 allows choices from among a set of instruments, Turkish rural development policy may require special instruments aimed at education, job creation and income in rural areas (see Chapter 5). We make no assumptions about how specifically funds are used, but expect that Turkey *will* use the funds available for EU rural policy.

Structural policy

Structural policy in Turkey has focused mainly on infrastructure and – for rural areas – on large irrigation projects. Infrastructure often needs a long lead time for development. This type of project does not directly address the need to find employment for large numbers of young people with low skills who will soon enter the job market. For a country like Turkey with large differences in regional development and large cities, there is a strong tendency to invest in urban centres because of scale effects (Krugman, 1991). However, development funding needs to be focused on rural areas in order to redress regional inequality within Turkey.

The potential availability of structural policy funds for Turkey is substantial but this would put a large burden on the EU budget. This budget pressure is the main reason for the European Commission to suggest reduced funding for structural policy in Turkey (European Commission, 2004b).

---

[3] Employment and income-generating policies, although not absent in EU rural policy (see for example the LEADER+ rural development initiative) are not central to EU rural policy. This is due to the history of these policies where a large part of the budget goes to environmental improvements in relation to agriculture, investments in farming and community development.

Food policy

Turkey adopted a food act in 1995 - succeeded by the Food Law 560 of 2004 - and the Codex Alimentarius in 1997. However, these adjustments require changes in the relevant institutions and considerable investments because many quality improvements demand new technology and facilities. The required changes could be facilitated by more foreign direct investment in the food processing and retail sectors. Most importantly, they require effective and rigorous implementation of food control systems.

During the pre-accession period, food quality issues will receive a lot of attention (for example, through twinning projects). Given the rather long lead time to 2015, implementing the *acquis* by 2015 does not appear to be impossible. However, even if the level of food safety and food quality improves, in particular in cities and in relation to the activities of supermarkets, this does not imply that sufficient guarantees can be given to allow free market exchange within the EU.

Animal welfare

Upon accession, Turkey will have to comply with EU rules for the transport and slaughtering of animals, and with regulations concerning, for example, animal housing and stocking densities for intensive livestock production.

Implications for trade policy, and animal and plant health

Trade

Several changes to Turkey's current trade situation can be expected if it becomes a full member by 2015. First, Turkey will adopt the common external tariff of the EU for every agricultural product. Currently, Turkey has some higher tariff bindings than those of the EU. If the Doha Round Agreement imposes smaller tariff cuts on developing countries than on developed countries, average tariff bindings for the EU will be reduced by more than those of Turkey, and so by 2015 Turkey may have higher bindings than EU common external tariff (CET) bindings for more individual products than at present.

Thus, the tariff gap between Turkey and the EU could actually increase in the period prior to trade harmonisation. This means that, if agricultural trade harmonisation between the EU and Turkey occurs in 2015, the fall in the maximum allowed tariff protection for these more highly protected products in Turkey will be relatively greater than it appears today. However, *actual* prices will fall only to the extent that Turkey's applied tariffs are higher than the EU's applied CET tariffs *and* fully determine the price gap between domestic and world market prices. Currently, this is not always the case.

If current policy trends continue up to 2015, the largest price falls following trade harmonisation will be experienced in the livestock sector. Meat, eggs and butter prices are currently considerably higher in Turkey than in the EU, due to high tariff barriers and Turkey's 8-year-long import ban on red meat. These factors severely limit consumption of animal products. Although the current EU consumption levels of animal products are considered by some nutritionists to be unhealthy, even a doubling of meat consumption after their prices fall would raise Turkish per capita consumption to only about 50 per cent of the average EU-15 level. Assuming steady income growth between now and 2015, and bearing in mind that these products have a relatively high income elasticity of demand (particularly at such a low level of consumption), a strong expansion in animal product demand is likely to occur if trade protection falls and internal price levels are equalised. To the extent that this demand expansion is met largely by imports from the rest of the EU, internal prices should become stronger elsewhere in the Union and the price fall in Turkey will be slightly less than one might predict on the basis of extrapolating from current differentials.

Currently, Turkey sources a part of its cereals and oilseeds imports from North America. These imports are purchased at world market prices and represent a source of tariff revenue for Turkey. After joining the Union, there is likely to be trade diversion involving imported cereals, as they are replaced by cereals imported from elsewhere in the Union. Trade diversion of other agricultural imports is likely to be small.

Membership of the EU would imply that Turkey surrenders its right to a national trade policy. This means that, whilst remaining an individual WTO member, Turkey no longer negotiates independently in WTO multilateral negotiations, and its import and export regimes and protocols become those of the EU. During the pre-accession phase, one would expect Turkey's position on international trade issues to become closely aligned with that of the EU.

By 2015, Turkey must be in a position to operate the degree of controls for goods on its external borders that is required to implement all EU measures for goods that cross its frontiers. Apart from a small border in the north-west with Europe, Turkey's borders consist of coastline, and land borders in the south and east with Asian countries. Developing the infrastructure, administrative capacity and commitment for effective border control will be a challenging element in Turkey's adoption of the *acquis*.

The future of Turkey's participation in ECOTA[4] will depend on whether its trade agreement with other members of the ECO can be incorporated into some kind of preferential agreement with the EU. To the extent that the EU already has bilateral trade agreements with some of the countries concerned[5], we can expect a rationalisation of agreements on this front. Indeed, with Turkey inside the Union, other members of ECO become 'frontier states', and it becomes even more in the EU's interests to have stable long-term trade ties with them.

[4] The Economic Cooperation Organisation Trade Agreement (ECOTA) signed between Afghanistan, Iran, Tajikistan and Turkey in July 2003 (see Section 8.2).

[5] A Trade and Cooperation Agreement between Iran and the EU has been under negotiation since 2002. The EU signed a Partnership and Cooperation Agreement with Tajikistan in October 2004.

Animal and plant health

A significant part of the agricultural *acquis communautaire* concerns animal and plant health, animal welfare, hygiene standards and food safety. Within this body of regulations, the sanitary and phytosanitary (SPS) regulations relate specifically to the health of plants and animals, and set standards necessary for maintaining plants and animals in a healthy state, both in order to protect the plant and animal populations themselves, and also to avoid any consequences for human health that may derive from consumption of diseased or infested plant or animal material.

In discussing Turkey's adoption of the SPS *acquis*, it is necessary to distinguish three phases: (a) assessing the appropriate legislation and setting up the administrative infrastructure and frameworks required by the legislation; (b) implementation of the regulations with the rigour and expertise necessary for them to function effectively; (c) convergence of the actual levels of plant and animal health with those of the EU.

Chapter 10 has reported that Turkey is already making a good start with the first stage: the programme of legislation and training envisaged up to 2006 is a significant step in the right direction, and is financed through a combination of national and EU funding (NPAA, 2003). However, given the fragmentation of agriculture, the large number of farmers and their generally low educational level, it seems clear that a much greater number of trained personnel will be needed to breathe life into this legislation than has been allowed so far.

At the second stage, there are several issues. It has to be borne in mind that the creation of an infrastructure to manage the *acquis* is a longer and more difficult process than simply adopting the necessary legislation. Again, the fragmentation of the sector and the fact that a substantial share of output is disposed of in informal markets or by own consumption means that effective communication of the SPS *acquis* to producers, as well as monitoring and control of all the regulations, will be a difficult and daunting task. Moreover, governance issues are important. The training of inspectors and extension officers, and the efficient organisation of these services so that regulation can be implemented and farmers can participate, will require a huge effort.

The third stage, namely the convergence of the actual levels of plant and animal health with those of the EU, cannot happen overnight. Eradication of diseases and pest populations may take a very long time because biological processes with long cycles are involved. Moreover, the good functioning of an effective, participatory biosecurity system can also take some time to achieve.

Assuming accession in 2015 and adoption of the SPS *acquis* into the legislation, it is unlikely that the conditions for a single market in animal products, without border controls for SPS inspection, will be possible for the EU and Turkey for some years after that. A cautious line must be adopted here, as an expanded EU without internal checks on animal movements will have the international animal-health status of whichever country has the weakest status, even in the absence of reported trade flows of animals or animal products from that country to other parts of the EU.

Turkey will have to adopt strong, proactive eradication programmes, and strengthen existing programmes for disease eradication and control where a programme already exists, in parallel with increased border controls for illegal animal movements, which will have to involve more than simply enforcing strict border inspection on main road routes with neighbouring countries.

Implications for policies concerning environmental impacts of agriculture

Administrative capacity

By accession, Turkey will have to adopt EU legislation aiming to avoid adverse impacts of agricultural activities on the environment. Overall, Turkey's current alignment with the environmental *acquis* is limited. The adoption of a new Regulation on Environmental Inspection represents a positive step towards increasing Turkish administrative capacity to implement the *acquis* (European Commission, 2004a). Critical issues to be addressed in this respect relate to capacity development in the implementation, monitoring and enforcement of the environmental *acquis*. Here, Turkey may face difficulties because of society's low valuation of environmental amenities and resources. Lack of environmental awareness could also hinder the further development of the organic food sector. NGOs and public-private partnerships can be expected to play a prominent role in the conservation and protection of the environment and in easing the adverse environmental effects of agricultural activities. Such partnerships are already underway in some EU member countries.

Monitoring and enforcement will be one of the most difficult issues to address, but accession will serve as a catalyst in this respect. Furthermore, the use of agri-environmental indicators at European standards[6] will help to benchmark the integration of environmental concerns into Turkey's agricultural activities.

Natural resources and biodiversity

To withstand stronger competition in the domestic market, Turkish agriculture will have to become more efficient. Government measures such as strengthening water pricing institutions and improving farm extension services will not only promote efficiency but will also have positive environmental benefits. Thus, membership of the EU in 2015 could help Turkey to bring under control various adverse effects of agricultural activities on the environment, such as soil degradation and salination. For example, the Water Framework Directive could be fully implemented by 2015, the same year foreseen for current member states (Sözen *et al.*, 2003). Similarly, future policy instruments originating from the

[6] The IRENA project: Indicator Report on the integration of Environmental concerns into the Agricultural policy.

European thematic strategy on soil protection will be highly relevant for Turkey.

Accession will require the application of the environmental *acquis* regarding the protection of vulnerable natural habitats and species, and the provision of funds for biodiversity conservation. New legislation covering all biodiversity conservation activities needs to be prepared and international commitments be incorporated in this legislation.

Turkey will be able to draw on the support provided by the EU for the protection and enhancement of the environment, including funding from the second pillar of the CAP for land management and rural development actions related to Natura 2000 nature protection sites. Financial assistance from other EU instruments and from other international organisations will be crucial for continuing to fund environmental protection projects. Here, too, NGOs will play a key role in the protection and conservation of these important natural assets and in raising public awareness. Accession could also stimulate the growth of domestically funded activities to protect the environment, but this depends on the weight given to environmental protection in the agendas of stakeholders with conflicting priorities.

Regulations and behaviour

Adapting Turkey's national environmental legislation to European standards should pave the way for better protection and use of the environment. However, in areas where there is a strong behavioural element, it is likely that advances will be slow. The level of environmental awareness in the Turkish society is low relative to the EU-15 countries and environmental values are not well integrated into individual decision-making processes.

The recently adopted legislation to facilitate registration and financing of associations is a strong signal that the Turkish government wants more local participation in the formation of policies, including environmental policies. Currently, municipalities are in charge of implementing environmental law. The process of local participation, together with NGO support, is likely to gain momentum during the pre-accession period and after membership. On the other hand, if policy implementation is decentralised, government may lack reliable information on environmental problems that is mostly available only at local level.

With the introduction of the CAP single farm payment, environmental cross compliance will become mandatory, and will be linked to the use of farm extension advice. This may help to control those farming practices regarding chemical and water use that are detrimental to environmental protection The extension services will have a key role in providing farmers with training on environmentally friendly agricultural practices, which could be of great benefit in Turkey providing the necessary expertise is available. The CAP second pillar will also be important for helping Turkish farmers to improve allocative efficiency in resource utilisation. MARA's current promotion of organic farming indicates that this is an important sector which the Turkish government intends to develop.

12.4 Implications for agrifood chain development

Bottlenecks to improved competitiveness

Joining the EU's single market in 2015 would subject the Turkish food supply chain to competition from mature industries and highly efficient, well-organised companies in the rest of the Union. To be ready for this competition, the Turkish agrifood chain has to tackle a number of serious bottlenecks. The structure of the Turkish farm sector, the low share of upstream and downstream firms using modern technology and equipment, and the general over-capacity of companies in the food sector have all been described in Chapter 6. The product flows through the chain mainly pass via the open market. The wholesale market system is considered rigid and inefficient, and lacks quality improvement incentives and price transparency. Low profitability, fragmentation, weak integration and low quality awareness are characteristics of the Turkish agrifood supply chain.

Competition on the fairly saturated EU food markets occurs increasingly with respect to quality. In Turkey, consumer awareness of quality issues is limited to a more prosperous minority. However, this group will increase in size as incomes rise and consumers become more discerning. Public policies in this field are important, too. For instance, most retailers in Turkey do not yet emphasise consumer packaging and safety issues when they define quality standards, partly because safety standards are not clearly defined or efficiently enforced by the authorities (Codron *et al.*, 2004). Setting standards on food quality and food safety and enforcing the players to accept the rules of the game are important public responsibilities.

However, as well as income growth and public policies, the move towards competition on quality will be pushed by the rise of supermarket chains in the country. In fact, as examples from other parts of the world show (Reardon and Berdegué, 2002; Dries *et al.*, 2004, Codron *et al.*, 2004), the growth of the retail channels is a key factor for the development of the food industry and farming sector in the short and medium terms. Supermarkets pay increasing attention to quality as part of their strategy to gain market share from the traditional retail channels. If the Turkish agrifood chain wants to take part in the expected expansion of the modern retail sector, it has to match the quality of its supply with the quality demanded by the supermarkets. In setting their private standards, supermarkets normally take public standards as a minimum level. Complying with requirements set by the supermarkets in the coming years should help the Turkish agrifood chain prepare for possible EU membership.

Impact on the food industry

The increasing weight of large format supermarkets in the retail sector will boost the trend towards further consolidation in the food industry. The major driving force is the quality issue. Supermarkets in Turkey increasingly set conditions in terms of prices and quality of the supply offered by the food industry, as well as in terms of other product attributes such as appearance, product diversity, convenience, safety, and so on. Supermarkets also demand that suppliers comply with requirements for packaging and delivery times. Processors who want to be part of the supply chain have to adjust to these demands. This requires investment in production and process technology. Presently, most companies in the Turkish food processing industry use only basic production technology: only one out of six firms uses modern technology for production and quality control (Sirtioglu, 2004). Much investment in modern technology needs a minimum operational size to reach the break-even point. At present, in many branches a large share of the processing companies is too fragmented to make such investments. Small processing firms will *have* to invest in expansion, or merge with others or form alliances to gain economies of scale, or leave the business.

Consolidation in the food industry will also be driven by supermarkets' preference for dealing with a limited number of large suppliers to minimise transaction costs. Food processors may also want to expand in order to strengthen their bargaining position vis-à-vis the large retail chains. The present low-capacity usage in much of the food industry is detrimental to profitability and will inevitably lead to a restructuring of the industry.

Foreign direct investment (FDI) should play an increasing role in the restructuring and modernisation of the Turkish food industry. Basic conditions for attracting FDI are political and economic stability. As long as these two conditions are not met, foreign companies will remain cautious about investing in the country. Experience from central and eastern Europe indicates that the prospect of becoming an EU member can increase a country's attractiveness even if the date of accession is somewhere in the future. Since the 2001 crisis, Turkey's economic situation and the investment climate for foreign investors have improved. The new FDI law of 2003 (TÜSİAD and YASED, 2004) is also expected to encourage foreign investors.

Impact on the farm sector

The retail and food processing industry will pass on the more demanding requirements with respect to food quality and safety to the farming sector. These demands may push many small farmers out of the market when they find it hard to comply with the requirements (Reardon and Berdegué, 2002; Berdegué *et al.*, 2003). Small farmers often cannot make the necessary investments, because they have insufficient own resources and face problems in getting external credit. Moreover, especially in the case of unprocessed, perishable products such as fresh fruit and vegetables, large transaction costs make it more costly for retailers to deal with many small farmers rather than with a few larger suppliers. However, small and medium farmers *can* have a future in modern retail chains. Investments by retailers and/or food processors and vertical coordination with suppliers appear to be crucial in this process (see Berdegué *et al.*, 2003; Dries and Swinnen, 2004). Examples from elsewhere show that a farm assistance programme offered by retailers or food processors may be an important instrument for giving farms access to inputs such as knowledge and techniques, and enhancing their output in terms of quality and quantity.

Vertical contracting may exclude small farms. However, the equity implications of such integrated chains are a justification for public policy involvement. Areas for government initiatives could include stimulating the emergence of alternative marketing structures (e.g. cooperatives), promoting associations of (small) farms to increase their bargaining power vis-à-vis the agribusiness companies and reducing transaction costs for companies dealing with small farms.

12.5 Implications for institutions and their functioning

Real progress in aligning the institutional framework of Turkish agriculture to that of EU agriculture will be needed if Turkey becomes a member in 2015. Preparations observed so far have concentrated on adopting the legislative requirements of the EU *acquis* as quickly as possible. A large number of laws and regulations have been passed, but with little attention to their effective implementation. Implementation demands a radical change in the mindset of not only bureaucrats and policy makers but also of market participants.

Competition rules

Major difficulties and ambiguities still characterise the operation of the Turkish Competition Authority (European Commission, 2004a). Although the adverse effects of state enterprises on the development of competitive agricultural markets are recognised, the privatisation process has so far been slow. Moreover, the manner in which privatisation has been achieved has not always been transparent or above suspicion of bias. In addition, inconsistent decisions appear to arise from weak administrative and expert capacity in the field of competition.

Land, labour and water institutions

Increasing yield has always been a primary goal of agricultural policies in Turkey. It is a goal that has implications for agricultural resource use, technology adoption and diffusion, and natural resource management. At present, institutional arrangements that concern the efficient use of land, labour, water and the environment are progressing. EU membership would act as a catalyst for more efficient use of these resources through improved land institutions (rules and mechanisms for using pasture land and irrigated land), water institutions (water pricing mechanisms and mechanisms for effective distribution of water across regions), labour institutions (social security schemes specifically designed for agricultural workers), and environmental institutions (mechanisms for sustainable use of environment).

Duality in agriculture (traditional versus modern) and regional differences in agricultural technology use (land, labour, and capital-saving technologies) comprise the two key characteristics of Turkish agriculture. These characteristics imply that competition, resources, technology and institutions may also differ both across regions and across types of agriculture. Uniform application of new institutions required by the *acquis* will be extremely demanding. Therefore, a challenge for Turkey is to start addressing these issues soon, because basic institutions should be in place before accession. The duality issue will be critical especially after membership, as it holds implications for the functioning of competitive agricultural markets under the CAP.

Labour market institutions have been brought closer to EU standards, but developments regarding agricultural labour remain inadequate. Currently, a large informal sector in agriculture (more than 90 per cent of workers in agriculture are unregistered) undermines the provision of social security and the enforcement of minimum wage regulations (OECD, 2004). The presence of this huge informal sector further hampers the enforcement of product market institutions, including food quality standards and intellectual property laws. EU accession would force Turkey to resolve these problems because its labour market institutions would become those of the EU. However, the presence of a large informal economy in agriculture must be regarded as a serious handicap in adopting the *acquis*.

Regarding land and water property rights, the regulations are already in place. However, the implementation of a land rental market and water use contracts still needs to be improved to comply with the *acquis*. As for intellectual property rights in traditional knowledge, ample scope exists for progress, although in 2003 significant progress was made to protect new varieties and plant breeders' rights.

Until the recent agricultural policy reform, the actors involved in the agricultural policy process were all under the direct control of the government. Even after the restructuring of institutions, the influence of government remains strong in the policy process. This makes it difficult for agricultural producers to demand changes regarding social dialogue and social protection. Turkey will be expected to offer social safeguards to its large group of agricultural producers in line with the *acquis*. Adopting the *acquis*, however, does not guarantee improved production and welfare. Success would depend strongly on how effectively relevant regulations and laws on social dialogue and protection are enforced and whether farm investment is stimulated.

Agricultural research, extension and knowledge institutions

Turkey's farming sector will have to become much more efficient after membership as it will encounter stronger competitive pressure from farmers in the rest of the EU. Increasing the capacity of the Turkish farming sector would be greatly facilitated by progress in an agricultural knowledge and information system, appropriate priority setting for agricultural research, adequate farm extension services and improved linkages between farm and off-farm sectors. Critical in this process will be the recognition of knowledge as an engine of growth and the preparation of the required legal framework for protecting intellectual property rights. At present, almost all farmers regard knowledge as a public good, impeding the further development of the intellectual property rights system. Changing the mindset of farmers with a small parcel of land and limited access to credit and technology seems to be a key challenge for Turkish agriculture. Under these conditions, the public sector naturally becomes the major producer of agricultural knowledge and the promoter of institutions for the adoption and diffusion of this knowledge. For the pre-accession period, it is realistic to expect that the public sector will continue to be the key actor in agricultural research.

Participatory approaches to natural resource management are increasingly applied in EU member countries. The general tendency in Turkey, however, has been to apply command-control mechanisms without consulting stakeholders such as farmers' organisations, NGOs, processors, input suppliers, trade associations, etc. A shift from the command system to a participatory approach may significantly strengthen the capacity of the agricultural sector to employ agricultural resources more effectively. For example, through participatory approaches, farmers can gain familiarity with rules of common-resource use (such as rules that relate to the use of pasture land open to common use, or rules governing the use of irrigation water). Such a move would also pave the way for the adoption of the *acquis* in areas that relate to technological and institutional innovations in agriculture.

In theory, involving farmers in participatory activities could be promoted by the extension service. At present, however, Turkish agriculture does not have a properly functioning extension system, although this is crucial for helping Turkish agriculture to respond to competitive pressure from the EU. Moreover, Regulation (EC) 1783/2003 requires member states to offer farm advisory services to farmers, and in 2010 the Council may also make farmer participation in farm advisory activity compulsory. Thus, the latest CAP reform has increased the priority given to agricultural extension. A good extension system would contribute to Turkey's competitiveness by supporting farmers with advice on appropriate cropping patterns and new farming technologies (OECD, 2004). Turkey must redress this situation in order to promote an enabling environment for agricultural innovations and their uptake by farmers.

Environmental institutions

No progress has been made concerning the integration of environmental protection into agricultural policy. By contrast, with respect to the environmental impact assessment regulation, the requirements of the EU *acquis* are largely met (European Commission, 2004a). The challenge for Turkey will be to improve its administrative capacity for the implementation of EU requirements.

Since Turkey's overall recorded levels of environmental degradation are low compared to developed countries and because environmental regulations have financial and technological implications, Turkey might be tempted to delay the enforcement of the environmental regulations adopted so far. This would be counter-productive. Environmental damage can be slow to reverse and in many local areas it is already emerging. Most important, since Turkey would have to impose and effectively operate farmer cross compliance regulations on becoming an EU member, it is essential to have appropriate institutions with experienced staff in place by the time of accession. Adopting the environmental *acquis*, understanding its implications for Turkish agriculture and developing the required institutions could be done over 10 years if it is tackled promptly.

Education

With respect to adult literacy, Turkey not only lags behind the level expected from EU members but also shows significant gender inequality (see Chapter 5). The existing gender inequality seems to originate largely from patriarchal characteristics of Turkish society. There are marked regional differences in literacy rates. There is time between now and 2015 to close the regional gaps by targeting educational spending in regions that are lagging behind. However, closing the gender gap requires more than investment because its roots are embedded in the social fabric of the society. Regarding university education, quality varies significantly across regions. The challenge for Turkey before and after 2015 would be to close the gap between genders and between regions, and make significant progress in the stock of human capital. There is a strong role still to be played by more widespread and improved nation-wide education services.

Institutions relating to foreign investment

With significant improvement in economic and political institutions before membership in 2015, the current low levels of FDI are likely to increase. Expectations are high especially in the vegetable oils and fats sector, and in the retail food industries, but success depends on how quickly and effectively new rules are internalised by market participants. A priority area for improvement is the administrative capacity of the legal system and competition-related institutions (Dutz *et al.*, 2003; OECD, 2004). For example, the lack of stable, permanent, reliable laws and regulations comparable to those of the EU invites corruption and bribery. Another priority need is the adoption of time-consistent policies. For example, the lack of a political culture that respects decisions and commitments made by previous authorities reinforces unpredictability and increases country risk (TÜSIAD, 2004).

It is not unrealistic to expect a significant increase in FDI inflow up to 2015 if Turkey takes measures to boost confidence by removing obstacles in these areas. Turkey has already missed major FDI opportunities because of these institutional weaknesses. In particular, Turkey's customs union with the EU did not lead to an increased FDI inflow, due mainly to the lack of an enabling environment for investment with clear and unified rules of conduct. Expectations arise now with respect to Turkey's EU membership, but success is largely dependent on the signals given to foreign investors during the pre-accession period.

FDI inflow is expected to play a critical role in promoting technology transfer and hence economic growth. Technology transfer becomes attractive when appropriate skills are available to service and exploit it. Investment depends on adequate levels of human capital. The presence of an effective intellectual property rights system is also essential for attracting FDI. Currently, the legislation on the protection of intellectual property rights is in place, but its implementation is poor. Furthermore, some of the laws and regulations are not yet in harmony with those of the EU (TÜSIAD, 2004).

Food quality

EU membership would mean that Turkey has to comply fully with the General Food Law of the EU regarding standardisation of food regulations. If current trends continue up to 2015, Turkey should not encounter serious difficulties in the adoption of the Law; however, enforcing it as it is intended to operate will be more onerous. A key difficulty relates to the current governance structure of food safety and quality control. Local authorities assume responsibility for food inspection, which might lead to different practices across jurisdictions, while the EU requires a central authority responsible for all standards. The danger with decentralisation is that it can create incentives for interest group activities.

Quality and functioning of institutions

EU membership would bring with it both opportunities and obligations for Turkey. EU membership offers Turkey the opportunity to become the modern, Western-style country that it has the ambition to be. A corresponding obligation is that as an EU member Turkey would be expected to adopt and fully implement the EU *acquis*. These two challenges are closely related to each other in that, in order to meet them, it is imperative for Turkey to address weaknesses in quality and enforcement of its institutions.

Institutional reform needs to begin with government. For example, regarding institutions that relate to FDI, a large gap exists between Turkey and EU countries. According to a World Bank survey, 92 per cent of investors ranked complexity and non-transparency of government regulatory policies as a serious constraint to business operations (OECD, 2004). Various chapters of this book refer to duplication and lack of clear strategic and operational guidelines for various institutions, as well as failure of communication between them.

Improvement is also required in budget and tax collection institutions (OECD, 2004), bearing in mind that large budget deficits were the key source of Turkey's recent financial crises. With the unregistered economy currently estimated at about 50 per cent of total GNP, Turkey is losing a significant amount of tax revenue (OECD, 2004). An informal economy of this size constitutes a serious obstacle to sustainable growth of the Turkish economy. This is an example of how ineffectively designed rules and regulations create a problem of adverse selection. The regulations concerning the labour market and taxation, characterised by high employment protection, substantial labour tax and social contribution wedges, create incentives for a large informal sector and hence a shrinking tax base and lower social contributions (OECD, 2004).

Corruption and bribery are widespread especially in the areas of customs, public procurement, taxation, the municipalities, the courts and the implementation of incentive schemes. Hospitals and the police, especially the traffic police, are areas where so-called '*petite corruption*' is common. According to the Corruption Perceptions Index of Transparency International, Turkey's place among 91 countries is somewhere in the middle in 2000 and 2001 (Şenatalar, 2002). Equally critical is the widespread tax evasion due to the application of value-added tax. Typically, consumers prefer not to have receipts especially when their expenditure involves large sums. The growing civil society participation and a networked media are two effective means for curbing not only corruption and bribery but also all kinds of irregularities observed in the private and public sectors. For example, the privatisation process that started after the 1980s came under such suspicion of bribery that the government started to broadcast auctions of state-owned firms on live television (Stafford and Pizzo, undated).[7]

Poor infrastructure of the justice system is also a concern. There are many indications that courts are not equipped with adequate physical, human, and financial resources, and that judicial services are slow. EU twinning projects and other international cooperation may bring useful contributions in this area. Regulatory rules as well as their actual enforcement need to be improved. The biggest problems are found in the administration of licensing regulations, land-planning and zoning decisions, environment protection regulations, tax accounting rules, corporate law and bankruptcy law (OECD, 2004).

The implications of these weaknesses for Turkey are obvious. The institutional framework for agriculture is at an early stage of development. In addition, for the framework to be operational, the quality and enforcement of rules, laws and regulations must be improved before membership in 2015. Moreover, current levels of corruption, bribery, and discriminatory application of regulations are not conducive to boosting economic growth, quite apart from the issue of institutional discipline that the EU imposes on member states. The 10-year period up to 2015 may be long enough for the most important changes in formal institutions and formal institutional rules that are required for membership. However, the 'bedding in' of these changes is likely to take longer. As for the informal rules (people's expectations of how rules work in practice, of the impartiality of formal rule systems and of the true antisocial nature of corrupt practices), these changes will probably take even longer and represent a long-term challenge to Turkey as a nation and prospective EU member.

---

[7] The Governance Indicators for 2002, published by the World Bank (Kaufman *et al.*, 2003), show Turkey's indicators for Government Effectiveness, Regulatory Quality, Rule of Law and Control of Corruption as –0.20, 0.08, 0.00 and –0.38, respectively, on a scale of –2.5 (lowest) to +2.5 (highest). For comparison, these indicators are 0.91, 1.15, 0.82 and 0.80 for Italy, 1.76, 1.59, 1.73 and 1.82 for Germany, 2.14, 1.87, 1.83 and 2.15 for the Netherlands, 1.03, 1.47, 1.3 and 1.33 for Portugal, and 0.78, 1.21, 0.90 and 0.60 for Hungary. For further comparison on Control of Corruption, the indicators for Poland, Bulgaria, Romania and Ukraine are 0.39, -0.17, -0.34 and –0.96, respectively.

12.6 Economic consequences

Budget flows

The net budget effect of Turkey's membership is a transfer from the EU-27 to Turkey.[8] This transfer has to be financed from the EU budget. The various budget headings are: CAP (first pillar), separated into market and price support, and direct income support, CAP (second pillar budget: rural development policy), structural policy funding (including cohesion funds), other budget items (administration, education, research, security and justice, environment, fisheries, etc.) and, in the reverse direction, Turkey's budget contribution to the EU.

Payments to Turkey from the EU budget, and Turkey's contribution to the budget, were presented and discussed in Chapter 11. Following the methodology explained there, Turkey's net budget receipts for 2015 are calculated at EUR 10.9 to 18 billion (at 2004 prices) and EUR 13.6 to 22.5 billion (at 2015 prices, assuming inflation of the euro at 2 per cent per year). The lower figure in the range corresponds to an absorption capacity for structural funds of 2 per cent of GDP, whereas the upper figure is for absorption at the rate 3.5 per cent of GDP. Even at the lower absorption rate, structural and cohesion spending make up over one half of gross budget receipts. Given the very low income level of Turkey and even under the assumed real annual growth of GDP of 5.2 per cent per year, 25 of Turkey's 26 NUTS II regions would qualify as Objective 1 areas in 2015. Turkey would also be eligible for cohesion funds.

Other consequences

Net budgetary flows are only one aspect of the consequences of EU entry for Turkey. Often this aspect is neglected altogether because of a stronger focus on expected benefits such as higher economic growth, more opportunities for international trade, greater transparency in governance matters and lower corruption, etc. These effects can be more important in the long run, but in the first years after accession and after full implementation of structural policy, budget transfers will also play an important role for economic growth. Future budget transfers can be used by acceding countries to justify important reforms that are necessary to bring rules and regulations to the level of the *acquis communautaire*. This process is facilitated by pre-accession funding, which has not been included in our analysis. An important challenge is to design programmes for structural and cohesion spending and rural development that address some of Turkey's particular weaknesses such as low levels of human capital, poor opportunities for non-agricultural employment in rural areas, and low levels of health and quality of life in rural areas.

Longer-term effects for the economy arise from a more open trade position, a more stable economic growth path, a reduction in the financial burden of foreign

---

[8] Part of the change will also be that (lower) tariff revenues will largely go to Brussels instead of to the Turkish budget. This shift has not been quantified.

debt because of increased stability and – above all – a clear direction for policy change. By comparison, the budgetary flows, defined in advance by objective rules, are the most predictable consequence of accession. Other less tangible aspects, particularly those that depend on or consist of changes in informal institutions, depend on the ability and willingness of Turkish people themselves.

12.7 Selected studies on Turkish accession

We have reviewed a number of studies that use general and partial equilibrium models to assess the expected consequences for Turkey of EU accession. Of these studies, only four are considered relevant (with respect to choice of time horizon and focus on agriculture) in the context of this book (see the Appendix to this chapter). Drawing on the main findings of these studies, this section discusses effects of accession in four areas.

The first area concerns the quality and functioning of institutions. This is an important area since the effects of enhanced institutions would be observed everywhere in society, promising much larger gains than the gains from improved agricultural institutions only. It has been stressed that, due to the ineffective functioning of institutions, the poor quality of regulations and the high level of corruption, Turkey has so far not been able to take advantage of existing opportunities nor to create new opportunities in the areas of FDI and trade. The second area relates to the effects of accession on welfare and income distribution. The third area concerns the competitiveness of agricultural products. The question of competitiveness immediately follows from the information presented in Sections 4.5 and 8.3. Which agricultural sub-sector(s) are likely to remain competitive after accession, and which may be threatened with collapse? Finally, the fourth area, already discussed in Sections 3.3 and 7.3, relates to harmonisation with EU's food safety standards.

Discussion

- *Significantly reduced corruption promises higher growth than access to the EU internal market.* An important issue explored in the literature is the effect on growth of institutional reforms expected to be triggered by the prospect of Turkey's membership. The reforms undertaken so far suggest that possible EU membership has already started to work as a catalyst for improving the institutional framework of Turkish economy in general, and of Turkish agriculture in particular. However, the effects of these reforms on the quality and functioning of this framework and hence on growth remain to be seen.

In the CPB study, Lejour *et al.* (2004) assess separately the macro-economic impact of three aspects of Turkey's EU accession: removal of non-tariff barriers upon entry to the single market, reduced corruption, and labour migration to EU-15 (for a description of the model, see Appendix). Of these three different features of accession, it is the reduction in corruption (the *ad hoc* assumption that Turkey moves from 64th to 25th place in the Corruption Perceptions Index) that brings the greatest gains to Turkey. Trade increases by 57 per cent, which in turn leads to an increase in GDP of 5.6 per cent and in consumption of 8.9 per cent. Reduced corruption would also have a positive impact on Turkey's trade partners. For example, consumption in Romania would rise by 1.1 per cent because of cheaper imports from Turkey and welfare in the EU-15 would increase by the equivalent of USD 8.5 billion. By contrast, the reduction in non-tariff barriers (NTBs) upon accession to the EU internal market would lead to predicted increases of just 0.8 per cent in GDP and 1.4 per cent in consumption.

These results call for several comments. There is, in fact, considerable correlation across countries between the capacity of administrative bodies, the rule of law, the quality of institutions and control of corruption. These aspects are all interrelated, and it is very difficult to isolate the effects of corruption from the effects of poor quality institutions, weak administrative capacity, and so on. Institutional reforms, which cover a much larger domain than corruption, could have been modelled more broadly by including changes in the capacity of public administration, in rule of law and regulatory organisation. Alternatively, it may be that the estimated impact on trade of lower corruption in this model also picks up the effect of these other highly correlated institutional features, in which case its interpretation as showing the effects of reduced corruption alone is too narrow. In any case, it is a weakness of the model that the positive effects of reduced corruption in Turkey are limited to trade effects only. The positive externalities from less corruption would be widespread across all the sectors of the economy concerned, and hence welfare gains are likely to be underestimated.

- *With EU membership, agricultural producers (farmers) lose but food consumers gain, implying income transfer from rural to urban sector. Within producer and consumer groups, membership reduces income inequality.* Grethe (2004) finds that, compared to the status quo, producers lose but consumers gain, both when trade is completely liberalised and when agriculture is brought into the customs union (CU) with the EU. Under these scenarios, the total loss in producers' surplus amounts to about EUR 2.8 and EUR 1 billion respectively, while consumers record welfare gains of about EUR 3.5 billion and EUR 1.5 billion respectively. As a result, the net gain from full trade liberalisation is EUR 0.7 billion, whereas the full CU scenario yields a net gain of EUR 0.5 billion. When budgetary effects of tariffs, export subsidies and producer premiums are taken into account, total welfare gains amount to EUR 667 million (about 2.3 per cent of projected agricultural production value, or 0.4 per cent of projected GDP) with liberalisation and EUR 482 million under the CU scenario. Cakmak and Kasnakoglu (2003) find similar results. With membership, the degree of producer protection and the general price level decrease, while imports of cheap livestock products increase, which in turn increases consumers' welfare.

Grethe's findings further suggest that liberalisation reduces intra-sectoral income inequality in absolute terms. Full liberalisation reduces income inequality among farmers, as large wealthy farmers receive most of the gains in producer surplus resulting from current price support. On the other hand, liberalisation leads to a more equal distribution of real income relative to the status quo due to lower food prices. Bearing in mind that producers are mostly located in rural areas, their deteriorating economic situation, together with consumers' welfare gains, can be interpreted as an income transfer from the rural to the urban sector.

- *With EU membership, fruit and vegetables remain competitive but cereals and livestock products are uncompetitive.* In Grethe's study (2004), in both the full liberalisation scenario and the agriculture-in-the customs-union scenario, Turkey appears to be a net exporter of fruits and vegetables and a net importer of cereals and processed products (mostly animal products). Under both scenarios, Turkey also remains a net exporter of plant products as a total, including fruits, vegetables, cereals and other crops. These findings are consistent with those of Cakmak and Kasnakoglu (2003) (see Appendix). They show that with membership the volume and value of the production of cereals decline, while those of vegetables, fruits and nuts rise. They further report large imports of livestock products. Grethe finds that, under both the liberalisation and full customs union scenarios, Turkey is expected to be a net importer of agricultural products, although it is a net exporter in the base period and the status quo scenario.
- *With either full trade liberalisation or simply including agriculture in the customs union with the EU, considerable changes in the regional distribution of producer surplus would occur.* Grethe (2004) indicates that, in relative terms, farmers in the northeast and the Black Sea regions suffer the highest loss in producer surplus because of the high share of sugar in the northeast and tea in the Black Sea region. Losses amount to about 13 per cent of production value. Farmers in the Aegean, Mediterranean, and southeast regions, however, suffer the smallest loss because of their high share of fruit, vegetables and cotton, which still remain competitive. For animal producers, the absolute welfare loss is more equally distributed among regions.

The findings of Cakmak and Kasnakoglu (2003) further indicate that the South-eastern Anatolia Project (GAP) region would benefit the most from EU membership since the positive effects of an expansion of irrigated land are relatively easily captured with membership. It is the only region that enjoys an increase in agricultural production value. In the case of membership with area compensation payments, almost all the payments accrue to the Central Anatolia region, which is predicted to experience a 9 per cent fall in revenues following accession.

- *Harmonisation with the EU health and safety standards yields absolute welfare gains for Turkey, whereas for the EU, gains are conditional on agreements regarding border frictions and certification rules.* Zahariadis (2002) finds that Turkey records welfare gains under all the five scenarios he studied. On the other hand, the EU gains in the border and certification scenarios only, implying a possible cooperation between Turkey and the EU under these two scenarios. With mutual agreement on certification rules, Turkey is expected to increase exports of animal products and processed food considerably. As suggested in Chapter 10, the question of certification for Turkey's meat and dairy products is likely to be problematic for some years. Regarding standardisation, Turkey gains from the adoption of EU standards, while the EU loses because the new standards lead to a reorganisation of production, and possibly to investment in improved technologies. The new standards will also make Turkish products more easily substitutable for EU products, thus enhancing their demand in both Turkish and EU markets. In fact, Turkey's increasing trade surplus under the standardisation scenario can be attributed to increasing exports (especially to the EU) relative to imports.

12.8 Conclusions

This chapter is based on the main trends described in Chapters 2 to 11. The working hypothesis is that Turkey joins the EU in 2015. The discussion has evolved around two questions: What are the implications for Turkey of adopting the *acquis* by 2015? What are the expected consequences for Turkey of entering the EU in 2015? Here we summarise the key conclusions.

Adaptation of the formal institutional framework for agriculture is underway, and the Turkish administration has been addressing a wide range of issues in order to incorporate the *acquis*. Competition, property rights, and education and research institutions are largely in place, but difficulties still arise especially in implementation. Similar improvements and difficulties are also observed in the organisation and implementation of rules and regulations that concern agricultural resources, technology, production and trade. A massive upgrading of the research and extension system is needed by 2015 in order to fulfil certain aspects of the *acquis*, and to help Turkish agriculture exploit the opportunities of EU entry.

Regarding economy-wide institutions, tax collection, the functioning of the judicial system, and credibility and time-consistency of public policies were and remain the key areas to be improved. Parallel evolution of informal rules and expectations is important if new formal institutions are to function as intended, and as they do in other EU member states.

The supermarket industry has begun moving along a stable path, promising considerable improvements in food quality by 2015. However, for a stronger and more competitive food supply chain, the Turkish farming sector with its large share of semi-subsistent and fragmented farms needs to be re-structured.

The changes in Turkish agricultural policy due to ARIP should help Turkish agriculture adjust to the realities of the CAP. The incomplete privatisation of state enterprises, however, still jeopardises the continuing alignment of Turkish agricultural policy to the CAP. The establishment of a rural development strategy, currently lacking in Turkey, is especially important in the light of the poor competitiveness of the farm sector. Turkish food policy legislation is progressing in line with the requirements of the General Food Law of the EU. However, in practice, safety standards are not clearly defined or efficiently enforced by the authorities.

In trade policy, Turkey will adopt the common external tariff of the EU for agricultural products in or before 2015. Agricultural trade harmonisation between the EU and Turkey will bring tariff reductions in Turkey for most products. The largest tariff adjustment is expected in the livestock sector. The challenge for Turkey is to develop the infrastructure, administrative capacity and commitment for effective border control in the run-up to EU membership.

The veterinary situation amongst Turkey's grazing livestock populations falls well below standards in EU member countries. Even with greater efforts to eradicate the most infectious diseases endemic in Turkey, biological processes and structural features of the sector constitute serious impediments to effective short- and medium-term improvement in the situation. It is unlikely that by 2015 a single market in animal products, without SPS border controls, can be operated.

Adapting Turkey's agricultural and environmental legislation to conform with the *acquis* should lead to better protection of the environment. Environmental cross compliance linked to the CAP Single Farm Payment, and CAP rural development measures, will encourage Turkish farmers to improve and diversify their use of natural resources. Where there is a strong behavioural element, advances will be slow until environmental awareness increases. Monitoring and enforcement will be one of the greatest challenges, but accession will act as a catalyst for improvement.

Research findings from the literature suggest various consequences of Turkey's accession. First, if EU membership leads to reduced corruption, this could be one of the strongest welfare-enhancing effects of accession. Second, fruit and vegetables are likely to remain competitive in the EU single market, while cereals and livestock products will suffer severely. Third, producers (farmers) lose but consumers gain, implying income transfers from rural to urban groups. Fourth, liberalisation and/or the inclusion of agriculture in the customs union with the EU may lead to considerable changes in the regional distribution of agricultural incomes. Lastly, harmonisation with EU food safety standards is expected to yield absolute welfare gains for Turkey.

A major challenge will be to design programmes for structural and cohesion spending that address some of Turkey's specific weaknesses, such as low levels of human capital, poor opportunities for non-agricultural employment in rural areas, and low levels of health and quality of life in rural areas.

References

Berdegué, J.A., Balsevitch, F., Flores, L. and Reardon, T. (2003) Supermarkets and private standards for produce quality and safety in Central America: Development implications. Report to USAID under the RAISE/SPS project, Michigan State University and RIMISP, July 2003.

Cakmak, E. and Kasnakoglu, H. (2003) The impact of EU membership on agriculture in Turkey. Conference on *EU-Med. Association Agreements and the EU's New Neighbourhood Policy Vision*, 6 December 2003, Marseilles.

Codron, J.M., Bouhsina, Z., Fort, F., Coudel, E. and Puech, A. (2004) Supermarkets in low-income Mediterranean countries: Impacts on horticulture systems. *Development Policy Review* 22 (5), 587-602.

Dries, L. and Swinnen, J.F.M. (2004) Foreign direct investment, vertical integration and local suppliers: Evidence from the Polish dairy sector. *World Development* 32(9), 1525-1544.

Dries, F., Reardon, T. and Swinnen, J.F.M. (2004) The rapid rise of supermarkets in Central and Eastern Europe: Implications for the agrifood sector and the rural development. *Development Policy Review* 22, 525-556.

Dutz, M., Us, M. and Yilmaz, K. (2003) Turkey's foreign direct investment challenges: Competition, the rule of law, and EU accession. Paper presented at the conference 'Turkey: towards EU accession', 10-11 May, Ankara.

European Commission (2004a) 2004 Regular report on Turkey's progress towards accession. SEC (2004) 1201, Brussels.

European Commission (2004b) Issues arising from Turkey's membership perspective. SEC (2004) 1202, Brussels.

European Commission (2004c) A description of the common organisation of the market in sugar. AGRI/63362/2004, Brussels.

Grethe, H. (2004) Effects of Including Agricultural Products in the Customs Union between Turkey and the EU: A Partial Equilibrium Analysis for Turkey. CeGE-Schriften, Vol 9, Center for Globalization and Europeanization of the Economy, Georg-August-Universität Göttingen. Peter Lang Verlag, Frankfurt am Main.

Kaufman, D., Kraay, A. and Mastruzzi, M. (2003) *Governance III: Governance Indicators for 1996-2002*. World Bank, Washington, DC.

Krugman, P. (1991) Increasing returns and economic-geography. *Journal of Political Economy* 99,483-499.

Lejour, A.M., de Mooij, R.A. and Capel, C.H. (2004) Assessing the economic implications of Turkish accession to the EU. CPB document No 56. CPB Netherlands Bureau for Economic Policy Analysis, The Hague, March 2004.

NPAA(2003) Turkey's national plan for adaptation of the *acquis*. *Turkish Official Gazette* No. 25178. 24 July 2003.

OECD (2004) *Economic Survey for Turkey 2004*. OECD, Paris.

Reardon, T. and Berdegué, J.A. (2002) The rapid rise of supermarkets in Latin America: Challenges and opportunities for development. *Development Policy Review* 20(4), 317-334.

Republic of Turkey (2003) Preliminary national development plan (2004 - 2006). Ankara, December 2003.

Şenatalar, B. (2002). Public sector reform and combating corruption. In: Dunér, B. (ed.) *Turkey: the Road Ahead?* Swedish Institute of International Affairs, Stockholm.

Sirtioglu, I. (2004) Turkey retail food sector report 2004. *GAIN report*, Number TU4005. USDA Foreign Agricultural Service, Ankara.

Sözen, S., Avcioglu, E., Ozabaldi, A., Gorghun, E. and Orhon, D. (2003) European Union water policy – tasks for implementing 'Water Framework Directive' in pre-accession countries. *Journal of Environmental Science and Health*. A38(8), 1401-1410.

Stafford, R. and Pizzo, C.M. (undated) Informed questions paper: Turkish economic issues. National Defense University, Washington, DC.

TÜSIAD (2004) Investment environment and foreign direct investments in Turkey. Working Paper (TS/SEK/04-06) prepared for Investors Advisory Council Meeting, 15 March, 2004, Istanbul.

TÜSIAD and YASED (2004) FDI attractiveness of Turkey: A complete analysis. TÜSIAD, Istanbul, February 2004.

Zahariadis, Y. (2002) Deep integration in the EU-Turkey Customs Union: A preliminary analysis based on the GTAP model. Paper prepared for the *DESG Annual Conference in Development Eco*nomics, 18-20 April 2002, University of Nottingham. Sussex. European Institute, University of Sussex, UK.

Appendix: Summary of four studies using simulation models

Study	Model characteristics	Scenarios	Simulation results
Lejour, de Mooij and Capel (2004)	• CGE model for the world economy • 15 industries: agriculture, raw materials & energy, 8 manufacturing, 5 service • 6 EU regions, NMS, Bulgaria, Romania, Croatia, Turkey, 4 regions in rest of world • Projection path until 2025 • Gravity equation estimated from a cross-section of 38 countries for 2001 based on the GTAP v6 database • Gap between potential and actual trade translated as tariff equivalent barriers.	• Baseline scenario: Current Turkey-EU relations until 2025: customs union in industrial products, limited integration to the internal market, neither full membership nor further integration in other respects. Turkish real growth rate GDP 5.6% per year; GDP per capita grows by 4.5%. NMS and AC-2 enter the EU in 2004 and 2007, respectively • Scenario 1: Access to the EU internal market (elimination of NTBs) • Scenario 2: Membership triggers institutional reforms in Turkey (i.e., less corruption and more transparency) • Scenario 3: Migration owing to free movement of labour • Simulations assume that Turkey enters the EU in 2010	Accession benefits Turkey, with little effect on EU-15, NMS and AC-2. • Scenario 1: private income (i.e. welfare) in Turkey rises by USD 4.4 billion, GDP expands by about 0.8% in the long term, the largest impact in Turkey is on Textiles & Wearing Apparel at the expense of production of these sectors in other European countries. • Scenario 2: Potential gains are larger than those under Scenario 1, if Turkish institutions function comparably to those in Portugal. Welfare increases by USD 28.2 billion, while GDP would expand by 5.6%. • Scenario 3: Per capita income in Turkey will rise, while it falls slightly in the EU. If migrants are primarily unskilled, also wage inequality in the EU-15 is likely to rise. • Turkey benefits most under scenario 2.
Cakmak and Kasnakoglu (2003)	• Non-linear mathematical programming model of the agricultural sector • 50 commodities produced and 4 regions for production effects • Base period – average 1997-99 • Projections for 2005	• Out-EU: no EU membership • In-EU: EU membership in 2005 • In-EU 1: In-EU with CAP and EU prices, without compensatory area payments • In-EU 2: In-EU plus compensatory area payments • In-EU 3: In-EU 1 plus technological improvement in Turkey's livestock production	• Out-EU: increase in total surplus in 2005 compared to base period. • In-EU 1: an additional 1 per cent increase in total surplus. Producers' surplus decreases by 16%, consumers' surplus increases by 12%. • In-EU 2: slightly lower welfare compared to In-EU 1 because payments are not included in surplus calculation. • In-EU 3: gives the highest surplus, followed by In-EU 1 and then In-EU 2. • Overall, membership benefits consumers but hurts agricultural producers.

Grethe (2004)	• Static partial equilibrium model • Regional farm supply of plant and animal products; 12 crops, 17 vegetables and fruit, and 5 animal products • Income elasticities of demand from the 1994 expenditure surveys • Projections for 2006	• Base period - the average of the 1997-99 data for crops, of the 1998-99 data for animal products, and of the 1996-98 data for prices • Status quo (reference) scenario – unchanged agricultural policies at the end of the projection period 2006 (i.e., market intervention kept at the 1996-98 level) • Liberalisation - Turkey abolishes all market policies, e.g. tariffs, export subsidies, premiums • CU - agriculture included in the customs union (CU) with the EU	Compared to the status quo, producers lose but consumers gain under both the liberalisation and the CU scenarios. Net gains (consumers' welfare minus producers' surplus): EUR 0.7 billion in the liberalisation and EUR 0.5 billion in the CU. When tariffs, export subsidies, producer premiums are considered, total welfare gains are EUR 667 million (0.4% of projected GDP) for the liberalisation and EUR 482 million for the CU scenario, relative to status quo. Under all scenarios, Turkey is a net importer of cereals, processed products and animal products, and a net exporter of fruits, vegetables, and plant products as a whole. With respect to total agricultural products, under the liberalisation and CU scenarios, Turkey tends to be a net importer, while under the base period and status quo a net exporter.
Zahariadis (2002)	• Multi-sector, multi-region CGE model • For baseline, GTAP v5 dataset for 1997 • 9 regions and 20 sectors including 4 agricultural sectors • GTAP v4 dataset for 1995: data on pre-customs union industrial tariffs and estimates on regulatory costs	• Simple CU (baseline): Customs union (CU) established in 1996 • Border: abolish bilateral border/frictional costs by 1996 • CU 2001: Turkey adopts Free Trade Areas in industrial products with EU's preferential partners by 2001 • Standards: Turkey adopts EU technical regulations by 2001 • Certification: Turkey and the EU adopt a mutual recognition agreement. Products certified in one country freely move to another with no duplicative requirements	Turkey gains from liberalisation under all five scenarios. The total welfare gain is the strongest under the simple CU, followed by the Certification scenario. The EU appears to lose in the standardisation scenario due to the absence of product substitutability, while gaining from the abolition of border and the possible elimination of certification costs. Gains for Turkey are high from the adoption and enforcement of EU standards and certification procedures. For the EU, however, gains mostly stem from the improvement in mutually agreed certification procedures.

13

Opportunities, Threats and Challenges

Alison Burrell

13.1 Introduction

By 2015, it will be nearly 60 years since the Treaty of Rome launched the realisation of the European project. The founding member countries of what is now the European Union shared a centuries-long history of close, not always friendly, neighbourliness and had recently emerged from a common experience of war and destruction on an unprecedented scale. Their common and complementary economic priorities, their cultural empathy and above all an urgent desire to found joint institutions that would safeguard peace within Europe were the driving forces behind the construction of what is now the European Union.

In the intervening years, these institutions have multiplied and their competencies have grown. The areas in which member countries have agreed in successive treaties to adopt common legislation have been progressively extended, as member countries discovered more areas in which harmonisation would be mutually beneficial. From the early years, member countries defined their common project as one in which all regions would move forward together. Maintaining a certain degree of regional cohesiveness was one of the objectives. Whether this was essential to the success of the common project could be debated. However, it was a political choice and the redistributive policies accepted by sovereign member countries, in which budget transfers take place between them according to objective rules, are evidence of members' commitment to the project and to each other. The idea of the 'level playing field' as regards trade between countries has also been an important principle behind sectoral policies and rules for competition. Agriculture and, more recently, rural development and food are areas where policies have been strongly shaped by these principles.

In a technical sense, the growing body of common EU legislation developed over these decades by member countries has made the entry of new members increasingly difficult, the more so for countries whose economic and social evolution has not been running in parallel with that of the Union. However, every enlargement opportunity is historically unique and has to be examined on its merits. Two fundamental prerequisites for successful enlargement are the potential for mutual economic and political benefits, and strong political willingness and commitment.

In considering the issue of Turkey's potential membership, the chapters in this book have focused on the areas of agriculture, food, rural development and structural policy. They have examined Turkey's current situation in these areas in terms of its institutions, its compatibility and complementarity with the existing EU, and its performance. In particular, authors have sought to identify areas where adopting and implementing the EU's *acquis communautaire* might pose problems for Turkey, or where the EU might have to modify the *acquis* to accommodate the particular case presented by Turkey. A related question that has not been explicitly asked but that has, at least in some respects, been answered implicitly is whether the EU's agrifood *acquis* in fact offers the most appropriate set of rules and incentives for Turkey's agrifood sector and its rural population at their current stage of development.

Chapters 11 and 12 have taken Turkey's accession to the EU in 2015 as a working hypothesis, and have looked at the consequences, first for the member countries of the European Union and then for Turkey. This exercise is steeped in uncertainty. Uncertainties abound regarding exogenous developments in the world economy and the international policy context, and in relation to the agricultural, budget and structural policies of the EU. An equally important source of uncertainty concerns developments within Turkey in the coming years: what will be the impetus of the European Council's decision to begin accession negotiations with Turkey in October 2005? How will this decision, which Turkey has been awaiting for many years, strengthen economic stability, encourage foreign direct investment (FDI), accelerate institutional reform and stimulate economic growth? Despite these uncertainties, some issues have been clarified and various conclusions of a qualitative and quantitative nature can be drawn.

This chapter synthesises the findings of the book, in particular drawing on the conclusions and implications of the previous two chapters. Its aim is not to produce a comprehensive list of opportunities, threats and challenges, in relation to agriculture, food and rural areas, that would arise from Turkish accession but rather to shed light on some of the questions that existing EU member states are inevitably asking when considering the prospect of Turkey as a member. These questions include: how would Turkey's entry benefit the European project that has been under development for the last 50 years? How might member states individually and collectively gain in concrete terms? How might Turkey's entry threaten what has already been achieved, or what it is hoped will be achieved in the future? And what will be the most difficult challenges ahead, in the specific areas of interest addressed by this book, as negotiations bring Turkish accession closer?

13.2 Opportunities

Turkey's accession to the European Union offers certain opportunities for furthering the ideals and principles that have motivated the European project so far. First, it has the potential for increasing security and political stability in the south-eastern corner of the Union by incorporating a strong, committed member state on its south-eastern frontier. In practice, one can question whether Turkey currently has the capacity to fulfil this role in all its aspects. As was discussed in Chapter 11, controlling such long borders requires institutional and administrative capacities that still need to be developed. In the short term, therefore, the scope for illegal movements of goods, livestock and people across these borders could complicate the implementation of a whole range of EU policies with relation to animal health and biosecurity, migration, trade, anti-terrorism and security. Nevertheless, as a long-term aim, a strong stable EU member to the south-east, acting as a bridge for trade and political dialogue with the Middle East and Asia, must be considered an opportunity for the EU.

Of more immediate interest are the economic opportunities offered by the Turkish market for EU companies. This is particularly relevant in the agrifood sector where many factors combine to create strong opportunities for the profitable export of goods, services, know-how and capital to the Turkish market. These factors include Turkey's deficit in animal products, the underdeveloped state of food supply chains and the food retailing sector, the expanding population and the growth in consumer incomes that can be expected in the coming years with or without EU entry. In particular, growing demand for quality food products, which is expected to accompany rising incomes, implies rising demand for the value-adding know-how that will come with foreign direct investment (FDI). In addition, Turkey's entry into the EU would improve market access for EU agricultural exports to Turkey, and should stimulate improvements in the institutional, legal and commercial environment for FDI, which would be commercially interesting to EU firms. Stimulating trade, investment and general economic cooperation between countries in order to strengthen the connective tissue that binds them politically has been one of the driving motivations of the European project, and Turkey's accession would offer appropriate scope for its extension. This will be an important opportunity in the agrifood sector, given the barriers that have existed and still exist currently to foreign firms wanting to operate in this sector.

In recent years, EU legislation has focused increasingly on environmental protection and enhancement, and indeed is leading most parts of the world in its concern for the interaction between economic activity and the environment. As part of this trend, environmental objectives are increasingly recognised in agricultural and rural policies. Thus, environmental protection has become one of the defining principles of the European project, and of what the EU stands for. Turkey has a large territory that is rich in biodiversity, and it currently appears to have fewer environmental problems than many existing EU members. Turkey's adoption of the EU environmental *acquis* would promote the conservation of its environmental endowment, which has to be seen as preserving a public good of relevance to the whole Mediterranean region and to future generations. In the environmental domain, therefore, there is scope for extending and strengthening the application of another fundamental EU principle that helps to define the longer-term goals of the EU and how it aims to promote the common good, although this has not be quantified in economic terms.

13.3 Threats

This book has covered a number of aspects where potential threats to EU common interests in the area of agriculture, food and rural development have been identified. We have, for example, signalled the possibility that achievements already made in the areas of food safety and quality could be diluted by the entry of another member with current difficulties in this domain, that average standards of governance may be lowered by the incorporation of a country with very different institutions and a poor record in this respect, and that the risk of animal disease outbreaks in the EU as a whole may increase – or may be perceived as having increased, which also has negative consequences in trade terms. The extent and significance of most of these concerns are difficult to assess without making assumptions about how much progress can be made during the pre-accession period, how effectively Turkey could implement and enforce the *acquis* upon accession, and the speed with which Turkey's formal and informal institutions are likely to adapt to European norms. Past experience in western Europe with animal disease eradication suggests that more than ten years are needed to achieve full disease-free status without vaccination, but in this case quarantine measures with or without internal zoning could be used to contain the problem.

The question of potential migratory pressure from Turkey after accession has not been covered in this book, although elements relevant to that discussion are described in some detail in various chapters. Migration out of agriculture and from rural areas to urban areas within Turkey has been occurring for many years, as has migration from Turkey to mainly unskilled employment abroad. Many of the Turkish migrant workers abroad indeed have their origins in agriculture within less than one generation. Currently, Turkey still displays the features of a dual economy: relatively developed industrial sectors and urban centres alongside a very large agricultural sector in need of modernisation and a rural population that is poorly educated and remunerated at a very low level. Duality within the agricultural sector is also observed: commercially profitable agricultural holdings and enterprises, many of them export-oriented, co-exist with a large number of small-scale, semi-subsistence holdings. The policy adjustments and competitive pressure that Turkey would experience on entry would undoubtedly impact strongly on some subsectors of agriculture, and on some of the poorest regions in Turkey. In a society where there is no effective safety net for these people, they may well consider that remaining in their rural community is a better survival strategy than out-migration. It is also possible that better functioning job markets and social security systems elsewhere may create strong incentives for mobility.

We have not addressed the question of whether these developments would generate large flows of migrant workers to other parts of the EU, or whether the other EU member countries could absorb them. What we can say is that agriculture and rural areas are likely to bear the heaviest burden of post-accession adjustment. Moreover, it seems clear that large pockets of poor, uneducated and unemployed or underemployed people *anywhere* in the Union pose a problem, in economic and social terms, for policy makers.

The question of the budget cost of Turkish entry has been much discussed, and is perceived by some as a 'threat'. Others consider that, expressed per capita of the whole EU population, or as a share of total EU GDP, it is remarkably small. This view may overlook the fact that it would not be financed by member states in proportion to their population, and that as a share of the total EU budget, payments to Turkey would be substantial.

The estimated budget flows presented in Chapter 11, obtained using our own methodology, are within the range of the most recent estimates made by other researchers and organisations. The resulting figures are, of course, conditional upon the assumptions made about developments over the next ten years, which are clearly spelled out in Chapter 11.

One can argue, however, that the key question is not only how much it would cost and whether the cost is 'high' or 'low', but rather what the opportunity cost of using these funds to finance Turkish accession is in terms of other objectives and priorities of the Union. Member states have recognised the need to revitalise the somewhat sluggish economy of the 'old' EU, and the Lisbon agenda sets out a strategy for spearheading economic revival. Its focus on research and development, and on the enhancement of the knowledge economy, needs large resources to produce results and is in direct contrast to, and in *potential* competition with, the increased use of funds to improve structures and living standards in lagging rural regions. Although the Lisbon strategy was launched in 2000 for a ten-year period only, it is probably unrealistic to assume that by 2010, or even by 2015, the EU will have achieved its growth targets and will be proceeding more or less autonomously on a faster dynamic growth path. Moreover, alongside the Lisbon strategy's objective of encouraging older people to remain economically active, it is likely that provision by national EU governments for their ageing populations will also compete for funds with any increased redistributive claims from agriculture via the EU budget. In addition, as is shown in Chapter 11, transfers to Turkey arising from structural policy are likely to be at the cost of promoting the convergence of regions elsewhere in the existing Union.

Two questions are crucial in the discussion over the budget payments: what would be the impact on EU growth of Turkey's accession, taking into account *not only* any boost to trade and employment that may occur through market harmonisation and integration, *but also* the effect of draining budgetary resources out of the economies of some existing members to a new member country? And, under the existing rules, would the budgetary transfers going to Turkey be spent in a way most likely to improve growth and living standards in that country in the medium and long term?

Regarding the first question, various model simulations in the literature suggest that the macroeconomic impact of Turkish accession, via increased trade, will be small. This is not surprising, since EU-Turkey trade (apart from agrifood) was already liberalised under the customs union. The specific effect of financial flows depends on the relative size of the multipliers in the regions from which financial resources flow and the regions in Turkey where they are spent. There is no hard evidence on these magnitudes. However, it should be stressed that the multiplier effects of spending in Turkey from the EU structural fund depend crucially on the good design and efficient implementation of the projects that qualify for EU financing.

On the second question, there is no guarantee that the additional resources that would become available – as pre-accession financing, or post-accession from the agricultural budget (mainly direct payments to farmers) or from the structural and cohesion funds – would be used to finance the kind of developments that would put Turkey on a fast medium-term growth path. It is debatable whether the spending on pre-accession programmes to facilitate the adoption of the *acquis*, designed as it is for countries at a higher stage of economic and social development, will address what should be the main priorities for Turkey's development and success in the medium term. As is argued in Chapters 5 and 12, much greater investment in education, training and job creation is needed if Turkey is to perform well as an EU member, and for its economy to succeed within a competitive single market. Direct payments to farmers may well end up, at least in greater part, in higher consumption. Finally, although there are very few studies on this issue, the success of structural programmes within EU-15 in promoting sustainable growth in the economies of remote and poorly diversified lagging regions has not been impressive.

Therefore, it is impossible to predict whether the budget flows from the rest of the EU to Turkey after accession will be positive, neutral or negative for growth. The above discussion indicates some of the main factors involved, which should be taken into account when decisions are made regarding the allocation of these financial resources.

13.4 Challenges

To make a success of Turkish pre-accession and accession, many challenges need to be addressed and successfully dealt with. Here we focus on three that arise directly out of the material presented and discussed in various chapters of this book.

The first challenge concerns the institutional arrangements, both formal and informal, that govern public and business life in Turkey. For the EU *acquis* to be implemented correctly and to produce the results intended, it is imperative that this challenge is openly discussed and taken into account in negotiations, training programmes and institution-building initiatives focussing on EU accession. As regards agriculture, food and rural development, the intricacies of policy measures, with their already proven scope for error and mismanagement, in conjunction with the large number of Turkish farmers and food processors, and their fragmented nature, mean that good institutional performance will be essential if the *acquis* is to be implemented in the way it is intended.

The second challenge relates to the large educational deficit that characterises the Turkish population in general, including younger age groups, and particularly in rural areas and in agriculture. No acceding country in the past (with the possible exception of Portugal) has had such a large differential with existing member states in this respect as Turkey has today. The high proportion of the population in the younger age groups and the acknowledged low standard of educational provision mean that many young Turkish people are denied their potential for contributing to the economic success of their country. The various options for helping Turkey to deal with this problem, whether Turkey is outside or inside the Union, are not addressed in the chapters of this book. Although the educational and human capital deficit is particularly relevant to agriculture and rural areas, Turkey's educational deficit is a horizontal issue that needs to be addressed by programmes developed and implemented jointly by policy-making groups having various competencies and responsibilities (including education, agriculture, labour, social affairs). We have not examined what the optimal choices might be to address this need. What does emerge, however, from the material we present is that Turkey's future success, not only economically but as a mature and fully modern society, depends crucially on making very substantial improvements in educational provision and outcomes. This is relevant whether Turkey remains outside the Union, or whether it joins the Union in the foreseeable future.

As well as developing the value-creating potential of Turkey's growing, active population, a third challenge is that of harnessing it. This challenge involves a whole set of inter-related issues, including the need for labour mobility, more job creation in the formal economy, and greater opportunities for female employment. Two current indicators: the very large share of economic activity in the less secure and untaxed informal economy, and the high unemployment rate among better educated young people, are particularly worrying signs of labour market imperfections that are likely to prevent Turkey from achieving fully the balanced and secure growth that could otherwise result from the stimulus of EU negotiations and pre-accession spending. We stress this priority because the agricultural sector in Turkey has traditionally acted as a reservoir of labour, providing manpower to feed the growth of other economic sectors. Turkey's negotiations and accession to the EU, and the modernisation of the agriculture and food sectors that will ensue, will continue this trend by making large numbers of unskilled, and disproportionately uneducated, workers available to the rest of the economy. Indeed, the success of Turkey's agrifood sectors within the Union depends on this restructuring process occurring.

At the same time, Turkey will need additional, skilled labour to sustain the high growth rates that could be feasible in both the pre- and post-accession periods. Matching the supply and demand of the skills required by the economy during this challenging period means not only a huge effort in terms of training provision, but also a major reform of labour market institutions and provision of the social infrastructure for relocated workers. The budget resources that Turkey would receive under the EU structural policy, and pre-accession funding opportunities, would ideally be strongly focused on this priority.

Our book aims to highlight the ways in which Turkey's situation, as regards agricultural production, the food sector and rural development, differs significantly from that of existing member countries. Turkey's accession would bring a new set of challenges that have not yet been confronted, on this scale and in this combination, in any previous enlargement. Our assessment indicates that new strategies, going well beyond what has been done in previous enlargements and specifically designed to meet the particular case of Turkey, will be needed to optimise the accession process and to manage the inevitable – and necessary – structural adjustments that will be set in motion.

Appendix
Summary of Conclusions

Chapter 1: Introduction

- The book offers an overview of Turkey's agriculture, food industry and rural areas.
- Turkey's accession to the EU in 2015 is taken as a working hypothesis. The consequences of accession for both the EU and Turkey are explored on the basis of available literature and current trends, using an approach that places strong emphasis on long-term processes and institutional developments.

Chapter 2: Turkey's economy

- Turkey's economy is about half the size of the combined NMS, but GDP per capita is much lower and closer to that of the AC-2 countries. The comparison of Turkey's income with that of EU member countries is influenced by the units used to measure GDP. The use of PPS (Purchasing Power Standards) reduces the income gap between Turkey and EU member countries.
- Turkey's long-term growth rates are relatively low and have been heavily influenced by negative growth rates in years of economic crisis.
- The current account trade deficit increased sharply in 2003 to 3.7 per cent of GDP. Interest on foreign debt is a significant element in the current account, whereas worker remittances have decreased in importance.
- The distribution of household income is very unequal in Turkey. Rural incomes are less unequally distributed than urban incomes because there are fewer high-income households in rural areas.
- Turkey has much higher regional income inequality compared to other large EU countries. Reducing these regional income gaps will be very difficult because of the dependence of low-income regions on agriculture.

Chapter 3: The institutional framework of Turkish agriculture

- Turkish society has evolved on the basis of the guiding principles of Atatürk's Kemalist ideology, as enshrined in the Constitution of 1923.
- Turkey is a secular sovereign state, whose economy (except for agriculture) is largely open to foreign competition. However, some of the key economic institutions are still partly state-owned or state-controlled, and the benefits of the social security system are limited principally to those working in the

'formal' economy.

- Competition legislation of the late 1990s is partly compatible with that of the EU, but does not cover public sector undertakings and needs to be applied with more rigour.
- Land property rights are well internalised in principle but are not always well defined in practice. Land surveys and land registration are still incomplete, now covering over 75 per cent of agricultural land.
- Labour legislation exists on minimum wages, and to protect workers' conditions in line with ILO and EU standards, but there is inadequate protection of part-time and fixed-term employees. About half of Turkey's workforce operates in the unregistered economy with minimal legislative protection.
- Institutional arrangements concerning labour and farming contracts, water use rights, land transaction rules and environmental impact regulations are weak and not adequately enforced.
- The education system offers eight years of compulsory schooling. However, quality is variable, and enrolment rates are well below 100 per cent, particularly for girls.
- The national farm extension system has performed inadequately for several decades, and is partly responsible for the technological backwardness of Turkish farming.
- In Turkey's food safety and quality legislation, 93 per cent of Turkish standards are now based on European and international standards, while over 90 per cent of EU standards have been adopted as Turkish standards.
- No credible sanctions apply in the case of bureaucratic mismanagement. The bureaucracy has not yet internalised such principles as accountability and transparency.
- On becoming an EU member, Turkey would face the challenge of adapting its informal and formal institutions. Informal institutions are more difficult to change and slower to adapt than formal institutions. Some decades would be needed before EU institutions could perform in Turkey as effectively and efficiently as they do in long-standing EU member countries.

Chapter 4: Agricultural production

- Agriculture accounts for 12 per cent of Turkey's GDP, 34 per cent of employment and 11 per cent of merchandise exports. About 7 million people work in Turkish agriculture, about the same number of agricultural workers as in the entire EU-15.
- Relative to non-agricultural employment in Turkey, agricultural employment is characterised by a high labour force participation rate, low recorded unemployment levels, and high rates of unpaid family labour, particularly amongst females.
- Total agricultural area was somewhere between 35 and 41 million hectares in 2001, of which about 27 million hectares was under crops, fallow, vegetable gardens, orchards and vineyards.

- Much of Turkey's agricultural production originates from the coastal regions, which are well suited to fruit and vegetable production.
- The value of Turkey's agricultural production in 2002 was EUR 29 billion (one-tenth of EU-15 output value). The crop sector in Turkey accounts for a much larger share of output value (77 per cent) than in the EU (55 per cent). Fruit and vegetables together account for 43 per cent of total output value in Turkey, but only 15 per cent in the EU.
- Field crops have the largest share (35 per cent) of Turkey's agricultural output. Livestock products are less than 25 per cent of total output, although livestock output may be under-recorded in the official statistics by up to 30 per cent.
- Fruit production has increased by 55 per cent since 1980. Vegetables are dominated by tomatoes and watermelon, whereas grapes, citrus, stone and pome fruits dominate fruit production.
- Grazing livestock numbers have been falling for two decades, and red meat production has remained constant. Poultry numbers have increased by over 300 per cent over the same period.
- Turkey has only very limited agro-ecological potential for increasing agricultural area. However, adopting more area-specific farming practices, improving pastures and extending irrigation would increase the productivity of existing farming systems.
- Real agricultural prices fell during 1999-2001, but have picked up again in the last two years. Producer prices for most commodities in Turkey are higher than in the EU, with the exception of sheep meat, sugar, tobacco, cotton and milk. However, wholesale prices for dairy products are higher than in the EU, indicating an inefficient dairy processing sector.
- Average income per employed household member in Turkish agriculture is less than 40 per cent of the level for non-agricultural workers.
- Gross Value Added (GVA) per person in agriculture is one-eighth of the average EU-15 level, lower than the averages for the NMS and Bulgaria, but higher than in Romania. GVA per hectare is 45 per cent below the EU-15 average, but higher than in the NMS, Bulgaria and Romania.
- In 2003, Turkey exported (imported) EUR 4.3 (3.7) billion of agricultural and food products. Fruit and vegetables are the major export categories. Turkey regularly has a trade surplus in agricultural products.
- EU-15 member states are the destination for about 45 per cent of Turkey's agricultural exports.

Chapter 5: Turkey's rural population and agricultural workforce

- In 2003, nearly 40 per cent (27.3 million persons) of Turkey's civilian population was classified as rural. The rural population provided a workforce of 9.8 million. Agricultural workers numbered 7.2 million persons (34 per cent of the total work force).

- Labour force participation is higher in rural areas than in urban areas. The labour force participation rate is twice as high for rural women as for urban women. Measured unemployment rates are lower in rural areas than in urban areas.
- In rural areas, 56 per cent of working men and 89 per cent of working women are employed in agriculture. Four-fifths of rural women working in agriculture do so as unpaid family workers.
- The 15-24-year-old age group comprises 20 per cent of the population. Unemployment among well educated persons in this age group is much higher than for the adult labour force as a whole.
- The rate of illiteracy is 18 per cent among agricultural workers (28 per cent for female agricultural workers). Beyond primary school (which ends at age 11), school enrolment rates are lower in rural areas (particularly for girls) than in urban areas. A number of disincentives for rural children to obtain education have been identified.
- Spending on education, as well as levels of educational attainment, are low in Turkey relative to virtually all OECD and EU-25 countries.
- Improving education provision, quality, access and attainment levels for the rural population must become a top priority for Turkey in the coming years, together with improved access to the labour market for educated young people.
- There are large differences in quality of life indicators between urban and rural areas, and between 'west' and 'east'. Poverty is inversely correlated with education level. Even within each level of education, rates of poverty are much higher in rural areas.
- Turkey's performance as an EU member, and the success of its economy within a competitive market, depend crucially on the human capital of young Turkish people. However, because the *acquis* focuses on regulations to support the single market and to impose EU-level policies, there is a danger that the attention of Turkey's policy makers in a pre-accession phase may be drawn away from national education policy as a top priority area.
- The adjustment of the agricultural sector to the single market will put pressure on a large socio-economic group for whom there is currently no safety net. The creation of non-agricultural jobs in both rural and urban areas is needed, accompanied by liberalisation of the labour market and extension of the social security system to act as a genuine safety net.
- Low levels of literacy in vulnerable areas, and the particular age structure of the Turkish population, mean that new structural adjustment strategies will be needed that go beyond the experience gained in recent enlargements.

Chapter 6: Agricultural and food industry structure

- The industries upstream of farming are either dominated by a few large firms (fertilisers and pesticides), characterised by many smaller firms (animal feed) or public sector-dominated with increasing private sector involvement (seed). Only 40 per cent of seed used by farmers is produced by the 'formal' seed industry.

- The Agricultural Bank of Turkey, still publicly owned, now operates according to commercial banking guidelines. Agricultural credit subsidies have ceased, and credit provided to agriculture has declined since 2001.
- Farm structures are highly fragmented and semi-subsistence farming is an important characteristic of Turkish agriculture.
- The government had a dominant role in the agricultural cooperatives, which are active in purchasing, processing and selling many agricultural commodities in Turkey. The cooperatives are being transformed into independent organisations that must compete with private traders in the marketing of agricultural commodities.
- The wholesale market system for fresh products is still dominated by commissioners, appointed by the government. Market efficiency seems low. The system hinders the development of quality standards and low economic transparency limits opportunities for tracing products in the food chain.
- The Turkish food industry contributes 5 per cent of GNP, and accounts for 20 per cent of total production of the manufacturing sector. Its share in manufacturing industry export is 5-6 per cent and is in slight decline. There are over 100 thousand registered workers in the food sector, whereas the number of unregistered workers is unknown.
- In general, the food industry suffers from over-capacity. Although generally fragmented, there is significant concentration in a number of branches of the food industry. Market power does not seem to exist.
- Retailers offer relatively low quality standards, because consumer awareness of quality is still too low to warrant charging a premium. As the economy grows and quality awareness increases, the agrifood sector will face the challenge of meeting demand for higher quality standards all along the chain. Failure to meet consumer requirements may result in further import penetration.
- In the retail sector, supermarkets are rapidly replacing traditional stores. In 2003, modern food stores had a market share of 42 per cent. Foreign investment in the retail sector is rather limited. The new law on foreign direct investment, ratified in 2003, may encourage investment from abroad in the food sector.
- Processors purchase most agricultural commodities on the market: contract farming is not widespread in Turkish agriculture. Supermarkets, on the other hand, are shifting from buying fresh products at wholesale markets towards the use of more integrated channels in order to purchase guaranteed quantities and quality against competitive prices.

Chapter 7: Agricultural, food, rural and structural policies

- For many years, Turkish agriculture has been a political football in the competition to gain votes, rather than a sector benefiting from long-term policies to improve its efficiency and ability to adjust to societal needs. The main players have been the government, state-owned purchasing, processing and/or trading companies, the many product-specific agricultural sales co-operatives (hardly independent of government influence) and, more recently and indirectly, external organisations such as the World Bank and IMF.

- Farmers' representation via Chambers of Agriculture (semi-public institutions with little independence from the government) is weak, although there are also a few genuine farmer-controlled organisations and other independent NGOs.
- The countervailing power of consumers and taxpayers in the agricultural policy process has been very limited.
- The Agricultural Reform Implementation Project (ARIP) (2001-2005) is a radical change of direction for agricultural policy, which will help to prepare Turkey for the EU. Under ARIP, price support has been reduced, subsidies have been removed and direct income support has been introduced. Markets for some products, however, still enjoy high levels of trade protection.
- The institutional reform of State Economic Enterprises and Agricultural Sales Co-operatives, however, is proving more difficult. Steps are being taken, but so far there is no clear indication that a competitive private sector has emerged.
- A direct income support system for farmers has been introduced. Payments are paid at a flat rate per hectare, with a cap at 50 hectares.
- Since the introduction of ARIP, agricultural production has fallen by 4 per cent.
- Turkish and EU policies for table olives and olive oil are quite similar, except for the EU's per hectare premium. Current sugar policies in Turkey and in the EU are both highly protective and include a quota system.
- The private sector in Turkey has only just begun its involvement in the food safety standards of EUREPGAP, in the context of important Turkish export products such as fruit and vegetables.
- Food policy in Turkey mainly consists of measures to introduce international food safety standards. Domestic income developments have not been sufficient to increase demand for higher standards of food safety and food quality.
- Rural development policy in Turkey is more focused on large-scale investments in, for example, irrigation. Structural policy would be a new concept for Turkey, although the National Development Plan 2004-06 already adopts the same terminology as is used to frame EU structural policy.

Chapter 8: Turkey's foreign trade

- In 2003, Turkey's total imports and exports of goods were 29 and 20 per cent of GNP, respectively. The EU is Turkey's main trade partner. Agricultural products accounted for 11 per cent of Turkey's merchandise exports in 2002, and 4 per cent of imports.
- Since 1989, agricultural trade volumes have fluctuated around a constant level. The agricultural terms of trade improved in the later 1990s, but are now close to the level of the early 1990s.

- Fruit and vegetables represent over half of Turkey's agricultural exports, whereas the composition of agricultural imports is more diverse. One-third of agricultural imports are intermediate goods (textile fibres, hides/skins, tobacco, animal feed ingredients).
- Turkey has a strong positive balance on agricultural trade with the EU. A customs union between the European Union and Turkey came into force in January 1996. Agricultural products have remained outside the customs union, although trade preferences operate for agricultural product flows in each direction. As the EU had already granted preference to many of Turkey's agricultural exports, the customs union had no discernible impact on Turkey's exports to the EU.
- Turkey retains some very high tariff bindings for agricultural and food products. The tariff structure for these categories exhibits tariff escalation. Turkey has no bound allowance for domestic support to agriculture under the Uruguay Round Agreement, all domestic support having been declared as *de minimis* support. Export subsidies are used for a number of products.
- Turkey is a member of the WTO (where it has developing country status), and a party to various multilateral and bilateral regional trade cooperation agreements.
- At the WTO, Turkey has faced three SPS-motivated complaints, two of which appear to be unresolved. The 8-year ban on imports of red meat has been repeatedly challenged as an illegal use of SPS measures for protectionist purposes. Other complaints against Turkey's import regulations for agricultural products refer to frequent unnotified changes in import regulations, and cumbersome bureaucracy.
- In the Doha Development Round, Turkey follows the EU negotiating position as regards non-agricultural products, whereas for agriculture, its position is close to that of the "G-20" developing countries, who insist on large reductions in export subsidies and support by developed countries as a condition for further tariff reductions.
- A recent study indicates that, with full harmonisation of agricultural trade between Turkey and the EU, livestock prices in Turkey would fall significantly and domestic animal production would shrink. Consumers' welfare gain would be greater than producers' welfare loss. The same study concludes that Turkey could do much to improve its net trade position even without trade harmonisation, by internal restructuring and raising the efficiency of supply chains.
- Turkey's current pattern of self-sufficiency levels is the result of trade and market distortions, and in particular an over-protected livestock sector, to the possible detriment of human nutrition.

Chapter 9: Environment and agriculture

- The main environmental impacts of agriculture in Turkey are water and soil degradation, due to the overuse of water and chemicals. Fertiliser and pesticide use has decreased slightly in recent years. However, expansion of irrigated areas may stimulate water use, leading to more nutrient run-off and salination.

- In the last ten years, Turkey has adopted much new environmental legislation. The implementation of global and regional conventions, participation in international environmental fora and the goal of joining the EU have been major driving forces behind these reforms.
- Institutions dealing with agri-environmental issues are still poorly coordinated and there is a lack of effective implementation at local level. Regulations are the main policy tool and there are few economic instruments. There are few incentives for farmers to use environmentally friendly practices.
- Turkey has only just started to include environmental concerns in its agricultural and rural development policies, and there is still ample scope for further regulation and improvements to existing regulations.
- Although public opinion still gives low priority to the environment, civil society in Turkey is gradually becoming more involved in environmental policy making. Non-governmental organisations have a key role to play in increasing environmental awareness and public participation, and in advancing governmental policy. Recent changes facilitate the registration and financing of non-governmental organisations and their projects in the field of the environment.
- Typically, environmental investment is financed from the tight government budget. Bank lending for environmental projects is limited. Funding for these projects is mostly provided by international development agencies and other international donors. Most of these projects are scattered and of small scale.
- Organic farming has developed rapidly since the mid-1980s, but covers less than 0.5 per cent of the cultivated area. Production is driven by export demand and the sector offers potential for further growth, particularly if domestic demand is encouraged.
- Turkey is very rich in biodiversity. Many species and habitats are, however, at risk due to various kinds of agriculture-related developments, which include intensification of farming, land abandonment and the construction of large infrastructure projects.

Chapter 10: Animal and plant health in Turkey

- Turkey faces major challenges with respect to animal health. Some highly infectious animal diseases that have been eradicated in the EU remain endemic in Turkey. The situation is complicated by the fragmented structure of the livestock sector, Turkey's geographical location and its porous borders to the south and east. Operational shortcomings limit the efficiency of the veterinary services, political commitment to pursue effective control and eradication has been weak, and resources to do so have been limited.
- Of the 15 most infectious diseases identified by the World Animal Health Organisation, three diseases (foot and mouth disease, *peste des petits ruminants* and sheep and goat pox) have occurred in virtually every year since 1996. The last outbreaks of bluetongue and Newcastle disease were in 2000 and 2001 respectively.

- Turkey is also prone to outbreaks of anthrax and brucellosis. Rabies is present in the dog and wild animal population. Turkey has had no registered case of BSE, but the BSE risk has been classified as Level III.
- The most important zoonoses recorded in humans are anthrax, brucellosis, leishmaniosis and salmonellosis.
- The veterinary services are working towards harmonisation with EU legislation and are upgrading enforcement capacity. Although there has been progress with legislation, enforcement capacity is undeveloped, as is biosecurity awareness at every level of the livestock production chain.
- Even with effective implementation of the *acquis*, it will be many years before Turkey reaches disease-free status for all List A diseases. Until then, a single market in animal products with the rest of the EU will remain problematic. Zoning might be used to allow the country to acquire disease-free status on a region by region basis.
- 245 insects, 85 diseases and 75 weeds of an economically damaging nature have been reported in cultivated crops in Turkey. About 35 thousand tons of phytosanitary chemicals are used annually, valued at about USD 250 million. Biological control programmes are being developed for various open field and greenhouse crops.
- New quarantine legislation to bring Turkey closer in line with the EU has been adopted.
- The plant health situation gives less cause for concern. Export-quality fruit and vegetable products are already accepted on the EU market and around the world without difficulty.

Chapter 11: Consequences for EU-27 of Turkish entry

- Turkey's accession would mean that EU budget spending is less able to shift in favour of measures to increase competition and growth, and away from redistributive policies to support agriculture and rural regions.
- Because of Turkey's low per capita income, Turkish entry implies a reduction in annual average EU per capita income by about EUR 2520 (at 2004 values).
- Turkey's accession would add to the number of EU member countries that have difficulties in implementing EU requirements with respect to food safety, environmental, veterinary and phytosanitary standards.
- The boost to macroeconomic growth in EU-27 from Turkish accession would be low and could be cancelled out by high budget transfers from EU-27 to Turkey.
- To keep total direct income payments within the current agreed limits after Turkey's accession, these payments may have to be reduced for EU-27 member states.
- The accession of Turkey to the EU will lead to a large increase in EU borders, with attendant risks of smuggling of goods and people, and undermining of the single market. The initial and on-going costs of controlling these borders will be high.
- Before and after accession, Turkey will be a growing market for the EU-27 food

industry and retailing companies, regarding both exports and FDI.

- With Turkish entry in 2015, new regions with a combined population of 79 million people will become eligible for structural funding at a high rate. At the same time, 20 regions in EU-27 with about 33 million inhabitants will no longer be eligible for these funds.
- The annual net budget cost for the EU-27 of Turkey's accession in 2015 is estimated to be EUR 10.9-18 billion (at 2004 prices). The uncertainty of these estimates comes from the structural fund component, and depends on Turkey's absorption capacity for structural spending. These estimates are somewhat lower than the numbers indicated by the Centre for European Policy Studies and the European Commission.

Chapter 12: Consequences for Turkey of EU entry in 2015

- If new formal institutions are to function as intended, it is important that informal rules and expectations evolve in parallel with the new legislation.
- To develop a stronger and more competitive food supply chain, the Turkish farming sector, with its large share of semi-subsistent and fragmented farms, will need to be re-structured.
- In trade policy, Turkey would adopt the common external tariff of the EU for agricultural products. Given current tariff structures, agricultural trade harmonisation between the EU and Turkey by 2015 will mostly mean tariff reductions in Turkey. The largest downward tariff adjustment is expected in the livestock sector. The infrastructure, administrative capacity and commitment for effective border control must be heavily reinforced in preparation for EU membership.
- Veterinary standards amongst Turkey's grazing livestock populations fall well below those in EU member countries. It is unlikely that by 2015 a single market in all animal products, without SPS border controls, can be operated.
- The adoption of cross compliance conditions for direct payments represents an opportunity for Turkey to improve the environmental performance of agriculture, but will require good quality extension services and monitoring expertise that need to be in place by the time of accession.
- With accession in 2015, market and price support, and direct income payments to Turkish farmers, would amount to EUR 0.2 billion and EUR 3.4 billion, respectively (at the 2004 value of the euro). Budget payments arising from structural and cohesion policy would be between EUR 9.5 and 16.6 billion (2004 values).
- Access to structural policy funds will give Turkey the opportunity to tackle severe disparities in economic development across regions.
- An important challenge is to design programmes for structural and cohesion spending that address some of Turkey's specific weaknesses such as low levels of human capital, poor opportunities for non-agricultural employment in rural areas, and low levels of health and quality of life in rural areas.

Subject Index

cabi-publishing.org/bookshop
Browse Read and Buy
ANIMAL & VETERINARY SCIENCES
BIODIVERSITY CROP PROTECTION
HUMAN HEALTH NATURAL RESOURCES
ENVIRONMENT PLANT SCIENCES
SOCIAL SCIENCES
CABI Publishing
A division of CAB International
Online BOOK SHOP
Search
Reading Room
Bargains
New Titles
Forthcoming
Order & Pay Online!
MasterCard
VISA
AMERICAN EXPRESS
Crop Pollination by Bees
Keith S. Delaplane and Daniel F Mayer
A DICTIONARY OF Entomology
G Gordh and D H Headrick
Principles of CATTLE PRODUCTION
C.J.C. Phillips
Seeds
THE ECOLOGY OF REGENERATION IN PLANT COMMUNITIES
2nd EDITION
Edited by Michael Fenner
FULL DESCRIPTION
BUY THIS BOOK
BOOK OF THE MONTH
Tel: +44 (0)1491 832111 Fax: +44 (0)1491 829292